GOD MADE ME

Science/Worldview | Level 3

Author: Tamela Sechrist

Editors: Joshua Schwisow and R.A. Sheats

GOD MADE ME

Hi! I'm Thump the Pump, also known as your heart! God gave me an important job to do inside your body. I do my job without ever stopping, for as long as you live.

And I'm Puff, a puff of air! Air is wonderful. I like blowing it around where God wants it to go. When you see me in this book, you can remember that God made you and gives you the Breath of Life!

Your body is something that's always with you to show you that God is amazing!

Copyright © 2024 by Generations. All rights reserved. No part of this book may be reproduced in any form or by any means without permission in writing from the publisher.

Printed in the United States of America.

ISBN: 978-1-954745-62-9

Cover Design: Justin Turley
Interior Design: Sarah Lee Bryant

Published by:
Generations
19039 Plaza Drive Ste 210
Parker, Colorado 80134
Generations.org

Unless otherwise noted, Scripture taken from the New King James Version®.
Copyright © 1982 by Thomas Nelson. Used by permission. All rights reserved.

Scripture quotations marked "ESV" are taken from the English Standard Version of the Bible. Copyright © 2001 by Crossway, a publishing ministry of Good News Publishers. Used by permission. All rights reserved.

For more information on this and other titles from Generations,
visit Generations.org or call (888) 389-9080.

Table of Contents

Introduction 1

Chapter 1
God Made You the Way You Are 17

Chapter 2
God Used DNA to Make You 25

Chapter 3
God Knitted You Together 33

Chapter 4
A Time to Be Born 45

Chapter 5
God Gave You Blood 55

GOD MADE ME

Chapter 6
God Gave You a Pump 63

Chapter 7
God Gives You the Breath of Life 71

Chapter 8
Take a Trip with a Blood Cell 81

Chapter 9
Your Brain Is Amazing 91

Chapter 10
Your Nervous System 99

Chapter 11
Your Neurons 107

Chapter 12
Your Nervous System Works Automatically and on Purpose 115

Chapter 13
God Gives You Eyesight 127

Chapter 14
God Gives You the Sense of Touch 141

Chapter 15
God Gives You Hearing 153

Chapter 16
God Gives You Taste, Smell, and Other Senses 163

Chapter 17
What is Food? 179

Chapter 18
To Your Stomach 191

Chapter 19
Absorbing in Your Abdomen 201

Chapter 20
Your Large Intestine Finishes Digestion 211

Chapter 21
Your Kidneys Clean Your Blood 225

Chapter 22
Your Bladder and Its Tubes 233

Chapter 23
Your Body Has Many Parts That Purify 239

Chapter 24
God Purifies Your Cells 247

Chapter 25
God Gives You Strong Bones 255

Chapter 26
God Gives You Structure 263

Chapter 27
How Muscles Work 273

Chapter 28
Swift, Strong, Careful Movement 281

Chapter 29
God Heals Your Wounds 291

Chapter 30
Healing: Making Things New Again 297

Chapter 31
God Can Heal Your Illnesses 305

Chapter 32
Vigilant, Valiant Germ Fighters 313

Chapter 33
God Gives You Hormones 325

Chapter 34
Pituitary, Pineal, and Thymus Glands, Keeping Your Body on Schedule 331

Chapter 35
Thyroid, Parathyroid, and Pancreas Glands, Keeping You Diligent 337

Chapter 36
Hormones for Hard Times 343

Glossary 349

Image Credits 355

Introduction

. . . what may be known of God is manifest in them, for God has shown it to them. For since the creation of the world [God's] invisible attributes are clearly seen, being understood by the things that are made, even His eternal power and Godhead . . . (Romans 1:19-20)

Understanding God's creation helps us all understand God's attributes. (Romans 1:19-20) This science course for middle to upper elementary children is designed to bring to light the wisdom, power, and love of God that is evident in His creation. *God Made Me* presents the amazing way that the human body is wonderfully designed so every part works together in harmony. When God made mankind in His image, He made each structure and process to proclaim His glory with its intricacies, from the smallest organelle in a cell to the largest parts, such as the arms and legs.

Thump, the hard-working heart, and Puff, representing the breath of life, appear often in the course to add interesting facts or encourage praise for God's creation of the human body.

God Made Me is designed to be easy to understand and engaging for children. Kids learn in a variety of ways, and this course, with its companion activity book, provides several modes of instruction.

- **Reading:** The textbook is designed at a reading level appropriate for middle elementary students. Acknowledging that the study of the human body involves learning many new vocabulary words and focusing on multi-step processes, the text is kept at a simple level. This will enable the student to concentrate more on the science concept than on sorting out lengthy sentences or potentially unfamiliar non-science words. Fun analogies are sometimes used to help explain the way things work.

- **Seeing:** Beautiful pictures and fascinating illustrations are abundantly used to keep the attention of children. Diagrams matching the text concepts are used to reinforce learning.
- **Moving:** The *God Made Me* companion activity book is designed to add a variety of fun, physical activities that are not overly burdensome to the parent/teacher. Here, science topics are reinforced with a balance of observation, experiments, exercise, art, Scripture, cooking, and games. Occasional review exercises are provided as a way for the parent/teacher to check retention. Each activity corresponds to a section of the textbook and is numbered for easy reference.

How To Use *God Made Me*

God Made Me is divided into nine units of four chapters each. Each unit begins with a memory verse and hymn that the child will have a chance to work on in the activity book. This course contains 36 weeks of lessons with one chapter to be read each week. Each week's reading and companion activities are divided into three days.

Children learn best in small doses with time to internalize in between. New concepts are solidified as they play and are made permanent as they sleep. Because of this, the following weekly schedule is suggested:

- **Day 1** — Read the first section of the textbook chapter. Complete the corresponding activity in the activity book (as announced at the end of that section of text).
- **Day 2** — Break
- **Day 3** — Read the second section of the textbook chapter. Complete its corresponding activity in the activity book.
- **Day 4** — Break
- **Day 5** — Read the third section of the textbook chapter and complete its corresponding activity.

May God be glorified, and may you be richly blessed as you and your student study the pinnacle of God's material creation, the human body!

Tamela Sechrist
The Generations Curriculum Team
April 2024

UNIT 1
God Made You on Purpose

God planned for you to exist before you even began to be! Our memory verse and hymn remind us that God made you on purpose!

Your family met you when you were born. But your life began months earlier inside your mother, before she even knew you were there. And way before that, God knew you.

Memory Verse

Before I formed you in the womb I knew you. (Jeremiah 1:5)

Hymn to Sing: Let All Things Now Living

You can listen to this hymn by searching for "Let All Things Now Living children's choir" on the internet.

Let all things now living a song of thanksgiving
To God the Creator triumphantly raise.
Who fashioned and made us, protected and stayed us,
Who still guides us on to the end of our days.
God's banners are o'er us, His light goes before us,
A pillar of fire shining forth in the night.
Till shadows have vanished and darkness is banished
As forward we travel from light into light.

His law He enforces, the stars in their courses
And sun in its orbit obediently shine;
The hills and the mountains, the rivers and fountains,
The deeps of the ocean proclaim him divine.
We too should be voicing our love and rejoicing;
With glad adoration a song let us raise
Till all things now living unite in thanksgiving:
"To God in the highest, Hosanna and praise!"

"Let All Things Now Living" hymn text by Katherine K Davis © Copyright 1939, 1966 by E. C. Schirmer Music Company, Inc., a division of ECS Publishing Group. All rights reserved. Used with permission.

God made each person unique.

CHAPTER 1
God Made You the Way You Are

And the LORD God formed man of the dust of the ground and breathed into his nostrils the breath of life; and man became a living being. (Genesis 2:7)

You are Unique

God made people in a different way than He made animals. God breathed life into the man's nostrils with His own breath! The man's body became alive. God formed the first woman out of one of the man's rib bones. He gave life to her too. Even though God uses a different way to form peoples' bodies now, He still gives life to each person.

God made people so they would be different than the animals. He made people **unique** in these ways:
- We are made in God's image.
- We can know the difference between right and wrong.
- God gives us each a soul that will live forever.
- We can talk, sing, and praise God.
- We can create, invent, and think about complicated things.
- We can know and love God!

[God] teaches us more than the beasts of the earth, And makes us wiser than the birds of heaven. (Job 35:11)

Definitions

If something is not like anything else, we say it's **unique** (you-neek).

A **womb** (woom) is the special place inside a mother where God makes a baby grow before it's born. We say the mother is **pregnant** when there is a baby growing in her womb.

GOD MADE ME

Before I formed you in the womb I knew you. (Jeremiah 1:5)

Here's our memory verse again!

Our memory verse shows us that God knows each person **before** they even start to grow inside their mother. He knew you before you started to grow too! God knew just how He would make you look. He knew what color of eyes He would give you and whether you would be a boy or a girl. Are you good at drawing and math? Or are you so observant that you usually know where your little sister left her favorite blanket? God planned for you to be good at certain things.

God made you a unique person! There has never been anyone exactly like you. Find someone nearby and ask to see their left pointer finger. Now look at your own left pointer finger. Do they look the same? That's a tricky question because in many ways they do look the same. God gave fingers special jobs to do. That's why He gave everybody's left pointer fingers the same **structure** (the way it's put together) to do its **function** (job).

At the same time, there are small differences between the same structures on different people. Look at the left pointer fingers on the two of you again. Do you see a difference in the shape of the fingernails? Do the knuckles wrinkle differently? Can one of you "pop" your knuckle?

Even if you have an identical twin and your hands look the same, your fingertips will still look different. The tiny swirls on the skin of fingertips are formed by the different ways babies touch things when they are in their mother's womb. Your fingertip swirls are not like anyone else's! They are unique.

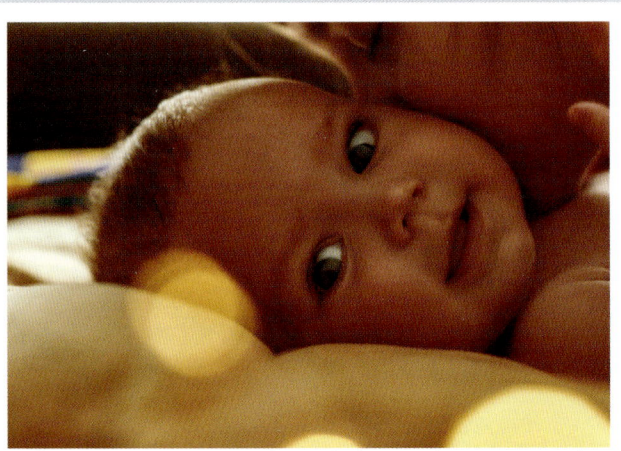

Before you were created, God chose who your mother would be. Then He put you in your mother's **womb**. All babies are special to God, even before He puts them in their mothers' wombs.

Even twins have unique fingertip swirls.

Time to do Activity 1 in the Activity Book!

You Came with Instructions

The first man was named Adam, and the first woman was Eve. When God made Adam and Eve, He put instructions inside their bodies. These instructions would tell their bodies how to automatically do what was needed for them to live.

After God made Adam and Eve, He wanted to put more people on Earth. These people would be formed in a different way than Adam and Eve had been. But they would need instructions inside their bodies too. God would give them (and all the people coming after them) instructions too.

What do these instructions do? These instructions tell your body how to work and how to grow. They might tell your body how to function, like when your heart needs to beat or when it's time for your body to stop growing. Others might give your body its structure by telling it how to

You never should laugh at someone because they look differently than you. God specially made each person. Let's not make fun of someone God created!

make your hair and whether to make it curly or straight. The instructions for millions of things were inside you when you were too tiny to be seen. But God put them there to make you exactly who He wanted you to be, with all your parts working well!

From the very first, your instructions said you would be a **human being**. You don't have to worry that you will ever grow antlers on your head or have pinecones instead of toes. Your set of instructions doesn't include antlers or pinecones. God gave you human instructions.

 Time to do Activity 2 in the Activity Book!

Zoom in to the Cell: The Home of Your Instructions

As we learn about your human body in this book, we will take a moment now and then to zoom in to the cell.

Microscopes help us see very small things.

People use magnifying glasses and hand lenses to make small things look bigger. But some things are too small to see even with a magnifying glass. To see very small things, people use microscopes.

Cells can't be seen with a magnifying glass, but you can see them with a microscope. To learn about cells, we will look at pictures and drawings that have been made using microscopes.

If you have a doll or an action figure, you can see that it has different parts. But each part is probably made of plastic. Some parts might be hard and some soft. Or the parts might be different colors. But if you look at the parts with a microscope, you will just see plastic.

CHAPTER 1: GOD MADE YOU THE WAY YOU ARE

> All of your cells contain DNA except for certain cells in your blood. These blood cells start with DNA where they are made. But by the time they are old enough to jump into your blood, their DNA (and their nuclei) have disappeared!

You are not made of plastic! If you look at your skin with a microscope, you will see that it's made of tiny structures attached to each other. If you look at your blood, you will see other tiny structures floating in liquid. If you are about eight to ten years old, your body is made of about 17 trillion of these tiny structures. They are called **cells**.

Let's look inside a cell now. Almost every cell in your body contains a copy of **all** your unique instructions. Even one cell in your tongue contains toenail-making instructions! Every cell also says that God made you human and tells you whether you are a girl or a boy.

Your skin loses about five dead cells every second. Each skin cell that lands in your bed, on your clothes, or in the dust you clean off the floor contains a copy of all the instructions about you!*

*When we look at dead skin cells with a microscope, we can't see any nuclei in them. Nuclei are the parts in cells that contain instructions. People used to think that dead skin cells didn't have nuclei and therefore couldn't have DNA (instructions). Now we know that when skin cells die, the nuclei are destroyed. But the DNA still exists in the cell.

Cells are kind of like little creatures. They are alive, they take in nutrition, they get rid of waste, and they have different

Definitions

A **cell** is the smallest living part of your body. Your body is made of various kinds of cells.

A cell's **nucleus** (noo-klee-us) is like the brain of the cell. The nucleus tells the cell what to do at the right time. It contains the DNA instructions. When we talk about more than one nucleus, we say **nuclei** (noo-klee-eye).

DNA is a large **molecule** that contains the instructions for your body.

A **molecule** is the smallest piece of something that still is that thing. The smallest piece of water is a molecule. If you broke a water molecule, it would break into smaller things that wouldn't act like water anymore.

GOD MADE ME

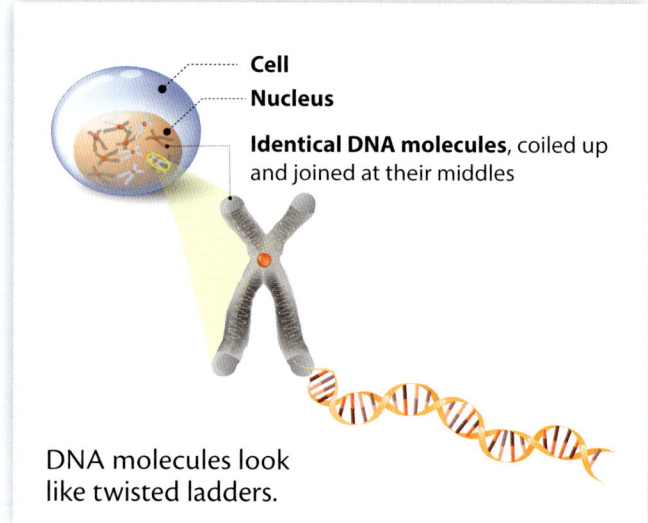

DNA molecules look like twisted ladders.

parts with different jobs. One part of the cell is the **nucleus**. The nucleus is like the cell's brain. It controls the cell by using its own copy of your instructions. The nucleus is where your instructions live most of the time.

Everything God created is made of **molecules** (mall-uh-kules). Even cells are made of molecules. Molecules are usually too small to see, even with a microscope.

The set of instructions in the nucleus of a cell is a set of molecules. Each of your cells has 46 instruction molecules. These instructions have a very long name: deoxyribonucleic acid. But we'll just use initials and call it **DNA**.

DNA molecules are very long. They need to be long so they can contain all your body's instructions. If you could hold up one of your 46 DNA molecules from one of your cells, it would stretch all the way from your head to your toes! These instructions need to be wound up like thread on spools when it's time for the cell to organize them.

Proteins are large molecules that have important jobs to do in your body. DNA tells cells how to make many **proteins**. Let's look at how DNA does this.

First, God has given your DNA the ability to know **when** proteins need to be made. Let's pretend you ate a lot of candy all at once. When you eat sweets, the sugar goes into your blood. But too much sugar in your blood can damage parts of your body. Your body has special cells that make a certain protein to take the extra sugar out of your blood and store it for another time. The DNA in these special cells knows when it's time to make more of the protein that takes extra sugar out of your blood.

Now, when it is time to make a protein, the DNA unwinds a section of itself a little so that the right instructions for that protein can be "seen." Then the DNA makes itself some little molecule messengers

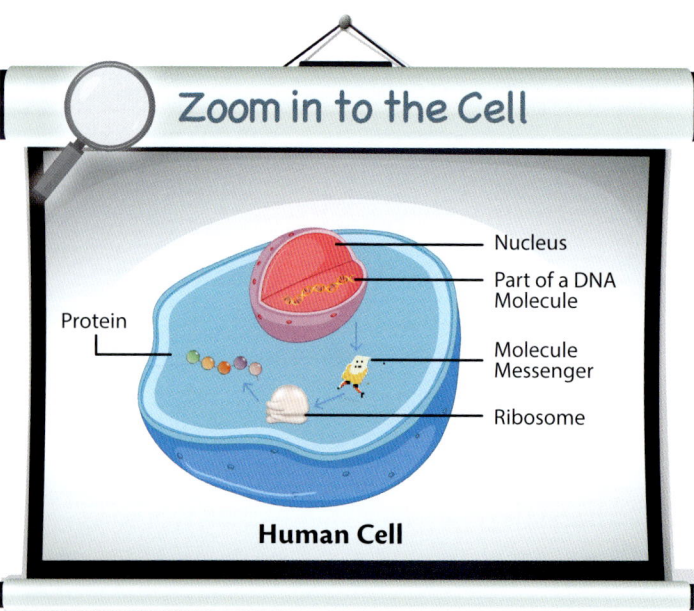

CHAPTER 1: GOD MADE YOU THE WAY YOU ARE

Death cap mushroom

buy and transplant in their area. These mushrooms pop up in late summer and autumn in wild areas and lawns. Eating just one death cap mushroom can kill a healthy adult. Its poison is deadly because it gets into cells and prevents DNA from making molecule messengers. Without the messengers, proteins can't be made. This can be dangerous!

to "see" this part of the instructions. Each messenger then leaves the nucleus and takes its message to one of the many **ribosomes** (rye-buh-sohms) in the cell's liquid. Ribosomes use these messages to find the right ingredients floating around in the cell. They grab the ingredients and make the protein according to the instructions.

The death cap mushroom is a very poisonous mushroom. It first grew above the roots of trees only in Europe. But now it has spread around the world by traveling on the roots of trees that people

Thump's Health Hint
If you find a mushroom, don't eat it! Not very many people can tell the difference between edible and poisonous wild mushrooms. It's best not to even touch wild mushrooms. If you touch one and then forget to wash your hands before touching your mouth, you could get very sick.

Prayer

God, thank You for making me the way I am. Thank You for making every person unique. You are so wise to make DNA instructions in each cell. They automatically control everything in my body to keep it working well. Thank You! Amen.

Time to do Activity 3 in the Activity Book!

CHAPTER 2
God Used DNA to Make You

And God said to them, "Be fruitful and multiply; fill the earth." (Genesis 1:28)

God's Plan to Fill the Earth with People

When God made Adam and Eve, He gave each of them their own set of DNA instructions. These instructions kept Adam and Eve's bodies working. But God also had another plan for their DNA. He planned to use Adam and Eve's DNA to create new people!

God created Adam and Eve as grownups. There had not been a baby on Earth yet. But God had a plan to fill the earth with people. His plan used babies! Where would the babies come from? From their mothers' wombs!

How did God give those babies their DNA instructions? Before the first baby began to grow in Eve's womb, God made a way for **half** of the baby's instructions to come from Eve and the other **half** to come from Adam. Adam and Eve's babies grew up and passed DNA instructions to their babies the same way. Ever since then, babies have been getting their DNA this way—half from the mother and half from the father.

You can often see that babies have DNA from their parents. Some children look like their mother and some look like

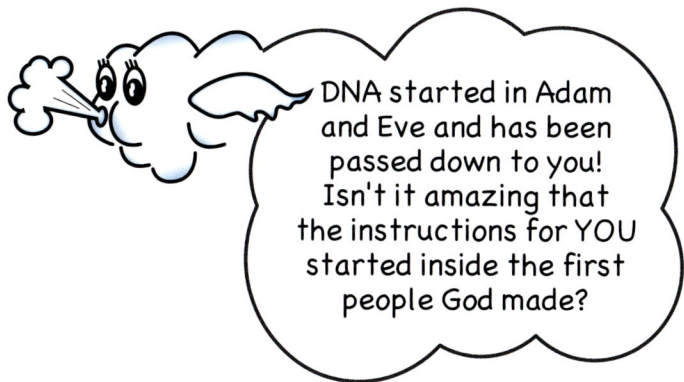

DNA started in Adam and Eve and has been passed down to you! Isn't it amazing that the instructions for YOU started inside the first people God made?

25

GOD MADE ME

Some children are **adopted**. This means that God has put them in a family with parents who are not the same parents that gave birth to them. These children have different DNA than their new parents. The new parents are so happy to welcome their adopted children. They love them very much!

their father. Sometimes they look like a mixture of both. Often children in the same family look similar to each other. But children from around the world can look very different. God has given DNA an amazing ability to make each person unique! Most of the differences between people happen because children get a mixture of possibilities from their parents' DNA.

Time to do Activity 4 in the Activity Book!

God Makes Variety

God is so wise! He made DNA for all living things—people, animals, plants, and microscopic life. Each kind of DNA is passed along to the next **generation** so the same kind of life will continue. Sunflower DNA will produce more sunflowers. Dog

Pigments are chemicals that give certain color to their surroundings. Certain pigments give paints their colors. Other pigments make freckles on skin.

CHAPTER 2: GOD USED DNA TO MAKE YOU

DNA will produce more dogs. But God is also creative! He made each type of DNA able to give **variety** within each kind of living thing.

Many parts of DNA help make our eyes able to see. But our DNA also has at least eight different parts that help make eye color. That's why there are so many beautiful eyes in the world. Let's look at DNA to see what might happen at one of these instruction places for eye color.

DNA has a place where its instructions say how much brown **pigment** goes into the colored part of someone's eye. If it says to make a lot of brown pigment, the person will have dark brown eyes. If it says to make no brown at all, the person could have bright blue eyes. (Other instructions could add other colors, making the eyes green, gray, or golden.)

Let's say a man and a woman get married. Their baby will get half of its DNA instructions from Mom and half from Dad. Let's see how the DNA instructions for brown eye pigment could work to give Baby either bright blue or dark brown eyes:

1. If Mom's DNA gives **blue** instructions (no brown pigment) and Dad's also gives **blue** instructions, Baby will have **blue** eyes.
2. If Mom's DNA gives **brown** instructions and Dad's also gives **brown** instructions, Baby will have **brown** eyes.

But what happens when the instructions the parents give are not the same? What if one DNA says "**blue**" and the other says "**brown**"? Amazingly, Baby will always have **brown** eyes! God has designed brown to be **dominant**. Think of dominance as though brown is selfish and always gets its way.

3. If Mom's DNA gives **blue** instructions and Dad's gives **brown** instructions, Baby will have **brown** eyes.
4. If Mom's DNA gives **brown** instructions and Dad's gives **blue** instructions, Baby will have **brown** eyes.

The DNA instructions for eye color are very complicated. But this example shows how dominant instructions work and helps us learn about DNA.

Here's a hard question for you: How is it possible for two **brown**-eyed parents to have a **blue**-eyed baby? (Puff has the answer on the next page.)

GOD MADE ME

For brown-eyed parents to have a blue-eyed baby, both of the baby's parents must have some "hidden" blue-eye DNA.

Maybe Baby has some grandparents with blue eyes. The parents' eyes are brown because brown is dominant, but Baby could still receive hidden blue-eye DNA from both Mom and Dad because of its grandparents' DNA.

 Time to do Activity 5 in the Activity Book!

Zoom in to the Cell

Your body has 46 pieces of DNA in the nucleus of each cell. Most of the time, your DNA is loose. It looks like a bowl of spaghetti made with 46 very long noodles.

When it's time to make new cells, the DNA changes. Each piece of DNA gathers itself together to take up less space. "Spaghetti DNA" could get tangled up as it tries to get into the new cells. To fix this, God made special proteins that wind up each piece of DNA. The DNA now gets packaged into 46 **chromosomes** (krome-uh-sohms).

We usually draw a chromosome as two strands joined at the middle. This is to show that the chromosome has made a copy of itself. DNA has to make a copy to give to the new cell. The two strands are identical and are called **sisters**. Even though there are now two strands, we still call the whole thing "a chromosome."

Your body cells have two of each kind of chromosome. One came from your mother, and one came from your father. Scientists have numbered each kind of chromosome. You have two of Chromosome 1, two of Chromosome 2—all the way up to two of

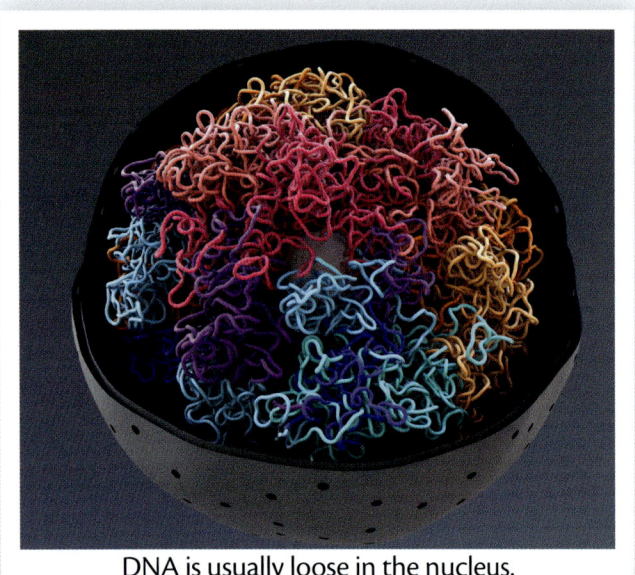
DNA is usually loose in the nucleus.

CHAPTER 2: GOD USED DNA TO MAKE YOU

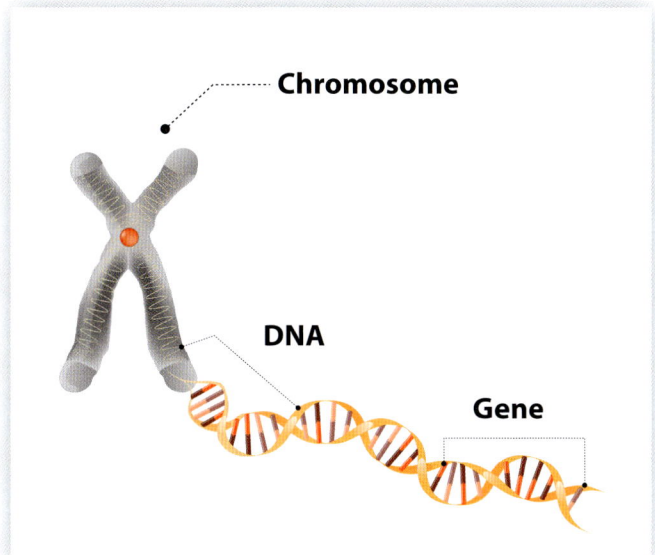

Definitions

Chromosomes are molecules of DNA wrapped around proteins. Chromosomes live in the nuclei of cells.

A **trait** is a certain feature about you. Green eyes or how tall you grow are both traits.

A **gene** (jeen) is a place on the DNA that gives instructions for a certain trait.

A **gamete** (gam-eet) is a cell with half of the DNA instructions. One gamete comes from the mother and another comes from the father. Their DNA joins together and a living being with complete instructions is made.

Chromosome 23.

Chromosome 15 has a certain place that helps decide the **trait** of eye color. That place is the same on both of your Chromosome 15's. It's also the same place on Chromosome 15 for everyone in the world. Each piece of your instructions has its own place. A single part of your instructions is called a **gene**.

The chart below shows the 23 pairs of chromosomes that people have in each cell. A pair of chromosomes includes one from the mother and one from the father. These chromosomes are shown at the time when they have made sister copies. In this chart, a chromosome pair would be two X shapes.

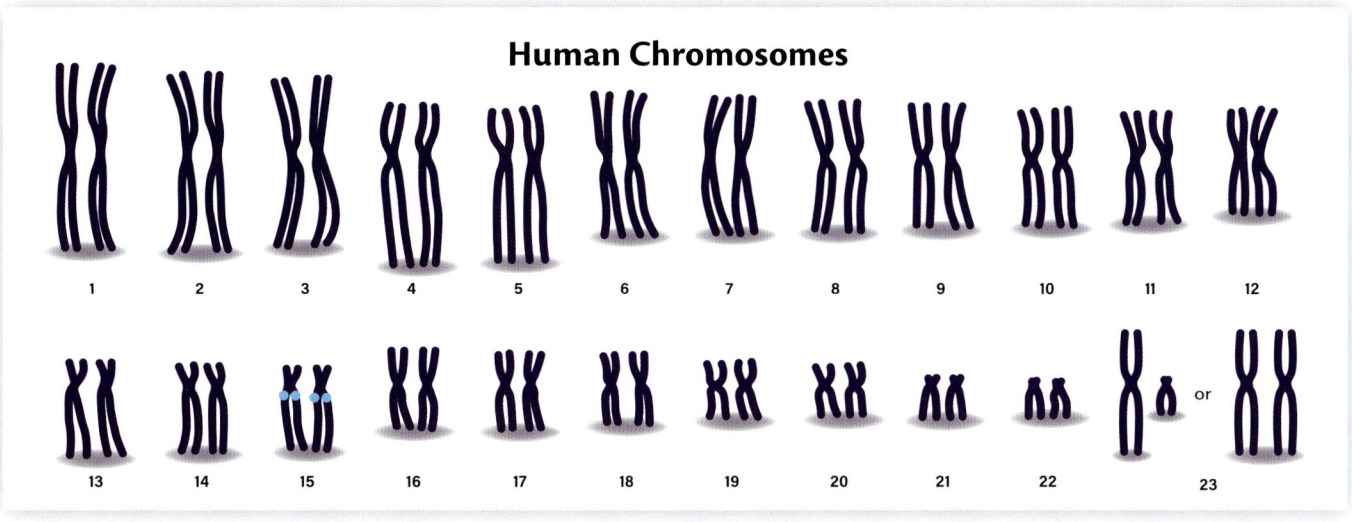

Human Chromosomes

29

GOD MADE ME

Chromosome 23 has two possibilities—one for a boy and the other for a girl. The blue dots on the Chromosome 15 pair is the location of the main gene for how much brown goes into eye color.

Let's learn about how a baby begins as a cell. Do you remember that a baby gets half his DNA instructions from Mom and half from Dad? God does this in an amazing way!

1. Mom's body makes a special cell (gamete) that has half of her DNA instructions. She got these genes from her parents.
2. Dad's body also makes a special cell (**gamete**) that has half of his DNA instructions. He got these genes from his parents.
3. These cells come together, and Dad's half of the instructions go into the nucleus of Mom's gamete.
4. In the nucleus, the chromosomes find their match. And a certain gene on Chromosome 15 gets busy deciding Baby's eye color!

You can protect your DNA by exercising, eating healthy foods, and avoiding unnatural chemicals.

Thump's Health Hint

God has given DNA molecules special structures on their ends to protect them. These structures work like the caps on the ends of shoelaces that keep them from unraveling. The DNA structures get shorter as people become older. When DNA structures get shorter, the person becomes less healthy. DNA gets shorter more quickly if a person smokes or has other unhealthy habits. You are helping your DNA last longer every time you take a brisk walk or eat colorful fruits and vegetables!

Prayer

Dear Lord, thank You for my parents and the DNA You gave me. It's amazing how You passed all our instructions through people. And You started with Adam and Eve! You picked out which genes I would get that would make me unique! Amen.

 Time to do Activity 6 in the Activity Book!

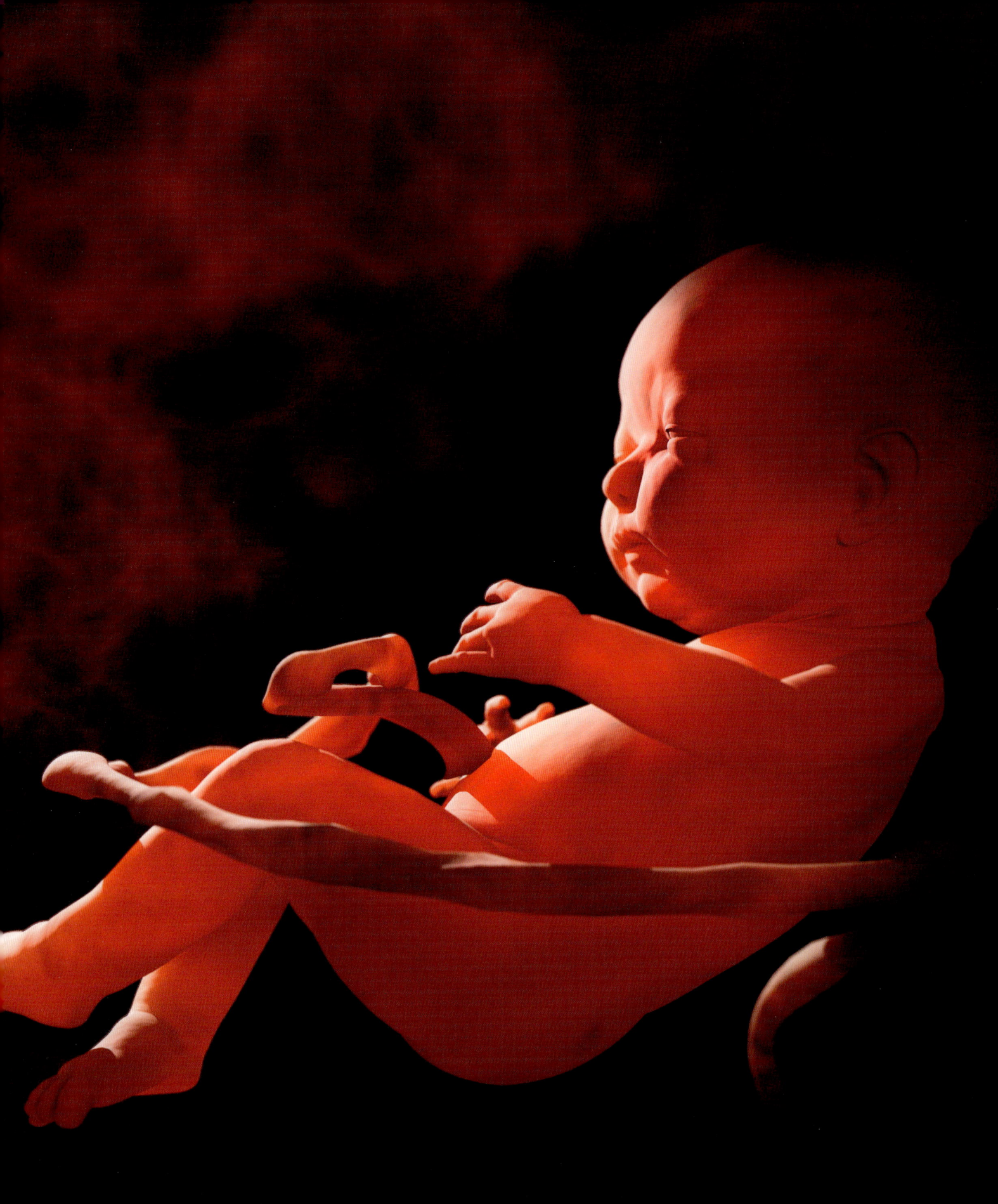

Child in the womb shortly before birth.

CHAPTER 3

God Knitted You Together

For You formed my inward parts; You knitted me together in my mother's womb. (Psalm 139:13) ESV

Do you remember that God specially **planned** you long before your body began to be? The verse above shows that, after you began to be, God also specially **formed** you. He carefully made your tiny inner parts and "knitted" you together in your mother's womb! He did this according to the DNA instructions He created for you.

Let's look at how you grew from one tiny cell into a complete baby before you were born.

Zoom in to the Cell

In the last chapter, we learned that you started from a special cell (gamete) that came from your mother. That cell had half of your DNA in its nucleus. When a gamete from your father added its half of your instructions to the egg's nucleus, you became a living being!

When you first began to be, you did not remain a single cell. About 24 hours (one day) later, you became two cells. Ever since then, your body has been growing by making more cells. When your body

Someone skillfully knitted this doll family. But only God has the wisdom to skillfully knit His complicated creations in the wombs of mothers.

33

GOD MADE ME

cells **reproduce**, they pass along complete instructions (not half) to each new cell. This kind of cell reproduction is how you grow and how your body heals.

Definitions

Something **reproduces** when it makes more living things of its kind.

Mitosis is the process God made for one cell to become two new cells by dividing. The new cells have the same DNA as the old cell.

In your body, one cell becomes two cells (reproduces) by dividing in half. Cells divide by a process called **mitosis** (my-toe-siss). God made mitosis so that new cells would get the same DNA as the old cell. New cells also need a nucleus, some liquid, and all the other tiny parts cells are made of. Let's have Thump show us how this happens on the next page!

Cells spend only a little time in mitosis. Most of their time is spent doing their other jobs.

God ceated birds and all living things to reproduce their kind! "So God created . . . every winged bird according to its kind. And God blessed them saying, 'Be fruitful and multiply [reproduce]. . . and let birds multiply on the earth.' " (Genesis 1:21-22)

CHAPTER 3: GOD KNITTED YOU TOGETHER

 ## Thump Teaches Mitosis

1. Before mitosis, the cell is busy doing the jobs that keep your body working well. The cell is also busy making an exact copy of all its DNA. This way, the two new cells made by mitosis will have the same DNA as the cell they came from. The first cell is also getting bigger and making more of its little parts to pass on to the new cells when it divides.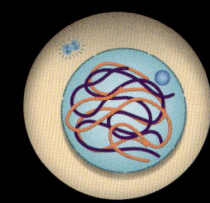

2. Mitosis starts when the DNA organizes itself. Instead of looking like a tangled spaghetti noodle, each piece of DNA and its copy (sister) now become shorter, fatter chromosomes. These matching sisters are joined at a spot (often near their middles) on their sides. This picture shows only four of the 46 chromosomes you have in your cells. The membrane (skin-like covering) holding the nucleus together now dissolves, and the chromosomes escape into the cell.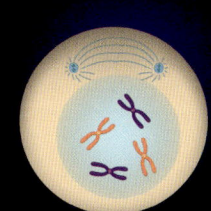

3. Now pretend the cell is like the earth. Two special structures have been moving to each "pole." Fibers are forming between the structure at the "north pole" and the structure at the "south pole." The fibers make a cage-like shape in the cell. The fibers also become attached to the chromosomes where the sisters are joined. Next, the chromosomes line up at the "equator" of the cell.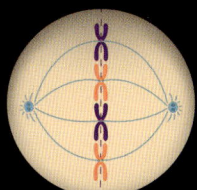

4. Then the sisters separate, and the fibers pull each sister to opposite "poles" of the cell.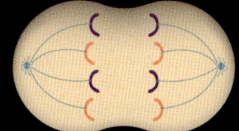

5. Next, two new membranes form around the two groups of chromosomes. This makes two new nuclei. The cage of fibers disappears. This is the end of mitosis.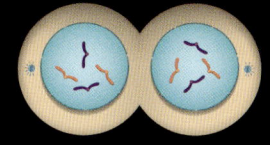

6. After mitosis, the membrane around the whole cell pinches inward until it separates the old cell into two new cells. Now their DNA looks like spaghetti again.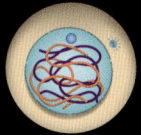

GOD MADE ME

 **Time to do Activity 7 in the Activity Book!**

Baby's Jobs in the Womb

Do you ever think about a baby when he or she is in Mom's womb? We often think babies are just waiting and growing with nothing to do until they are born. But God has given babies jobs to do. They prepare their mothers' bodies to welcome them, and they prepare themselves for their own future!

 How Baby Prepares Mom*

1. God makes our bodies able to fight things like germs when they get inside us. We recognize our own cells and leave them alone, but we treat other cells like enemies. Since a baby starts as a tiny creature different from Mom, Mom's body might try to fight it. But God solved that problem! As soon as Baby starts to be, he changes his mother's body in the area around himself so she won't think he's a germ to fight. The rest of her body is not changed. She can keep fighting real germs.
2. When Baby is about a week old, he attaches to the womb. Then Baby sends signals to Mom so no other babies will come at this time.
3. Baby sends signals to make Mom's womb grow bigger.
4. Baby sends a signal for Mom to make more blood. For every two drops of blood Mom had before she was pregnant, she will make one more while she is pregnant. Baby will need this extra blood to bring him food and oxygen.
5. When you turn a faucet to make more water come out, you are increasing the pressure of the water. Baby controls the pressure of Mom's blood during the whole pregnancy so he gets the right amount of blood.
6. Baby can make Mom's body turn food into energy more quickly. This is why pregnant mothers often feel hot when other people are chilly.
7. As Baby grows, Mom's skin will need to stretch. And some of her bones will need to spread apart to make room for the baby inside her. Baby makes a chemical called *relaxin* to make Mom's bone connections relax and her skin stretch.
8. Baby also sends signals for Mom to make milk so he will have something to eat when he is born!

Human Design: The Making of a Baby. Lecture by Dr. Randy Guliuzza. 2013. Dallas, TX: Institute for Creation Research, DVD.

CHAPTER 3: GOD KNITTED YOU TOGETHER

A mother's body gives a lot of its strength to care for the life of her unborn baby. After birth, a mother keeps giving of herself as she takes care of her baby. She gives and gives the rest of her life because she loves her child.

How Baby Prepares Herself

While God is knitting Baby together in Mom's womb, Baby begins to notice and learn things that will prepare her for life:

When Baby has lived in the womb for 11 weeks, she begins to push her feet against the womb when she happens to touch it. This helps her learn the stepping motion she will need for walking.

At 16 weeks, the Baby becomes aware of her body and where she is within the womb. She learns to control her motions according to what's around her.

At 24 weeks, Baby's senses develop. She can taste strong flavors that have gone from Mom's food into the liquid inside the womb. She hears Mom's heart beating and her stomach growling. She can recognize her mother's voice. She may turn toward a bright light shining on the outside of her cozy mother-home.

After six months inside Mom, Baby could live if she is born early. If early, she would need special care until her lungs start working well. But inside the womb, she will grow fatter. She will startle and

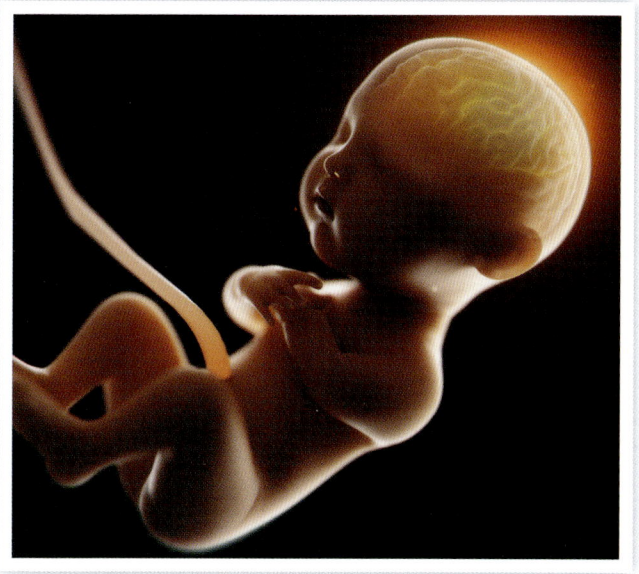

jump at sudden noises. She will sleep a lot and have dreams. She'll learn how to move her body and will practice breathing using the liquid around her. Baby will learn to suck and swallow so she will be ready to drink milk when she is born.

In the Bible, you can read about John the Baptist when he was still in his mother's womb. He had been growing there for about six months when something happened. That day, a visitor arrived and greeted his mother Elizabeth. This visitor was Elizabeth's cousin, Mary.

Mary had just been told by an angel that God had put baby Jesus in her womb. God put Jesus there even though Mary was not married. Baby Jesus was a special baby, so He started life in a different way— without a father. The Bible says Jesus would be called the Son of God. Mary was so amazed by this news from the angel that she rushed to see Elizabeth.

GOD MADE ME

When Mary arrived, she greeted Elizabeth. Suddenly, baby John the Baptist leaped in Elizabeth's womb! Do you think babies can know things before they are born? John did! Elizabeth was amazed. She told Mary that John had jumped inside her because he heard Mary's greeting. John heard the voice of the mother who was carrying Jesus, his Lord, in her womb. You can read about baby Jesus and baby John the Baptist in Luke 1 and 2!

 **Time to do Activity 8 in the Activity Book!**

How You Grew in Secret*

My frame was not hidden from You, When I was made in secret. (Psalm 139:15)

Sometimes parents can see pictures of their unborn babies. Doctors use a special camera called an **ultrasound** machine that allows parents to see pictures and videos of their baby at certain stages in the womb. Artists

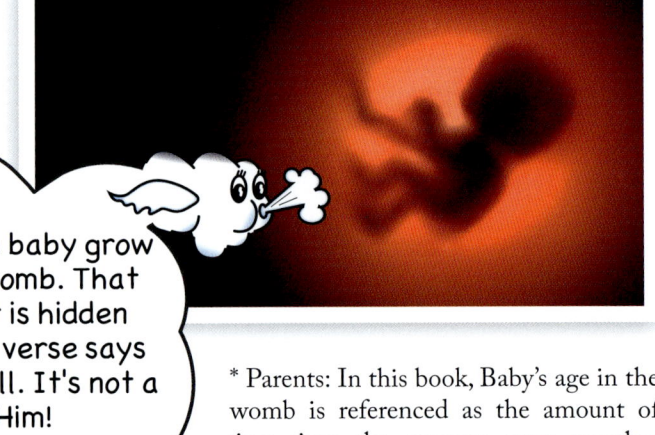

We can't watch a baby grow in his mother's womb. That wonderful sight is hidden from us. But this verse says that God sees it all. It's not a secret to Him!

* Parents: In this book, Baby's age in the womb is referenced as the amount of time since the gametes came together and life began. These ages may be two weeks less than other sources give for the same milestones. The reason for the difference is that, for practical purposes, these other sources often calculate Baby's age based on observable things happening in the mother's body two weeks before the gametes come together. The beginning of a baby's life is not observable to the mother.

38

CHAPTER 3: GOD KNITTED YOU TOGETHER

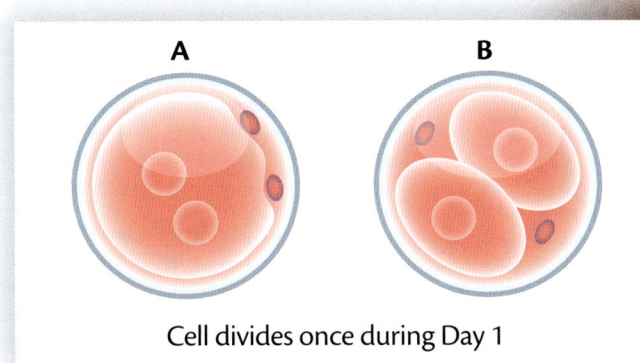

Cell divides once during Day 1

Day 4

have made realistic pictures of babies as they grow inside their mothers. Let's look at some pictures of babies at different ages:

Illustration A (above) shows the beginning of Baby's life.* Do you see the egg's nucleus and the nucleus from the father's cell? Their instructions are combining in the middle of the cell.

Only one day after Baby's life begins, the cell divides using mitosis. Two identical cells are made. (See B above.)

Over the next two days, the cells will keep dividing into matching cells using mitosis. Each of these cells has the ability to divide and make any kind of cell in the body. Cells that can do this are called **stem cells**.

At day six of Baby's life, the stem cells have formed a partly hollow ball. The group of stem cells on the *inside* of the ball will make the baby. The stem cells covering the outside of the ball now become attached to the womb. These *outside* cells will make the **placenta** (pluh-cent-uh). The placenta is a special structure that takes nutrition and oxygen from Mom's blood and transfers it to Baby's blood.

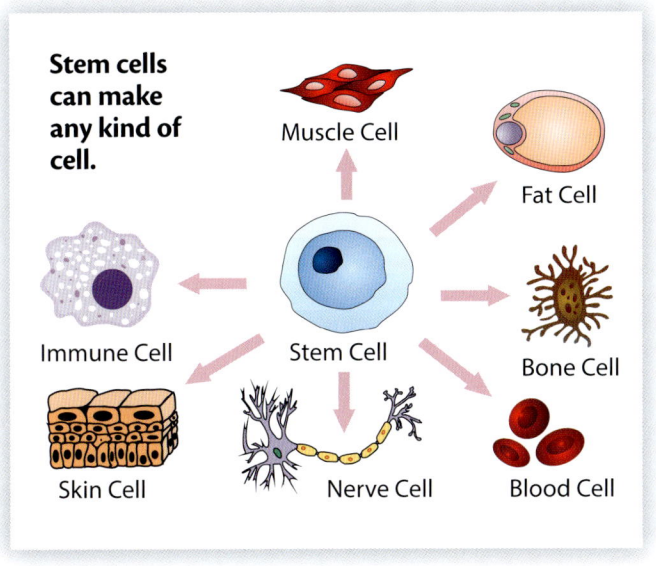

Stem cells can make any kind of cell.

*The beginning of life (when Mom's and Dad's DNA combines) is called **fertilization**. Fertilization is when a unique person begins. **Conception** is a word that once meant *fertilization*. But recently its meaning has been changed to mean any time from fertilization to a week later when the baby becomes attached to the wall of the womb. (From *Fearfully and Wonderfully Made* (Hebron, Kentucky: Answers in Genesis, 2021), 25.)

39

GOD MADE ME

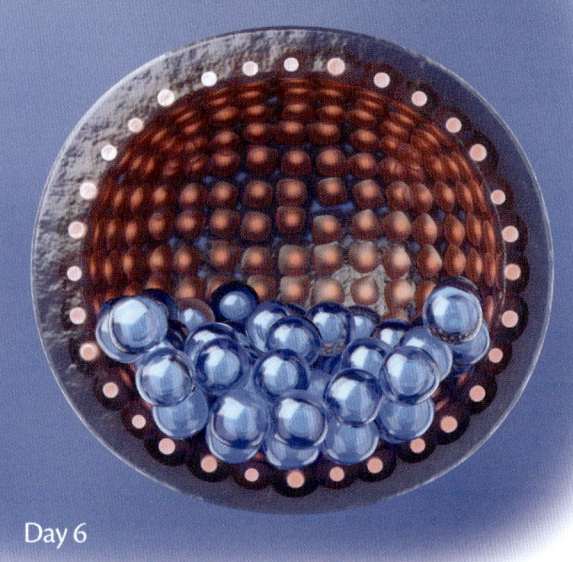

Day 6

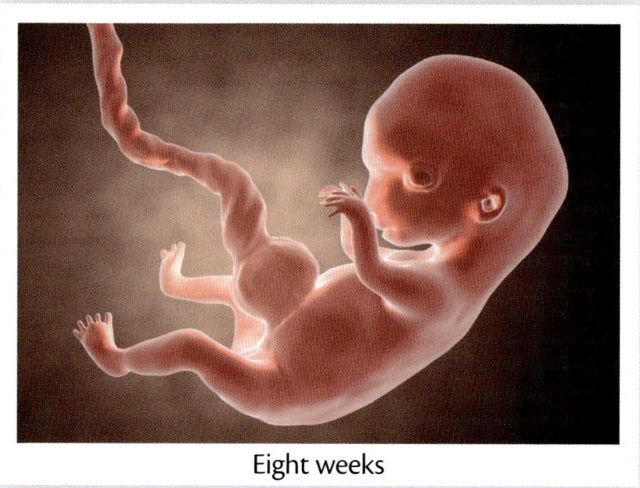

Eight weeks

At 22 days old, Baby's brain has been forming and his heart begins to beat with his own blood. In two days, his eyes and ears will begin to form.

By six weeks, Baby has hands, arms, feet, and legs.

By eight weeks, if we could see the baby, we could tell if it's a boy or a girl. Every organ is in place. Bones are hardening and Baby's ears can begin to hear!

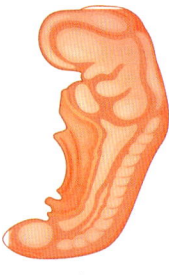

22 days

Week 11 of life: You can see Baby's **umbilical** (uhm-bill-uh-cull) **cord** attached to the placenta. Good things are transferred from Mom's blood to Baby's blood in the placenta. Then the good things travel to Baby through the umbilical cord. Wastes from Baby's blood travel back through the umbilical cord. They are transferred to Mom's blood in the placenta, and Mom's body gets rid of the wastes. Around this

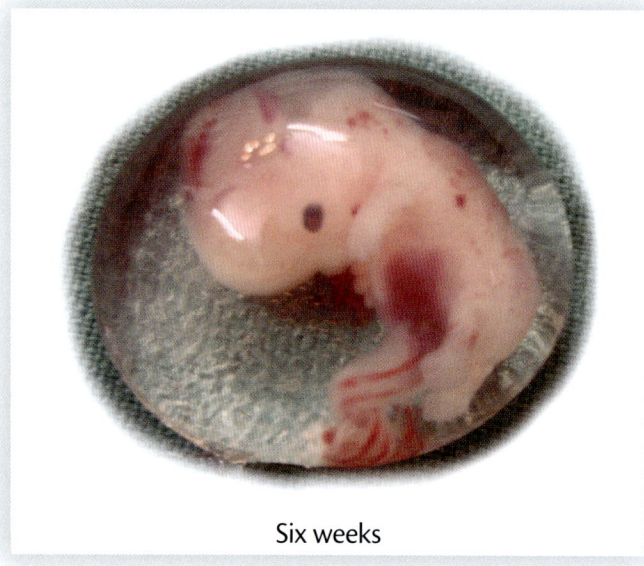

Six weeks

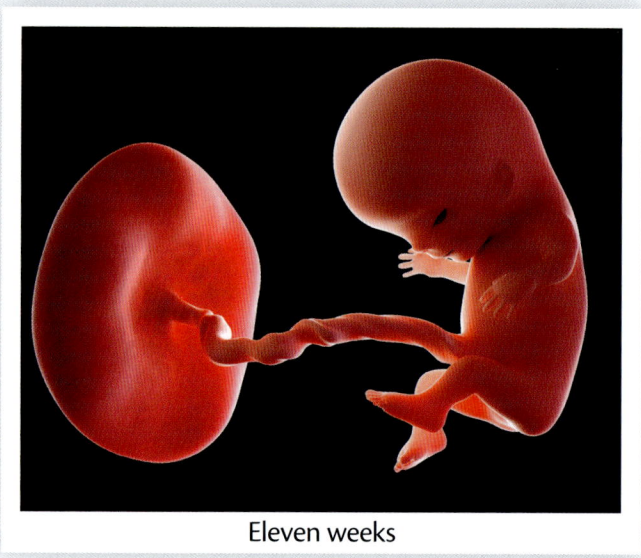

Eleven weeks

CHAPTER 3: GOD KNITTED YOU TOGETHER

time, Baby can suck his thumb, turn his head, frown, and hiccup. All organs are working, and teeth and fingernails are forming.

Week 20: By this time, Baby can move his joints, and Mom feels him kicking and stretching. He hears sounds from outside the womb and can even recognize his mother's voice! He tastes and swallows some of the liquid around himself, and his digestive system starts working.

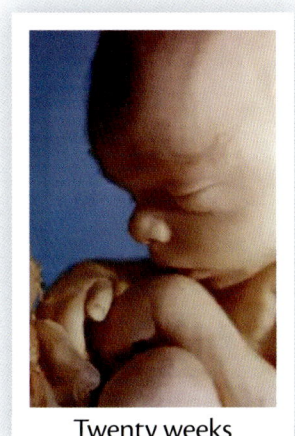
Twenty weeks

Week 26: Baby has started to grow hair on his head and fat under his skin. His brain is very active, and he can blink his eyes.

Weeks 37-41: Baby is ready to be born! His lungs have become ready to breathe air. He has been gaining about half a pound (.25 kg) in weight each week for the last two months.

As you do not know what is the way of the wind,
Or how the bones grow in the womb of her who is with child,
So you do not know the works of God who makes everything.
(Ecclesiastes 11:5)

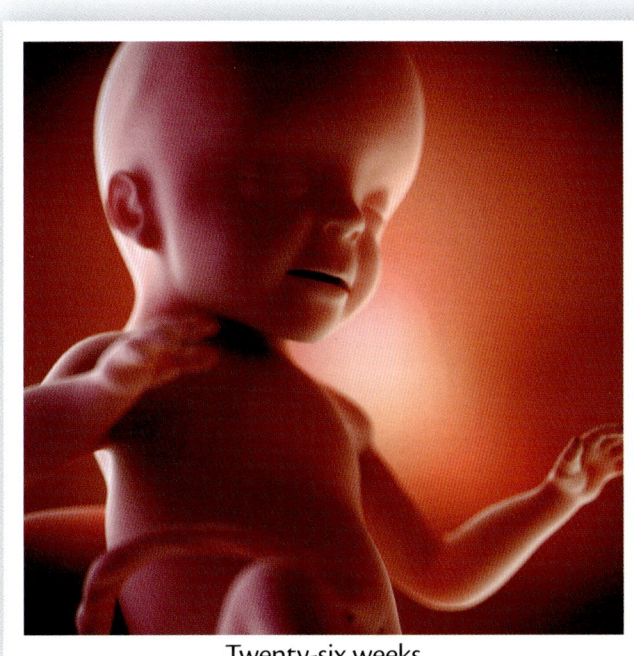
Twenty-six weeks

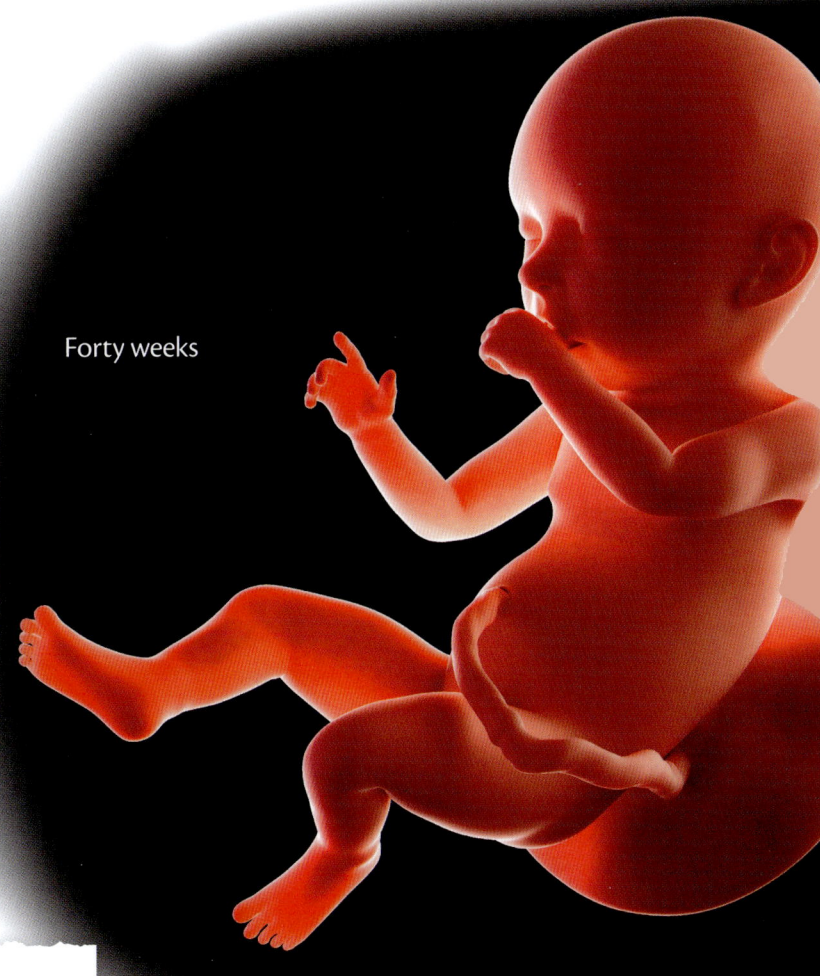
Forty weeks

GOD MADE ME

1 Month
Size of Red Currant

2 Month
Size of Cherry

3 Months
Size of Plum

4 Months
Size of Pear

5 Months
Size of Grapefruit

Thump's Health Hint

Mothers must keep their bodies healthy when they are pregnant. If they don't have the right nutrition, their babies may not become healthy children or adults.

- If a mom or dad are very overweight and eat a lot of sweets, they can get a disease that gives them too much sugar in their blood. This disease may change their DNA. Their children and grandchildren can easily get this disease because of DNA that doesn't act right.*

- Babies use a lot of good fats as their brains and eyes develop in the womb. These good fats are called omega 3 oils. Pregnant moms can get these fats by eating wild-caught salmon, walnuts, and flax seeds. Or they can buy supplements of omega 3 oils from a vitamin store. Since Baby needs so much omega 3, he will borrow what he needs from Mom's body. Mom may not feel very cheerful after her baby is born if Baby didn't leave her enough good fat.

Portha B, Grandjean V, Movassat J. *Mother or Father: Who Is in the Front Line? Mechanisms Underlying the Non-Genomic Transmission of Obesity/Diabetes via the Maternal or the Paternal Line.* Nutrients. 2019 Jan 22;11(2):233. doi: 10.3390/nu11020233. PMID: 30678214; PMCID: PMC6413176.

Prayer

It's wonderful how You knitted me together in my mother's womb, Lord. You skillfully formed my inward parts. I praise You! Amen.

CHAPTER 3: GOD KNITTED YOU TOGETHER

6 Month Size of Papaya

7 Month Size of Pineapple

8 Months Size of Cantaloupe

9 Months Size of Watermelon

Time to do Activity 9 in the Activity Book!

CHAPTER 4
A Time to Be Born

To everything there is a season, a time for every purpose under heaven: A time to be born . . . (Ecclesiastes 3:1-2)

A Big Change

You lived and grew for about nine months inside your mother. Your family waited a long time to meet you. They wondered if you would be born in the night or in the daytime, or maybe on Grandpa's birthday. Then, on the exact day that God decided, you were born! Your family changed in a special way when you became a part of them.

Things also changed for you when you were born. Before you were born, your little womb-room was filled with liquid. At birth, you went from living in water to living in air. You would never be able to live underwater again. Let's learn what happened to you when you made this change.

God planned the exact time you would be born! Let's see how your entrance into the world began.

- When you were in your mother's womb, your lungs were one of the last things to finish growing and be complete. When they were ready, your lungs sent a signal to your mother's womb: "Breathing will work now! Time for Baby to be born!"
- At birth, your body took the liquid out of your lungs and put it into your blood. Your body replaced the liquid in your lungs with oxygen to allow you to take your first breath.
- Before you were born, your blood received oxygen from the placenta. Your little heart took the blood from

GOD MADE ME

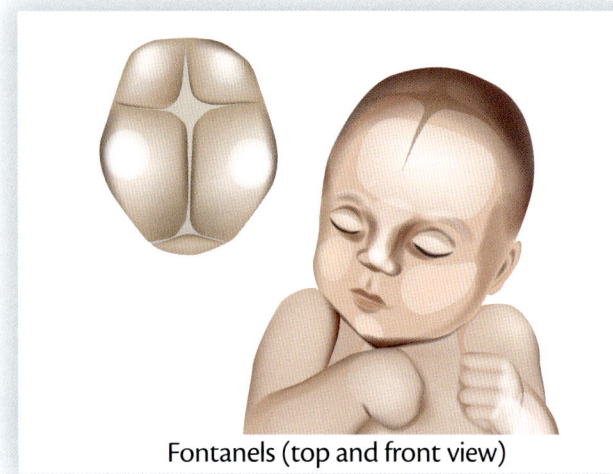

Fontanels (top and front view)

so they could supply your blood with oxygen. Your lungs have been doing this job ever since!

- In the womb, your head bones had not formed a skull yet. The bones were in separate pieces with spaces (**fontanels**) in between. This design allows a baby's head to change shape if it gets squeezed during birth. Sometimes you can see the skin on the top of a young baby's head moving up and down over one of these spaces. By the time a baby is one and a half years old, the bones have grown and closed these spaces.
- At the time you were born, the umbilical cord and placenta also left your mom's womb. Someone cut the umbilical cord and clamped or tied the end that was attached to you. (It didn't hurt you or your mom.) The piece of cord left on a baby will fall off in one to six weeks. All that's left is Baby's belly button!

the umbilical cord and pumped oxygen throughout your body. Your lungs weren't needed yet, so your blood went right past them. Your blood also took a shortcut through your heart.

- At your birth, the shortcut through your heart closed suddenly! This made your blood go to your lungs. But God had it all worked out. Your lungs had just begun to breathe air

Definitions

Your belly button is also called a **navel**.

Fontanels are the spaces between a baby's skull bones. We can sometimes see and touch these "soft spots."

Time to do Activity 10 in the Activity Book!

46

CHAPTER 4: A TIME TO BE BORN

Zoom in to the Cell

Often, the first thing your parents notice about you when you are born is whether you are a girl or a boy. What a wonderful surprise! Sometimes, though, parents like to know ahead of time so they can prepare clothes and a room for the baby. Parents can get this news early by asking their doctor to look at their baby in the womb with an ultrasound machine.

Of course, God planned for you to be either a girl or a boy long before you were in the womb. Let's see how He made this happen in the first cell that was you.

Remember, you have 23 pairs of chromosomes. These chromosomes are numbered by scientists. You have one chromosome of each number from your mom and a matching one of the same number from your dad. This is not quite true for Chromosome 23. Chromosome 23 is different from the other chromosomes because the one from the mom and the one from the dad don't match. Let's see what this means.

Your mom had two of Chromosome 23. They matched. They both looked like the letter X. We call them *X chromosomes*. You would have received one X from your mom in the gamete that became you.

Your dad also had two of Chromosome 23. They did not match. One looked like the letter X. The other looked kind of like the letter Y. We call that a *Y chromosome*. You would have received either an X chromosome from your dad or a Y chromosome from him.

If you received an X chromosome from Dad, you would have two X chromosomes for Chromosome 23. That means God made you a girl!

If you received a Y chromosome from Dad, you would have an X and a Y for Chromosome 23. That means God made you a boy!

GOD MADE ME

Time to do Activity 11 in the Activity Book!

Growing Up

> By You I have been upheld from birth;
> You are He who took me out of my mother's womb.
> My praise shall be continually of You.
> (Psalm 71:6)

This verse shows us that God brought us safely out of the womb and has upheld (or taken care of) us ever since. When you were first born, you could not take care of yourself at all. But God gave you parents to care for you.

Most animal babies are able to walk, swim, or fly soon after they are born. They can also take care of themselves even when they are still young. God made animal babies **develop** in the womb a little more than human babies. This is important because animal parents don't take care of their babies the same way people do. Human babies can't walk until they are about a year old. During that year, they need to be carried, fed, bathed, and have their diaper changed. Why are human babies born so helpless? The reason has to do with the size of our heads!

God made people different than

This baby donkey was just born!

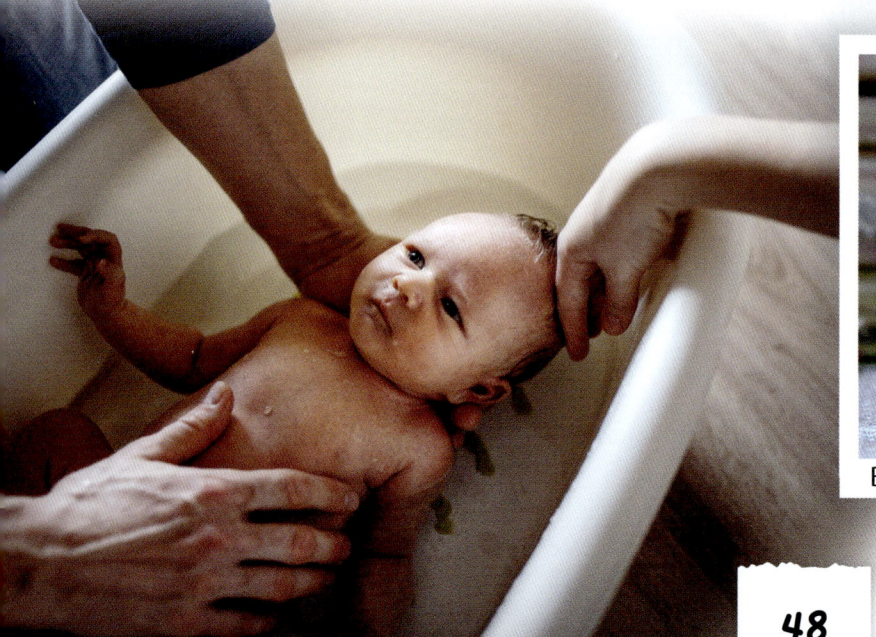

Baby birds can fly about three weeks after hatching.

A mother whale pushes her newborn calf to the water's surface for its first breath of air. After that, the calf can swim by itself!

animals. We have bigger brains and need bigger heads to hold our brains. Your head is a large part of your body. An animal's head is a smaller part of its body. Human babies also have bigger heads than animal babies. A baby's head and body would have to grow quite large if she stayed in the womb until she was able to walk! God knew that the pregnancy and birth of such a large baby would be too hard for the mother.

> God wants people to have loving relationships! We begin to understand love as our parents take care of us when we are too helpless to care for ourselves. This teaches us how much God cares for us too!

If a human baby wasn't born until she was as developed as a newborn calf, she would be born as a 21 pound (9.5 kg) toddler!

49

GOD MADE ME

Definitions

To **develop** (or **mature**) means to grow and become able to do more.

Before a baby walks, she usually crawls. Crawling helps her mind develop as she uses the different sides of her body. She is so excited to be able to go places and explore on her own. Exploring helps her learn.

Crawling also helps her neck bones form into the curve shape she'll need later in life.

Babies usually start walking by first crawling to a piece of furniture where they can pull themselves up to stand. Eventually, they take steps, holding onto the furniture and moving around it. One day, they get interested in something farther away and take a few steps toward it without holding on. They are usually a little surprised and happy with their accomplishment!

You are still getting taller and stronger. You are quickly learning new things. Even though you will someday stop growing taller, you will keep growing in many ways. Your hair and nails will grow for the rest

CHAPTER 4: A TIME TO BE BORN

of your life. Cells will die, and new ones will grow in their place. When you get a cut, new skin will grow. If you break a bone, new bone will grow to join the broken pieces together again. God has given you an amazing body!

Prayer

How amazing, God, that babies start breathing at just the right time! And their blood suddenly changes its route to keep babies alive as they are born. You bring them safely into the world. You have also kept us ever since! Thank You for helping me live and grow. Amen.

Time to do Activity 12 in the Activity Book!

UNIT 2
God Gives You Oxygen

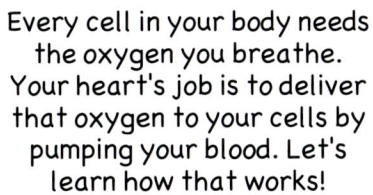

Every cell in your body needs the oxygen you breathe. Your heart's job is to deliver that oxygen to your cells by pumping your blood. Let's learn how that works!

God gave you the breath of life! He gives you life-giving oxygen each time you take a breath of air into your lungs.

Memory Verse

Let everything that has breath praise the Lord. Praise the Lord! (Psalm 150:6)

Hymn to Sing: All Creatures of Our God & King

You can listen to this hymn by searching for "All Creatures of Our God And King" on the internet.

All creatures of our God and King,
Lift up your voice and with us sing
Alleluia, alleluia!
Thou burning sun with golden beam,
Thou silver moon with softer gleam,

Chorus: O praise Him, O praise Him,
Alleluia, alleluia, alleluia!

Thou rushing wind that art so strong,
Ye clouds that sail in heav'n along,
O praise Him, alleluia!
Thou rising morn in praise rejoice,
Ye lights of evening, find a voice,
(Chorus)

Thou flowing water, pure and clear,
Make music for thy Lord to ear,
Alleluia, alleluia!
Thou fire so masterful and bright,
That gives to us both warmth and light
(Chorus)

And all ye men of tender heart,
Forgiving others, take your part,
O sing ye, alleluia!
Ye who long pain and sorrow bear,
Praise God and on Him cast your care.
(Chorus)

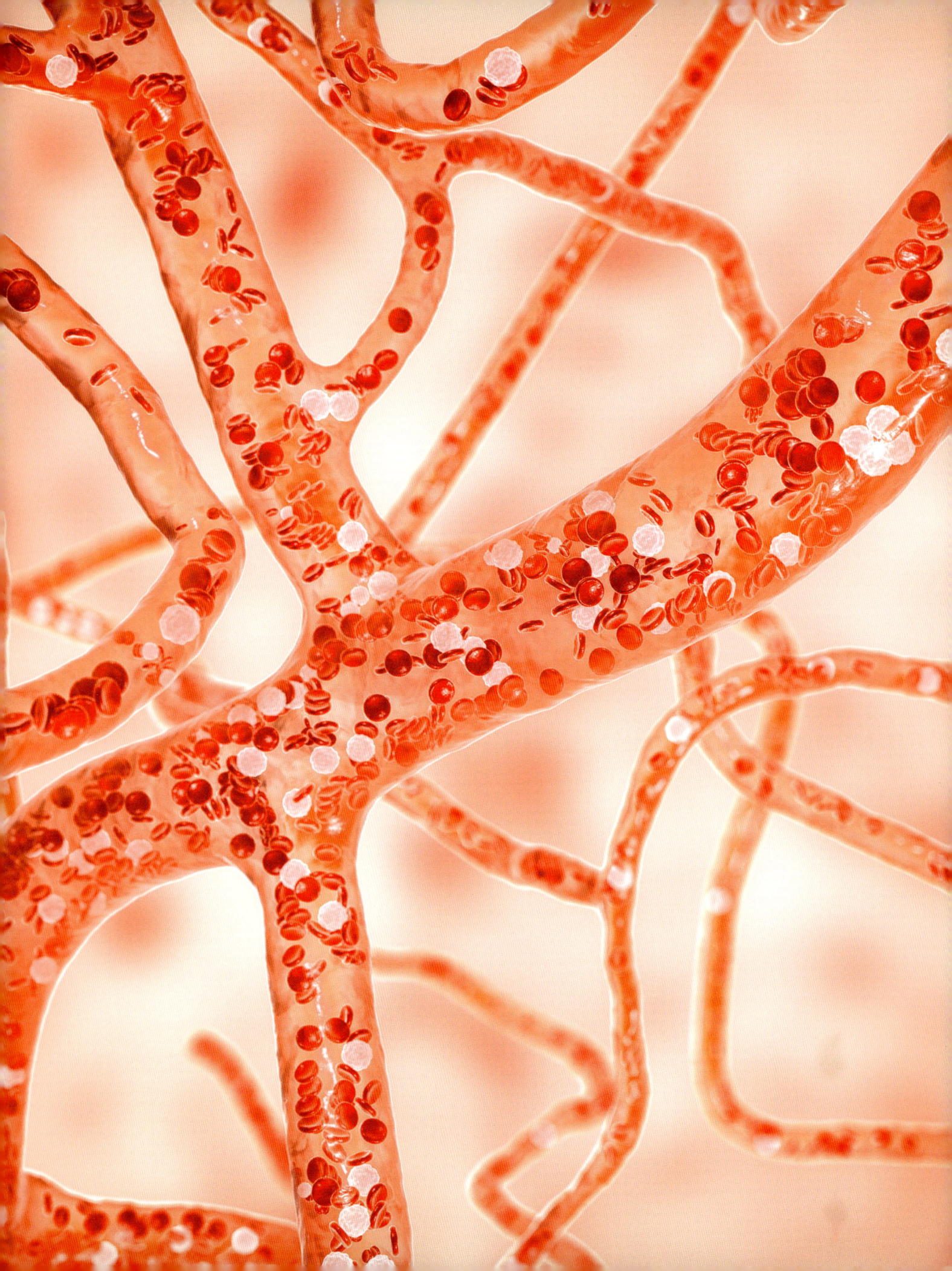

CHAPTER 5
God Gave You Blood

For the life of the flesh is in the blood. (Leviticus 17:11)

Blood Carries Oxygen

You started learning soon after God created you. You learned by noticing things in your mother's womb. You noticed and learned more after you were born. Then, when you could talk, you learned more by asking questions about the things you noticed. When you were old enough, you learned even more by doing your schoolwork. Do you like learning about the things God created? If so, you like learning science!

As we study science, we can see God's wisdom. We see that He makes things work well. We also see that He makes things work together.

Here are some things you may have learned in science about God's wisdom:

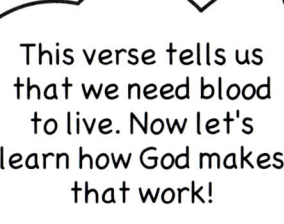

This verse tells us that we need blood to live. Now let's learn how God makes that work!

- Earth is the perfect distance from the sun. This makes life possible!
- Plants provide the right food and vitamins for people and animals all over the world.
- Air near the earth's surface has the perfect amount of oxygen gas for people and animals to safely breathe. Air that's farther away from Earth has different gases that protect us from harmful light.
- Plants make oxygen for people and animals to breathe. People and animals make carbon dioxide for plants to use.

GOD MADE ME

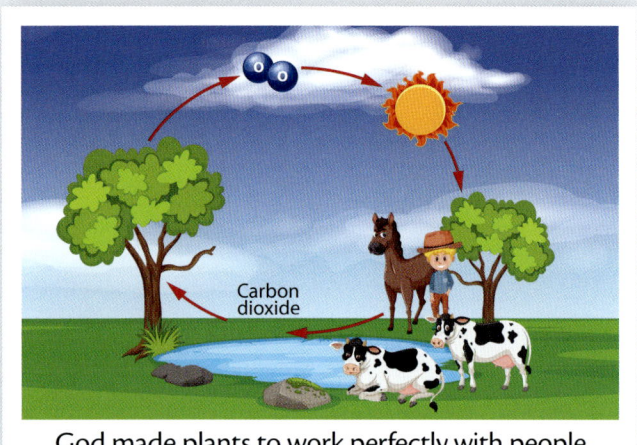

God made plants to work perfectly with people and animals.

Whenever you study God's creation, you will see the way He makes things work together perfectly. As we learn about the human body, you will see how all your parts were designed to work together perfectly too!

In this unit, we will learn how God formed your **heart**, **lungs**, and **blood** to do a special job. They work together to deliver oxygen to cells all over your body. To be alive, you need all three of these special things!

Let's start by learning about blood. Blood has many jobs to do in your body. In this chapter, we are going to learn about blood's important job of carrying **oxygen** to cells.

As your heart beats, it pumps your blood through little tubes inside you. These tubes are called **vessels**. Your blood vessels started to form only 18 days after you began life in your mother's womb. God made three kinds of blood vessels in your body. Thump will show you the three different kinds.

Definitions

Blood is the liquid that carries oxygen and other good things to the cells in your body. It also carries wastes away from your cells so your body can get rid of them.

Oxygen (O_2) is a gas in the air we breathe. We need oxygen for all the jobs our cells do. Plants make oxygen but they don't need it. They put it out into the air for us to breathe! Two oxygen atoms make the oxygen gas molecule, O_2.

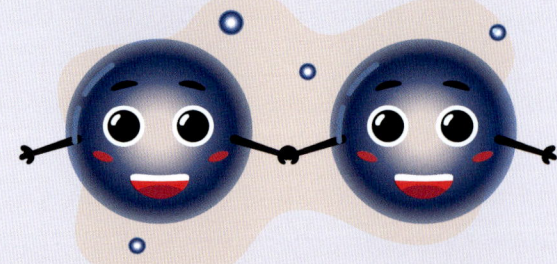

Carbon dioxide (CO_2) is a gas our bodies make. But we don't need it. Plants do need carbon dioxide. We breathe it out into the air for plants to use!

The sun gives the perfect amount of light and warmth for life on Earth.

56

CHAPTER 5: GOD GAVE YOU BLOOD

Your Blood Vessels

1. **Arteries** are blood vessels that carry blood away from your heart. Arteries near your heart are stretchy. They stretch to make room for the big squirts of blood that come along with every heartbeat. They return to normal size between beats. Arteries near your heart are large. But they divide into smaller vessels as they get farther away from your heart. The smaller arteries far from your heart don't need to be stretchy. Instead, they have **muscles** to help control how much blood goes through them. Muscles are groups of cells that can get shorter or longer to help control movement in your body.

2. **Capillaries** (cap-ill-air-eez) are the smallest blood vessels. Your smallest arteries branch into a lot of capillaries. Capillaries are where your red blood cells give away their oxygen. Almost every cell in your body is next to (or very close to) a capillary. Capillaries have walls that are only one cell thick. Oxygen from your blood goes through these thin walls to your body's cells. These cells use the oxygen to do their jobs. As they do their jobs, your body's cells make a waste called **carbon dioxide**.

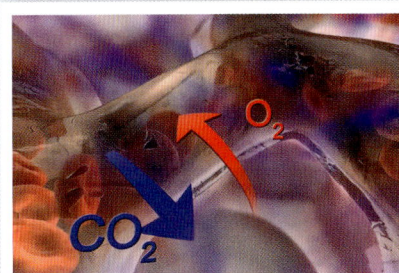
This capillary is trading gases through its thin wall.

Things happen in reverse when your blood flows back to your heart. The carbon dioxide waste goes from your body's cells into your capillaries. The capillaries take the waste to your veins.

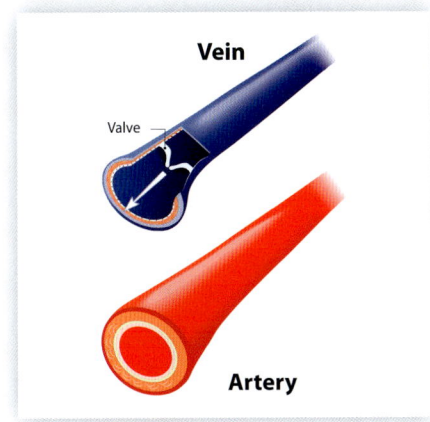

3. **Veins** are blood vessels that take blood back to the heart. First, capillaries come together to make your smallest veins. The small veins join with others to make larger veins. Veins become larger as they get closer to your heart. Veins don't need to stretch when your heart beats because they don't get the big squirts of blood that arteries do. Veins could have a different problem: gravity is always trying to pull the blood down. But God has put little **valves** inside the veins. A valve is a structure that keeps liquids going only one direction. The valves in veins are like trap doors that only swing open one way. After they swing open to let blood through, they slam closed so the blood won't run back the wrong way.

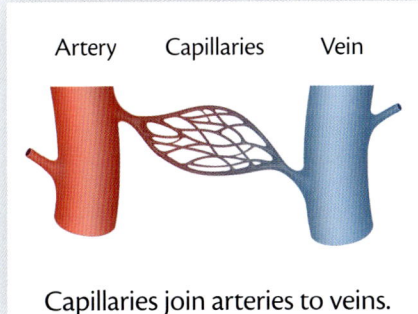
Capillaries join arteries to veins.

57

GOD MADE ME

Thump's Health Hint

When you move, your muscles squeeze your veins and help push the blood back to your heart. This is good for you. Get lots of exercise and try not to sit too long without moving!

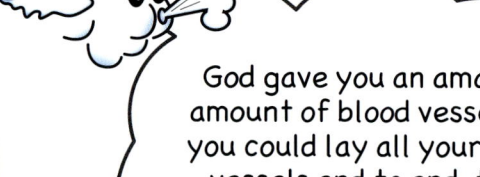

God gave you an amazing amount of blood vessels. If you could lay all your blood vessels end to end, they would go around the earth about 2½ times!

Time to do Activity 13 in the Activity Book!

What's in Blood?

You have probably seen your own blood. Have you ever had a paper cut, a nosebleed, or a scraped knee? Sometimes blood seems yucky or scary because it's coming out of our body when we think it's not supposed to. But blood is a wonderful liquid. God says that blood gives us life! Let's learn about the important things blood is made of.

If you could separate your blood into its different parts, here's what you would see (following page):

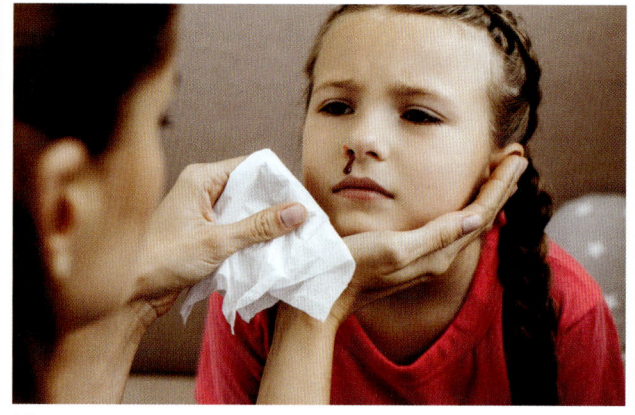

The Design of Blood

- White Blood Cell
- Platelets
- Red Blood Cells
- Plasma

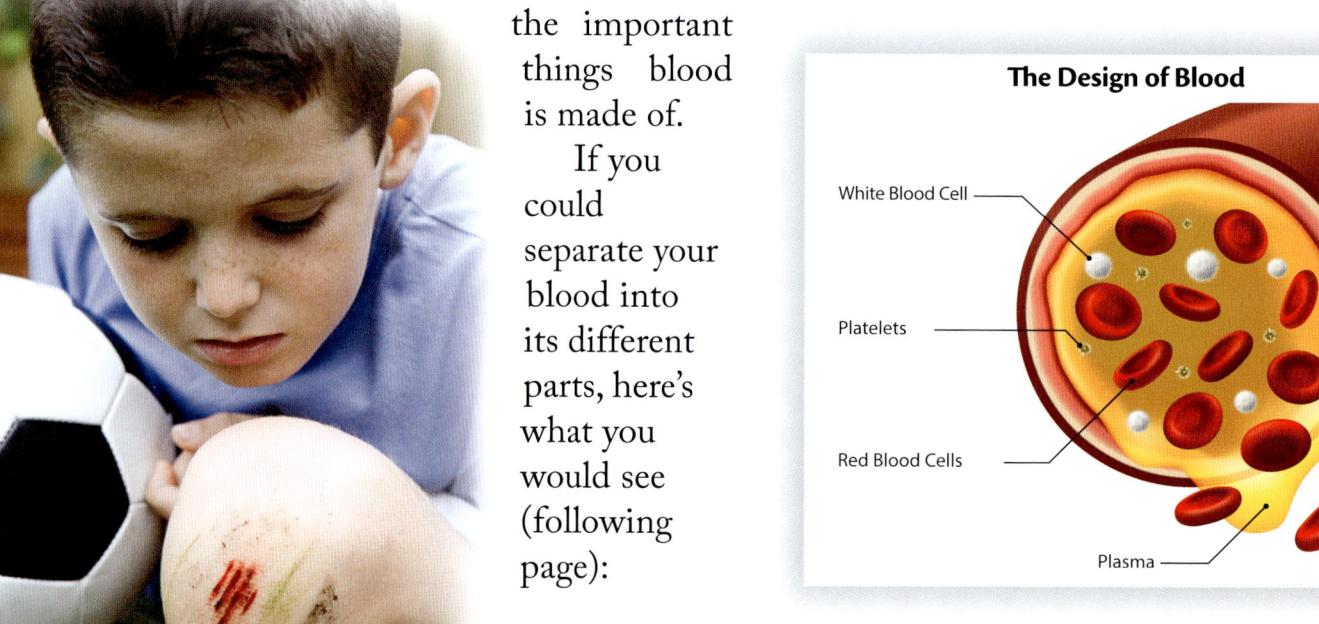

58

CHAPTER 5: GOD GAVE YOU BLOOD

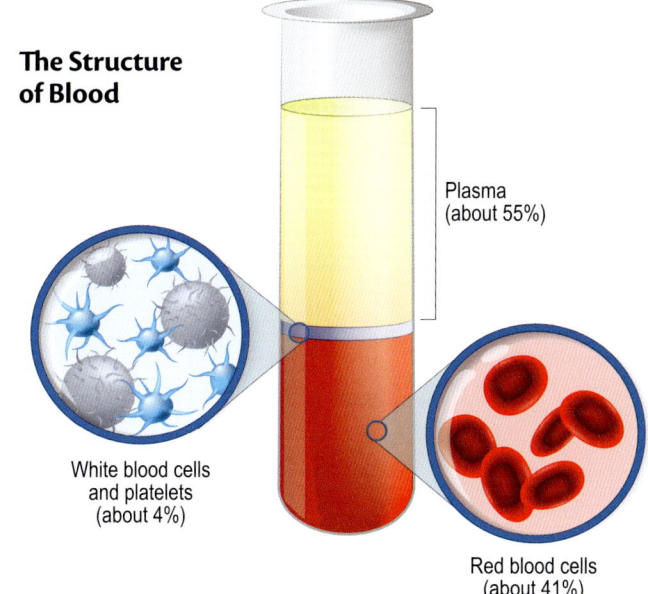

The Structure of Blood

Plasma (about 55%)

White blood cells and platelets (about 4%)

Red blood cells (about 41%)

 Plasma — a clear, yellowish liquid that contains all the blood cells. About half of your blood is plasma. Plasma is the liquid that carries the oxygen out of your capillaries and into your body's cells. Carbon dioxide gets back into the capillaries by riding in the plasma as it goes back into the capillaries.

 Red blood cells — flat, disk-shaped cells that give blood its red color. About half of your blood is red blood cells. These cells are red because of a chemical called **iron**. Iron is part of a molecule that can hold onto oxygen. When it has oxygen (O_2), the cell is bright red. When the molecule gets to a place in the body that needs oxygen, it gives the oxygen away. Then it picks up carbon dioxide (CO_2). Without O_2, the cell turns dark red instead of bright red.

 White blood cells — cells that help defend your body against germs.

 Platelets (plate-lets) — pieces of cells that help your blood to clot and stop you from bleeding too much. Platelets and white blood cells are only a small part of your blood. But they are important! We will learn more about them later.

Unless you are a horseshoe crab, your blood is never blue. Sometimes we think human blood is blue because veins look blue through light skin. But that's just how the dark red color looks when seen through skin.

Don't be confused by science drawings that have blue blood vessels. These drawings color certain vessels blue to show that their blood is low in oxygen. Real blood is dark red when it's low in oxygen.

 Time to do Activity 14 in the Activity Book!

GOD MADE ME

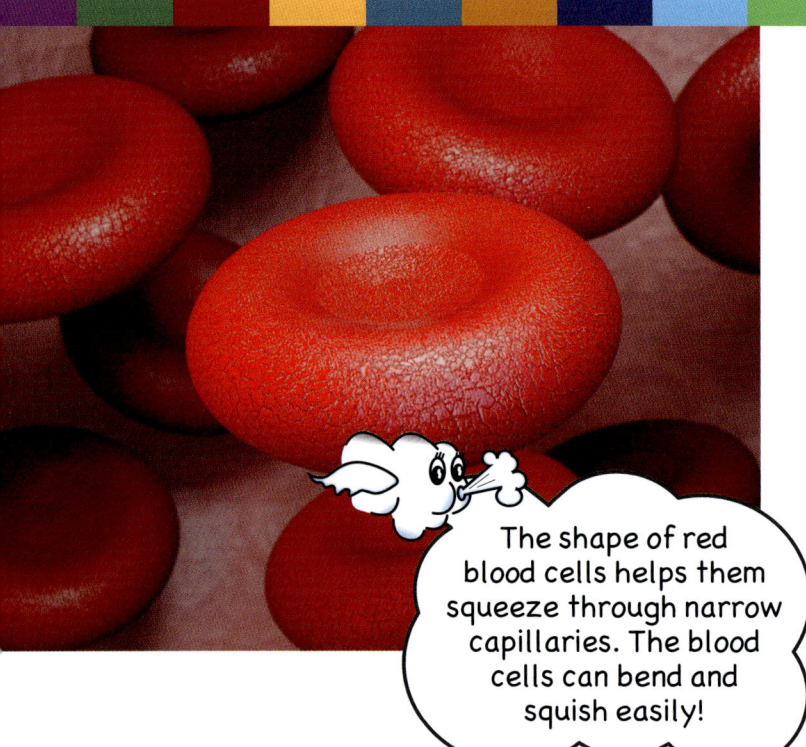

The shape of red blood cells helps them squeeze through narrow capillaries. The blood cells can bend and squish easily!

once they go into the blood, their nuclei are gone! In fact, they don't have any of the tiny parts that other cells have. They only have **cytoplasm** (sigh-toe-plazm) and **cell membranes**. Since a red blood cell's job is to carry oxygen and carbon dioxide, God has saved all its inside space for that job.

Each red blood cell lives for about four months. Every second, two million of your red blood cells die. But every second, two million new red blood cells are made!

Definitions

Cytoplasm is the liquid inside a cell.

A **cell membrane** is the skin-like covering that holds the cell together.

Zoom in to the Cell

Cytoplasm — Nucleus — Ribosomes — Cell membrane

Human Cell

The three kinds of blood cells inside you are made differently than most other kinds of cells in your body. They don't divide with mitosis. Instead, they are formed in the middle of some of your bones!

Blood cells grow inside something called **bone marrow**. As they are growing inside the bone, red blood cells have nuclei. But

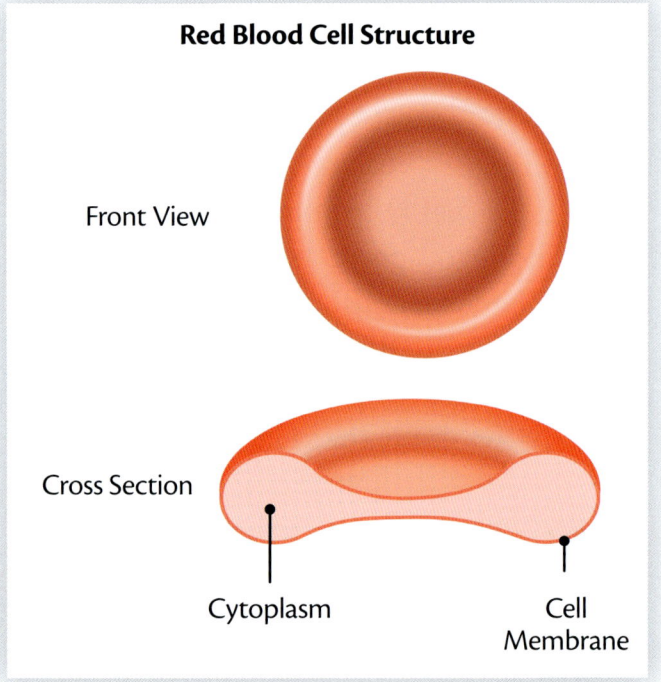

Red Blood Cell Structure

Front View

Cross Section — Cytoplasm — Cell Membrane

60

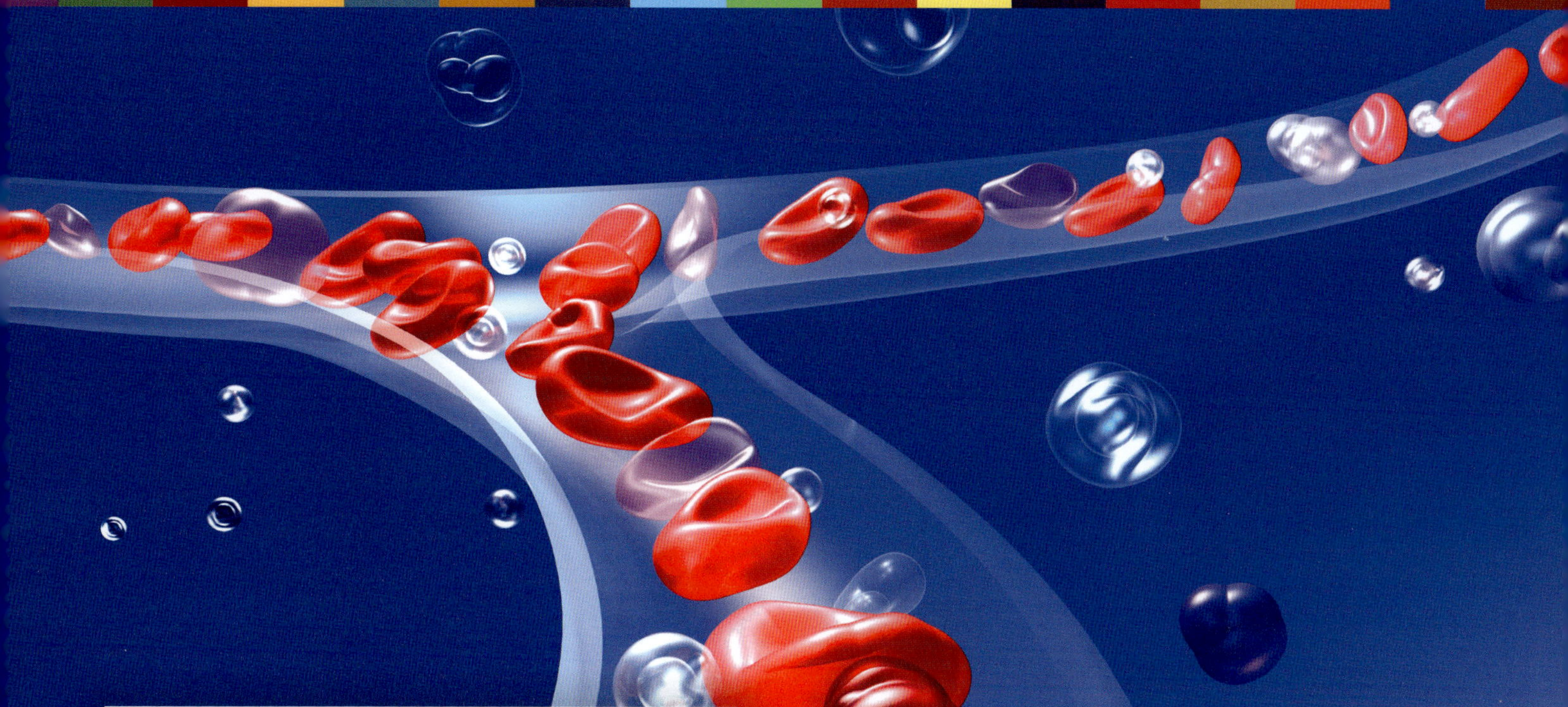

CHAPTER 5: GOD GAVE YOU BLOOD

Inside a Bone

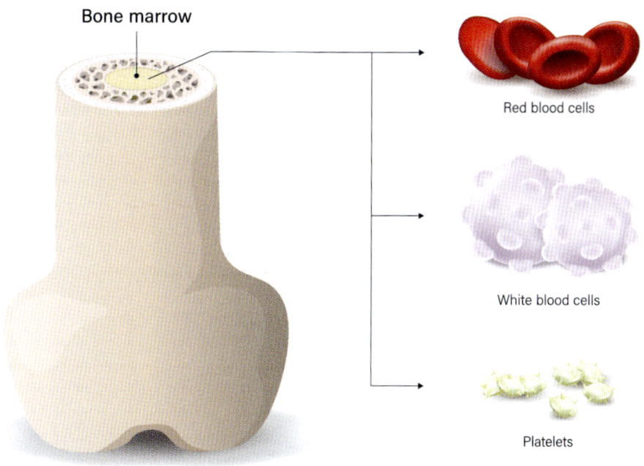

All types of blood cells are made in bone marrow.

Prayer

Lord, we praise You for giving us life through our blood. Our blood cells and vessels are perfect for taking life-giving oxygen to our cells. They also work perfectly to take away the carbon dioxide our cells don't need. Thank You! Amen.

Time to do Activity 15 in the Activity Book!

CHAPTER 6
God Gave You A Pump

Create in me a clean heart, O God, and renew a steadfast spirit within me. (Psalm 51:10)

How God Formed Your Heart Pump

What if you were a tiny, one-celled pond creature? If you were, you would be able to absorb all your oxygen from the liquid you were living in. Actually, when you were a tiny, new person in your mother's womb, you did absorb oxygen that way. But God knew you wouldn't stay tiny like a pond creature.

Three things happened when you were 18 days old in your mother's womb.

1. You were given blood vessels.
2. You were given your very own blood, separate from your mother's blood.
3. God started making your **heart**.

Four days later, you had become too big to absorb oxygen from the liquid around you. So, your new little heart began to beat! It began to **pump** your very own blood to your cells.

Doctors can use ultrasound machines to listen to a baby's heartbeat when the baby is only a few weeks old. When a baby is 18 to 20 weeks old, his heartbeat can be heard with a stethoscope like this one.

63

GOD MADE ME

Inside your body, you have things called **organs**. Each organ has a certain shape. Each one has a certain place to live inside you. And each organ has a certain job. Your heart was the first organ to begin working inside you.

But do you wonder how your heart became a heart? Why didn't it become your foot? If your body started as a ball of cells that were all alike, how did your body become a group of so many different parts? And how can it be that all the people on earth have their body parts arranged the same way? Only God knows the complete answers to these questions. But as scientists have studied, they have found out some very interesting things:

1. Do you remember learning that your first cells were all alike? They each had a copy of your DNA. They were called *stem cells* because each one had the ability to divide and make any cell in your body.
2. Do you also remember that your early stem cells formed into a partly hollow ball? It was made of outside cells and inside cells. At the right time, the inside stem cells suddenly began to move! God used mysterious ways to sort these moving cells into three layers. These three layers became you!
3. After the three layers were formed, God helped each stem cell know which layer it was in. Each cell also knew if it was on your right, left, back or front side.

Definitions

A tire **pump** is a machine that uses pressure to force air into a flat tire.

Your **heart** is a pump that forces blood to move through your body. We say your heart *pumps* your blood.

An **organ** in your body is something with a certain shape, a certain place, and a certain job to do. Your heart is an organ.

4. Now that your stem cells knew where they were, they started dividing to make certain body cells. Cells in the outer layer became your skin and brain cells. The middle layer became your muscle, bone, heart, and red blood cells. The inside layer became your stomach and lung cells.
5. How did your stem cells know that they should divide into cells that were different? Did their DNA

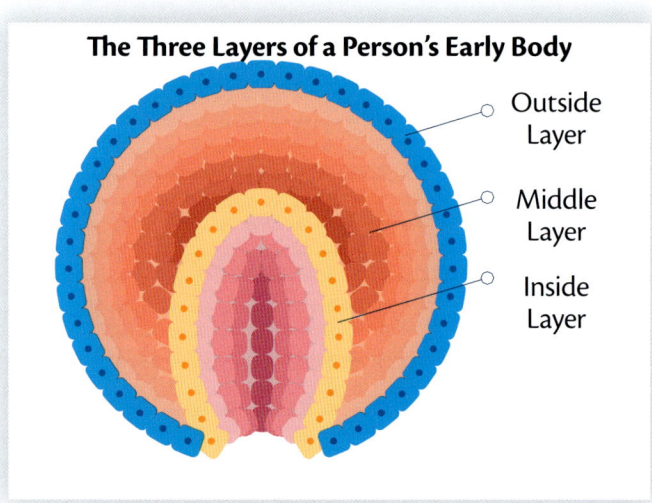

The Three Layers of a Person's Early Body

- Outside Layer
- Middle Layer
- Inside Layer

64

CHAPTER 6: GOD GAVE YOU A PUMP

change? No, they all still had the same copy of all your DNA. But another mysterious thing happened after the three layers were formed. If cells ended up in the right place to become your heart, heart-making instructions were switched on in their DNA. Those new heart cells still had DNA for all other instructions, but only the heart instructions were switched on. The same thing happened to all the other stem cells. The correct instructions were switched on according to where the cells were on your new little body! God made you work this way.

 **Time to do Activity 16 in the Activity Book!**

Get to Know Your Heart

Let's see what your heart looks like inside you now! Your heart lives in your chest. It's not exactly in the center. Your heart is a little to the left of the center of your chest.

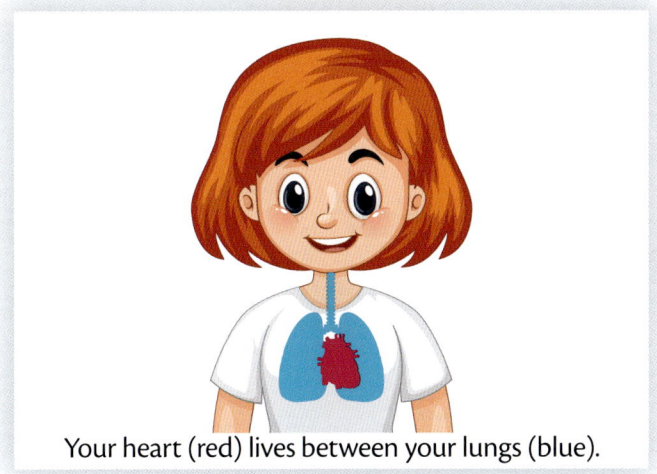

Your heart (red) lives between your lungs (blue).

"Then I will give them a heart to know Me, that I am the LORD; and they shall be My people, and I will be their God, for they shall return to Me with their whole heart." (Jeremiah 24:7)

Your heart pump lives deep inside your chest. In the Bible, God often talks about the heart. But sometimes He means something different than your heart organ.

When God talks about your heart, He means who you are on the inside. Your heart is where you love and care about the things that are important to you.

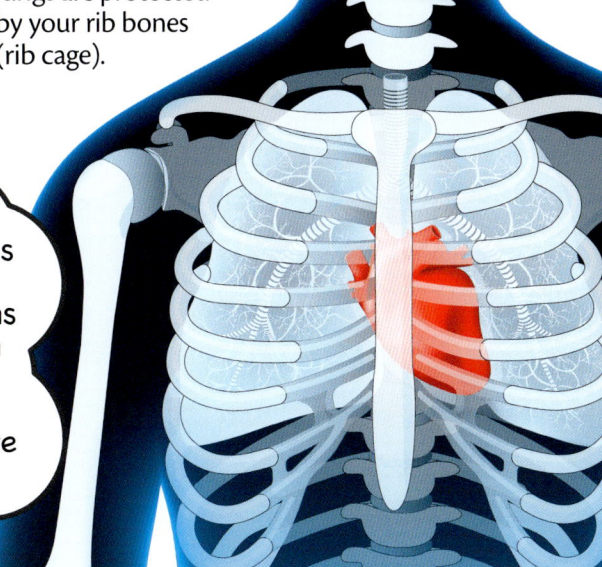

Your heart and lungs are protected by your rib bones (rib cage).

GOD MADE ME

If we could see the outside of your heart, we would see lots of big blood vessels attached to it. Some of these vessels are arteries coming out of the heart. Some are veins going in. These big vessels are doing their job for your *body*. They are carrying oxygen or carbon dioxide to your body's cells.

But we would also see smaller vessels fanning out on the surface of your heart. These vessels are arteries, veins, and capillaries doing their job for your *heart*. Just like all the cells in your body, your heart cells need to have oxygen. Without oxygen coming from these smaller vessels, your heart could not beat.

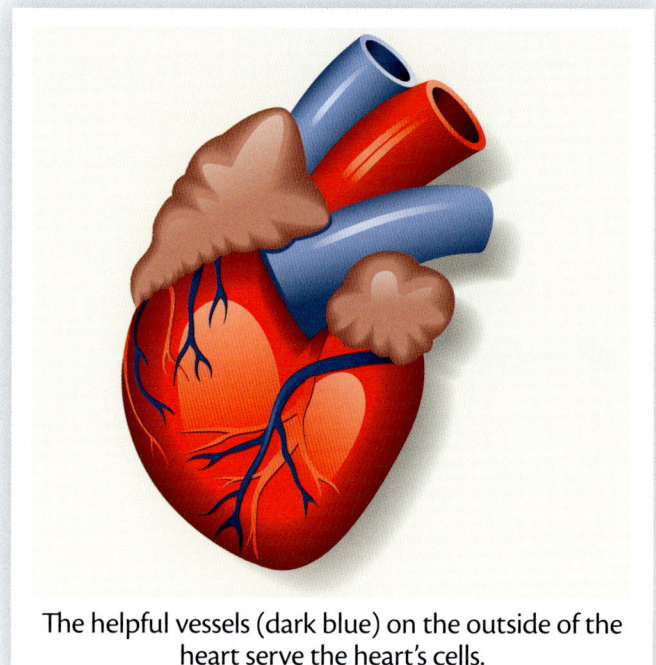

The helpful vessels (dark blue) on the outside of the heart serve the heart's cells.

Thump's Health Hint

Children like you usually have healthy hearts. The vessels that fan out over your heart are able to let plenty of blood through. Your heart cells get all the oxygen they need to do their job well.

But sometimes, as people get older, the vessels on their heart change. They can become narrow and not let enough blood through. Then the heart can't work properly, which is dangerous. Here are some things that will help your heart's important vessels work well.

- Eat healthy foods.
- Don't eat a lot of sweets or other unhealthy foods.
- Exercise often by doing activities that make you breathe hard.
- Don't smoke.
- Brush your teeth. The same germs that can cause sick gums can also harm your heart.
- Get plenty of sleep.
- Trust God. Our hearts can be healthier when we aren't worried and hurried.

Time to do Activity 17 in the Activity Book!

CHAPTER 6: GOD GAVE YOU A PUMP

Inside Your Heart

Your heart is made of muscle cells. These muscle cells are different than other muscles in your body. Usually, you tell your muscles what to do. Look at your legs. Now swing them back and forth. You decide when you want to swing your legs. Leg muscle cells need your brain to tell them to get to work. But your heart muscle doesn't need your help. Your heart squeezes automatically. You don't have to decide to make your heart pump. Its cells get electric signals to start each beat. God is very wise to make your heart do its job without you having to think about it.

Your heart is divided into two sides—the right side and the left side. Each side has a space on top where blood comes into the heart from somewhere else. The top space on the right side of your heart is called the **right atrium** (ay-tree-um). Blood from the body comes into the right atrium.

The top space on the left side of your heart is called the **left atrium**. Blood from the lungs comes into the left atrium.

There are two more spaces below the atriums. These spaces are called **ventricles** (ven-trih-culls). Blood comes into each ventricle from the atrium above it. Then the blood goes out of the ventricles to other places. The blood from the **right ventricle** goes to your lungs. Blood from the **left**

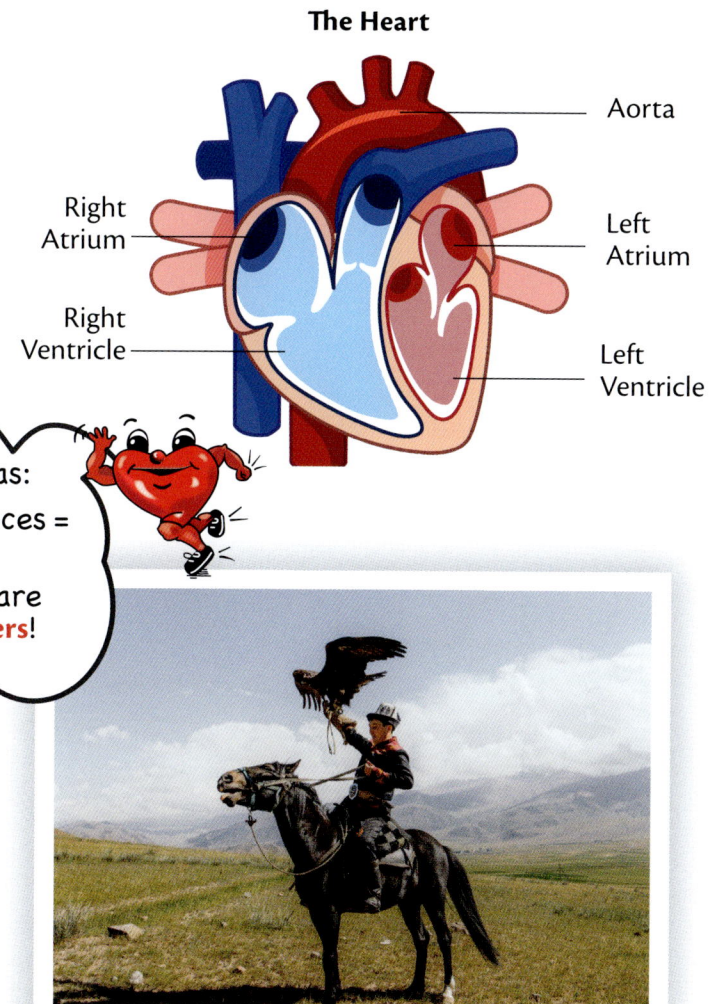

Your heart has:
2 spaces + 2 spaces = 4 spaces
These spaces are called **chambers**!

Humans, birds, and mammals all have hearts with four chambers.

Snakes and frogs have hearts with three chambers.

67

GOD MADE ME

A fish's heart has only two chambers.

Definitions

A **system** (siss-tum) is a group of things working together to do a certain job.

Your heart, blood vessels, and blood make up your **circulatory** (sir-kyou-luh-tore-ee) **system**. "Circulatory" is like the word "circle." We use this word because the blood keeps going around and around in your body like in a circle!

The "lub-dub" sound of the heart is caused by valves closing!

Human heart showing valves (green)

ventricle goes to your body.

Inside your heart are four special structures that keep blood going the right direction. These four structures are called **valves**. Just like with veins, these valves are doors that can only open in one direction.

God made your heart work **efficiently**. This means that it doesn't waste time or energy. In picture A (next page), you can see that both ventricles are squeezing different blood out at the same time. From the right ventricle, the blood is being squeezed through the blue vessels to the lungs. From the left ventricle, blood is being squeezed through the red vessel to the body. This red, hook-shaped vessel is the **aorta**.

In picture B, you can see that the ventricles are relaxed. The atriums are squeezing different blood into both ventricles at the same time. The heart has had a short rest between beats—*lub-dub*,

CHAPTER 6: GOD GAVE YOU A PUMP

rest, lub-dub, rest. During each rest, the right atrium receives blood from the body. The left atrium receives blood from the lungs.

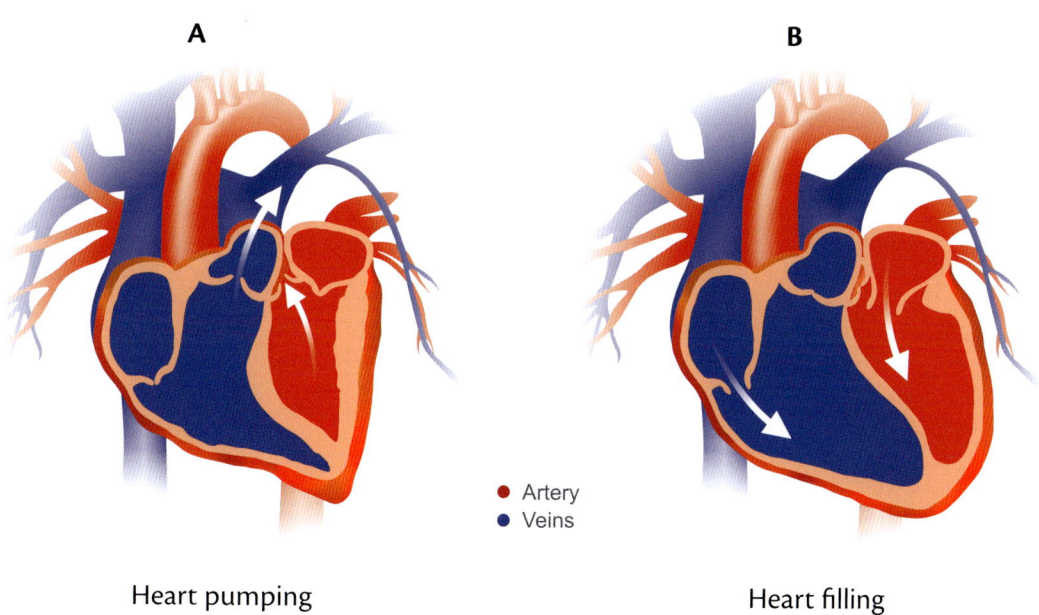

A Pumping Heart (seen from the front)

- Artery
- Veins

Heart pumping   Heart filling

Prayer

Dear God, thank You for forming our amazing hearts so early in our lives! Thank You that they have been beating ever since. We are happy that You care for our other kind of heart. You want our hearts to love You. Help us keep You as the most important love in our hearts. In Jesus' name, Amen.

Time to do Activity 18 in the Activity Book!

CHAPTER 7
God Gives You the Breath of Life

> He Himself gives to all mankind life and breath and everything. (Acts 17:25 ESV)

About 1,000 babies are born in the world every minute. Can you picture that many babies taking their first breath every minute?

We have learned that your heart was the first organ to start working while you grew in your mother's womb. The last organs to start working were your lungs. And they didn't start working until you took your first breath at birth! Isn't God amazing to give you two organs at such different times and make them begin working together at the perfect time?

Now we are going to learn about the system that works closely with the circulatory system. This system is called the **respiratory** (ress-purr-uh-tore-ee) **system**. One job of your respiratory system is to take oxygen from the air and

Inside Your Lungs

We have learned that one of the jobs of the circulatory system is to carry oxygen to all the cells in the body. The other job is to carry carbon dioxide away from the cells of the body. This oxygen and this carbon dioxide are carried in blood.

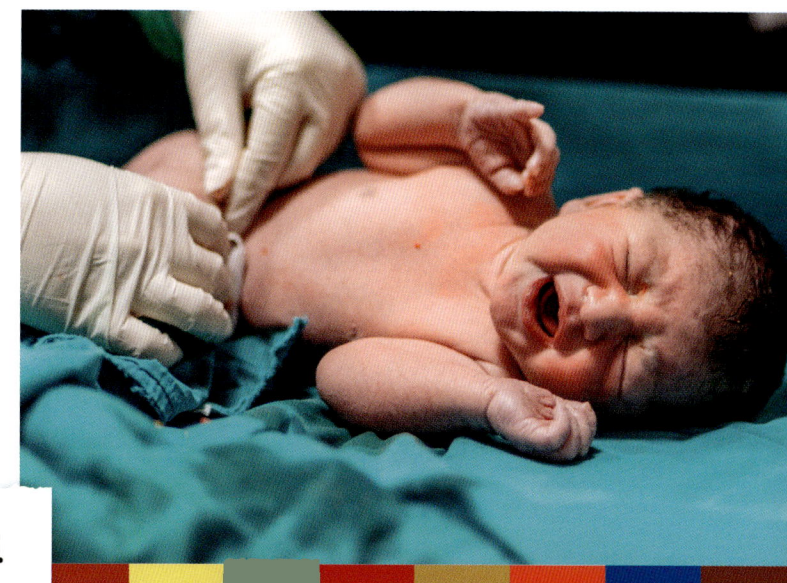

71

GOD MADE ME

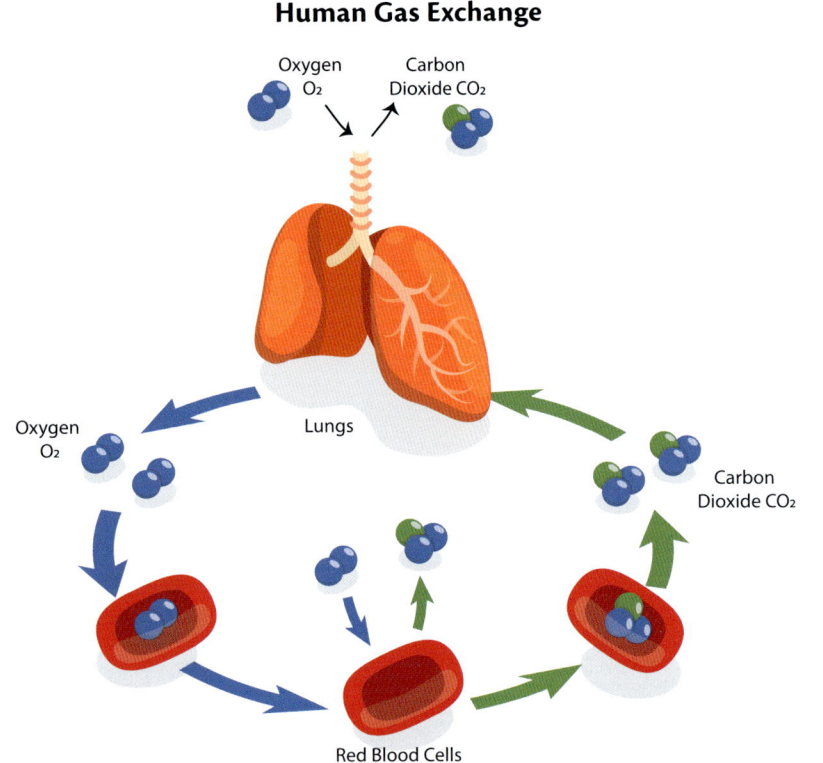

Human Gas Exchange

put it into your blood. Its other job is to take carbon dioxide from your blood and put it into the air. These jobs happen in your **lungs**. They happen while you are breathing!

Only part of the air you breathe is oxygen. Let's suppose you had 20 bottles filled with air. If you separated the oxygen out of the air, you would have four bottles of oxygen and 16 bottles of air with different gases. Your lungs don't use all the oxygen from the air you breathe. If you breathed in all the air from the 20 bottles, your body would not use all four bottles of oxygen. It would only use one of the bottles. The oxygen from the other three bottles would come back out of your lungs.

Take a big breath and fill your lungs with air. Now let the air out. When your lungs fill up with air, they are not like big balloons. They are more like sponges. Dry sponges have a lot of spaces that are filled with air. Lungs have a lot of spaces that can be filled with air too. These spaces are called **alveoli** (al-vee-uh-lie). They are very small spaces. Each alveolus is like a tiny balloon. It's lined with cells that can trade oxygen and carbon dioxide with nearby capillaries. Alveoli allow more places for lung cells to be. The lungs can do more work with alveoli than they could if each lung was like one big balloon.

CHAPTER 7: GOD GIVES YOU THE BREATH OF LIFE

Definitions

The **respiratory system** is your lungs and all the tubes and openings that air must pass through to get to your lungs.

The **lungs** are the main part of your respiratory system. Their job is to transfer oxygen and carbon dioxide between blood and air.

Alveoli are tiny sacks in your lungs. They are lined with cells that trade oxygen and carbon dioxide with the blood in nearby capillaries. One of these alveoli is called an *alveolus*.

Sometimes our indoor air isn't very fresh. It's nice to open the windows to let in different sounds and smells. This also lets out unhealthy things that may build up in a closed home. Let's look at ways to keep your indoor air healthy on the next page.

I love blowing around the world! I like to bring fresh air with me. When your windows are open, I can blow out the old air and bring in the new.

Libraries can hold a lot of books along their outer walls. But they can hold even more books by filling their rooms with shelves. Your lungs are like a library. Your alveoli are like library shelves. The hard-working cells in your alveoli are like books on the shelves. You have 300 million alveoli in your lungs. They hold a lot of cells just like library shelves hold a lot of books!

Alveoli pick up carbon dioxide from capillaries (blue). Alveoli also give oxygen to capillaries (red).

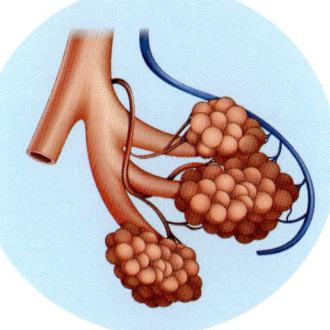

Alveoli

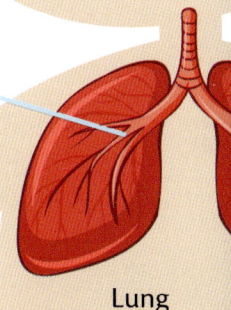

Lung

GOD MADE ME

Mold releases *spores* which are like seeds to make more mold.

To keep indoor air healthy:

- 🪟 Open your windows now and then to refresh your air. Always open windows to rid your house of smoke, new carpet chemicals, or paint fumes. Close your windows to keep out car or lawnmower exhaust when engines are running nearby.
- 🪟 Mold can grow in damp places. Some molds can release harmful chemicals. Watch for mold and mildew. Search out moldy smells. Check for water leaks and other damp areas so you can prevent mold.
- 🪟 Store lawnmowers, paints, and car chemicals away from the house.
- 🪟 Use natural, safe cleaning products. Don't ever mix products that contain bleach with products that contain ammonia. Mixing them will make a poisonous gas.
- 🪟 To get rid of bugs, try to find natural ways to control them instead of using chemicals.
- 🪟 Air out clothes that have just been dry cleaned. This will get rid of chemicals you shouldn't breathe.
- 🪟 Empty the trash often so germs, mold, and odors won't grow in your home.

 **Time to do Activity 19 in the Activity Book!**

Parts of the Respiratory System

The Spirit of God has made me, And the breath of the Almighty gives me life. (Job 33:4)

Isn't it wonderful that God's Spirit made you? This verse also says that His breath gives you life! We have learned about your lungs. Now let's learn about the other parts of the respiratory system!

CHAPTER 7: GOD GIVES YOU THE BREATH OF LIFE

🐑 Your **nasal (nose) cavity** is the first place air goes through when you breathe through your nose. It's lined with cells that make **mucus**. The mucus warms and moistens the air. It also traps dust you breathe in and keeps it from going into your lungs. Then tiny, finger-like **cilia** move the dust and mucus toward your throat. Soon you swallow and all the dust and mucus move out of your body in your waste.

🐑 Your **oral (mouth) cavity** is another place you can breathe through. It's nice to have another way to breathe when your nose is stuffy. You couldn't talk, sing, or laugh if you couldn't breathe through your mouth!

🐑 Your **epiglottis** (eh-pig-lot-tiss) is a very important structure. Since you eat and breathe through your mouth, your epiglottis helps keep these two jobs separate. As you breathe, your epiglottis is open to let the air get into your lungs. But when you eat, this flap closes off your breathing tube as you swallow. It keeps food out of your lungs!

🐑 Air must then pass through your **voice box** on its way to your lungs. Inside your voice box are two **vocal cords** that are stretched across the opening. You use these folds of tissue to make sound as you talk and sing. The sounds are made when air presses against the vocal cords and makes them vibrate. What an amazing creation our voices are!

🐑 After the voice box comes the **trachea**

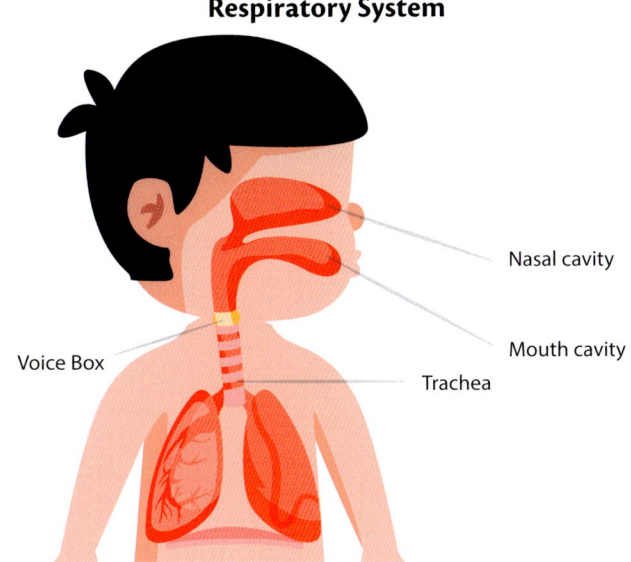

(tray-kee-uh). This "wind pipe" is held open by stiff rings of **cartilage**. Cartilage is tough stuff in your body that's softer than bone but hard enough to provide structure. (Your nose and ears also get their bendy structure from cartilage.)

🐑 Your trachea branches into two tubes,

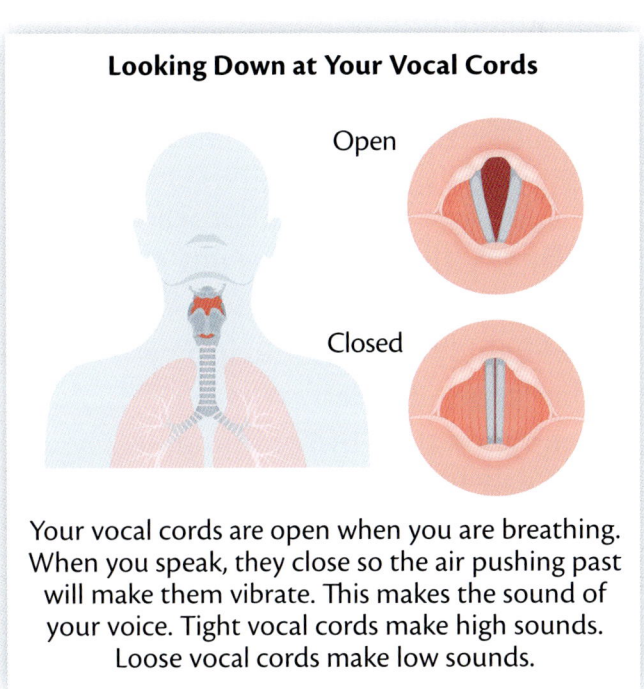

Your vocal cords are open when you are breathing. When you speak, they close so the air pushing past will make them vibrate. This makes the sound of your voice. Tight vocal cords make high sounds. Loose vocal cords make low sounds.

75

GOD MADE ME

one going to each lung. The tubes keep branching into smaller and smaller tubes until they lead to the alveoli.

- Your **diaphragm** (die-uh-fram) is a muscle below your lungs. It's shaped like an upside-down plate. It moves up and down to make you breathe.

While I live I will praise the LORD; I will sing praises to my God while I have my being. (Psalm 146:2)

This verse shows something you can do with your voice all your life!

 **Time to do Activity 20 in the Activity Book!**

How You Breathe

Mountain roads are built to go up and down. But sometimes it's easier for road builders to blast a tunnel *through* a mountain than to build a road over it. Sometimes children play a game when their car comes to a tunnel. They have a contest to see who can hold their breath all the way through. Usually the children figure out that taking a lot of deep breaths before entering the tunnel helps them hold their breath longer. In the game, children are controlling their breath two ways—by breathing deeply and by holding it.

Do you control your breath all the time? No, God has made breathing both controlled and automatic. Breathing usually happens automatically, without you having

CHAPTER 7: GOD GIVES YOU THE BREATH OF LIFE

to think about it. God knows you need oxygen while you're sleeping, playing, working, and learning. It would be hard to think about breathing too!

But there are many times when you *do* need to control the way you breathe. You need to hold your breath when you go under water. You also hold your breath a little and let it out slowly when you talk or sing.

Sometimes your automatic breathing acts differently. Think about the way your breath works when you laugh. How do you **inhale** (breathe in)? How do you **exhale** (breathe out)? Didn't God make laughter interesting and fun? Laughter shows that you are happy or that you think something is funny. It's amazing that people all over the world laugh when they are happy or when they think something is funny. What if you went next door and found your neighbors crying because something was funny? Or what if people in another country sighed at funny things and laughed when they were sad? No, God made humans so we would understand each other's **emotions**

You automatically sneeze and cough to clear your air pipes.

Definitions

Emotions are feelings we have in our minds. Some of our emotions are happiness, sadness, fear, anger, and surprise.

77

GOD MADE ME

anywhere in the world. We all laugh at funny things, cry when we're sad, and sigh when things aren't quite right.

Here's how breathing works:

1. Your body cells use oxygen to turn food into energy. When you run and jump, your cells need a lot of energy. This means they use a lot of oxygen and make a lot of carbon dioxide.
2. Special nerve cells in your brain and in certain arteries notice this. They notice your low amount of oxygen and high amount of carbon dioxide.
3. These nerve cells send a message to a certain part of your brain.
4. That part of your brain sends messages to your chest muscles and diaphragm.
5. Your lungs cannot pump air in and out of your body. But your diaphragm and chest muscles can. These muscles begin to work faster to make you breathe more often. When you inhale, your diaphragm pulls downward. At the same time, your chest muscles work to spread your ribs apart. This makes your chest get larger. Your lungs also get larger to fill the extra space. This makes air rush in to fill the extra space. The air brings in oxygen.
6. When your alveoli have traded oxygen for carbon dioxide, you exhale. This means your diaphragm pushes up and your ribs are pushed in. Air comes out! The carbon dioxide is in the air you breathe out.

Inhaling

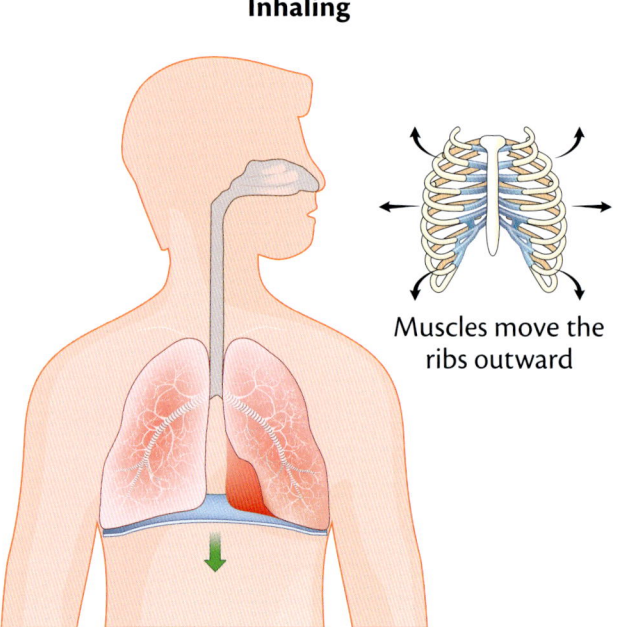

Muscles move the ribs outward

Diaphragm moves downward

Exhaling

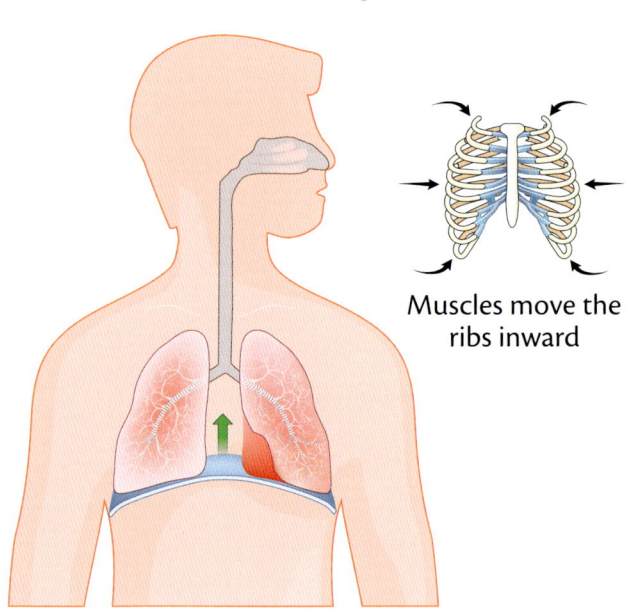

Muscles move the ribs inward

Diaphragm moves upward

CHAPTER 7: GOD GIVES YOU THE BREATH OF LIFE

While you're sleeping, the same special nerve cells notice that your body isn't using much oxygen because you are asleep. The nerve cells send signals to slow down your breathing.

**Let everything that has breath praise the LORD.
Praise the LORD! (Psalm 150:6)**

You breathe about 23,000 breaths of air in one day!

Prayer
Father, thank You for the breath of life! It's amazing how You made us able to use our breathing in so many ways. We Praise You that we can talk and sing, and laugh and cry. Amen.

Time to do Activity 21 in the Activity Book!

79

Tiger and Turtle staircase sculpture, Germany

CHAPTER 8
Take a Trip with a Blood Cell

We have been learning amazing things about the circulatory system and the respiratory system. Soon we will put together all the things we've learned. We'll follow a red blood cell on its trip through the circulatory system. But first, let's learn a few more things about the body God gave you.

The Lymph System

Your body has another system that works closely with your circulatory system. It's called the **lymph** (limf) **system**. The lymph system has several jobs. We will learn about one of these jobs now.

Remember, blood carries oxygen and other good things to your cells and carries carbon dioxide and other wastes away. These important jobs happen near your capillaries.

Capillaries have thin walls. Their walls are so thin that plasma goes in and out through the walls. Plasma is what carries good things out of your capillaries. Your nearby body cells absorb what they need from this plasma. Your body's cells also get rid of wastes by putting them out into this plasma. Most of the plasma then goes back into your capillaries and heads to your heart. But one tenth of the plasma does not go back into your capillaries.

The small amount of plasma that gets left behind is now called **lymph**. This

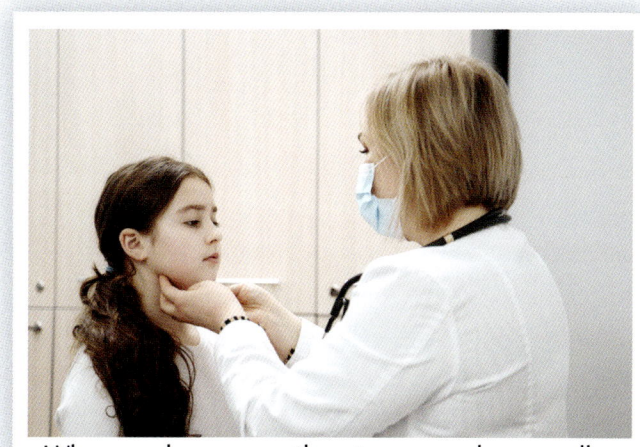

When you have a sore throat, you may have swollen lymph nodes. These can be felt from the outside.

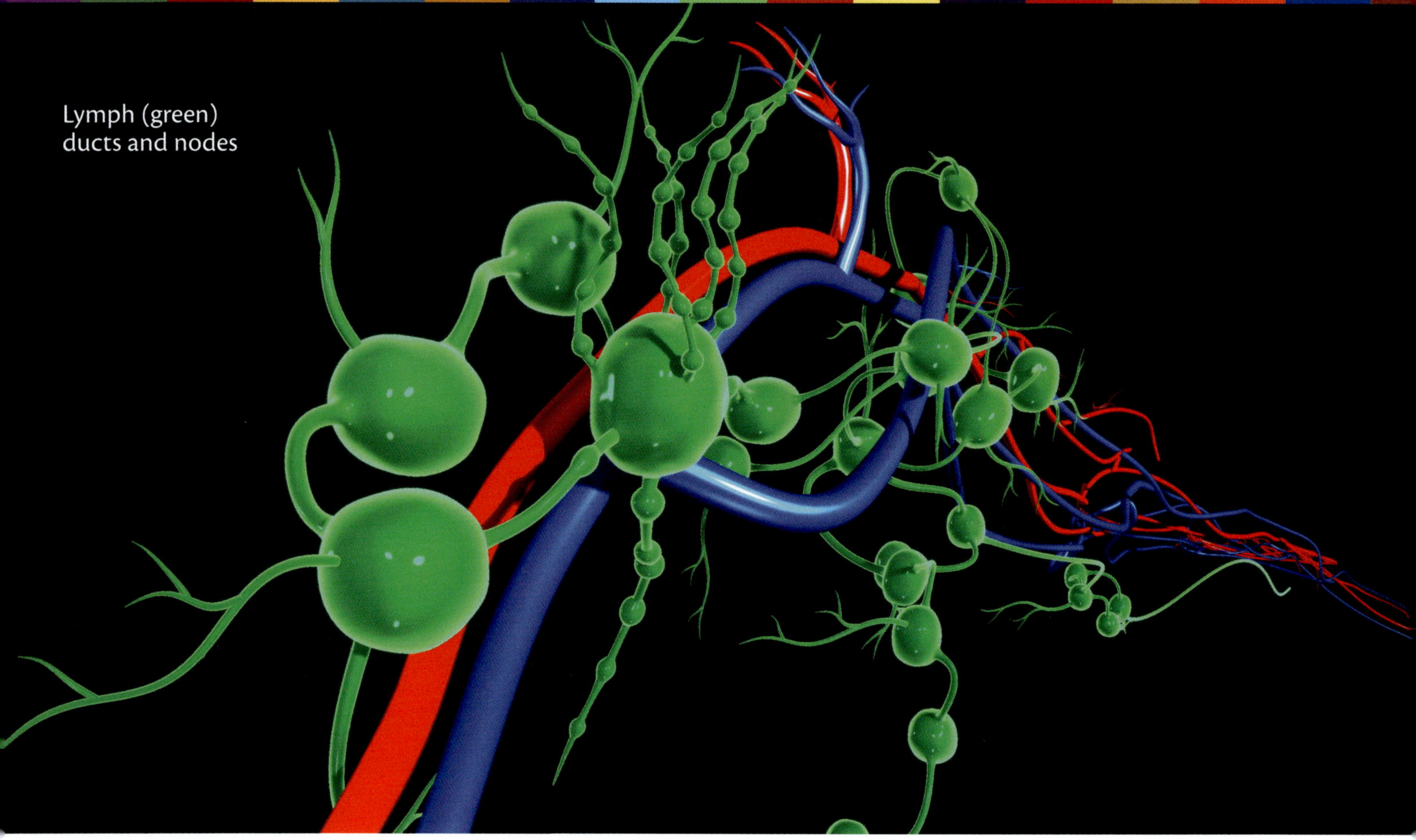

GOD MADE ME

Lymph (green) ducts and nodes

lymph picks up different wastes than the capillaries do. It picks up damaged cells, germs, and other waste.

Your lymph collects these wastes so they can be removed from your body or destroyed. This lymph does not go back into the capillaries. Instead, it drains into a nearby tube called a **lymph duct**. Lymph ducts are like blood vessels. They join with other lymph ducts to become larger ducts just like blood vessels join to become larger blood vessels. Lymph ducts have one-way valves (like veins do) to help your lymph travel in the right direction.

Thump's Health Hint

Lymph doesn't have a pump to move it through your body. Even though each lymph duct squeezes lymph along automatically, you can help!

- Drink enough water so lymph can move easily.
- Exercise so that your moving muscles will help squeeze lymph along.
- When you help with cleaning, use safe products so you won't breathe in harmful chemicals or get them on your skin. Your lymph system already has a big job to do. We shouldn't give it more wastes to get rid of.

CHAPTER 8: TAKE A TRIP WITH A BLOOD CELL

Thump's List of Ways Your Lymph System is Different from Your Circulatory System:

1. Your lymph system has no pump. Its ducts are lined with muscles that squeeze the lymph along. Your circulatory system has a pump: your heart.
2. Lymph ducts carry lymph into bean-shaped pockets called **lymph nodes**. Lymph nodes clean the lymph and put it back into the lymph duct. Blood vessels do not clean the blood. They deliver blood to organs (like the lungs) that clean out different wastes.
3. Lymph travels only one way—away from the body's cells. Blood travels both to and from the body's cells. Lymph ducts begin near your body's cells, join up with other ducts, and end inside your neck. This is where they dump the lymph into certain neck veins so the lymph can become part of your blood's plasma again.

You have lymph ducts and lymph nodes all over your body. Here we see the areas (green) where you have the most lymph nodes.

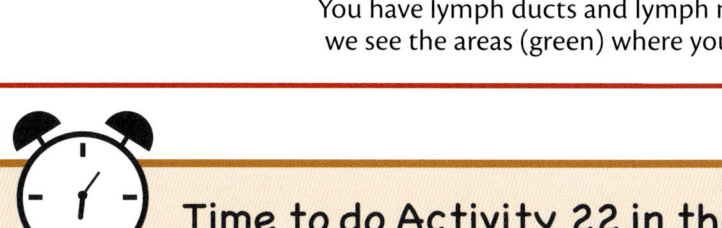

Time to do Activity 22 in the Activity Book!

Zoom in to the Cell

After working hard, you get to stop, rest, and maybe sleep. But your heart works hard your whole life without stopping. Aren't you glad God makes your heart pump to keep you alive even while you're sleeping?

Because your heart works so hard, it needs a lot of energy. Let's look at the structures in cells that make energy! These tiny structures are called **mitochondria** (my-toe-con-dree-uh). Mitochondria are

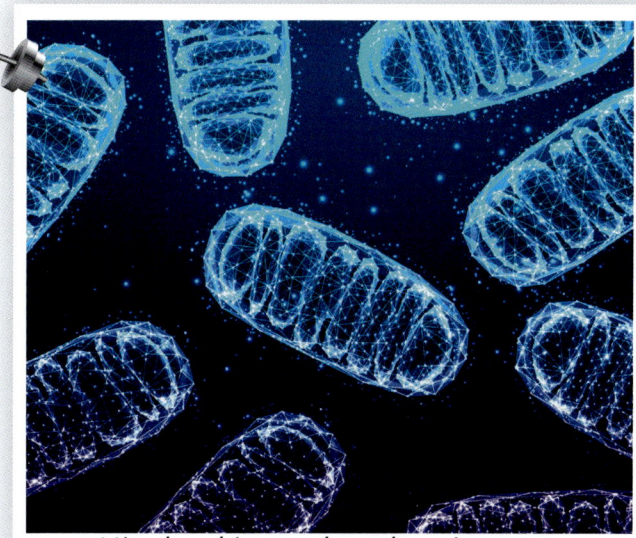
Mitochondria seen through a microscope

GOD MADE ME

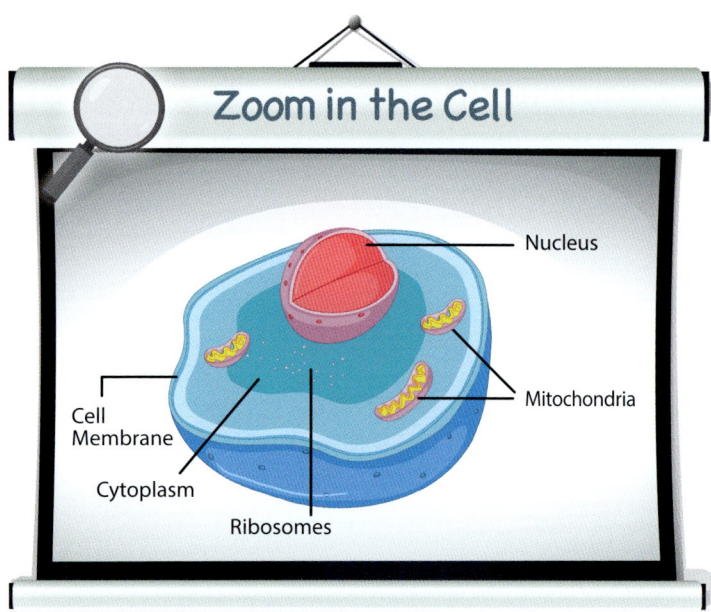

into carbon dioxide, water, and energy.

If this chemical reaction were to happen too quickly, the sudden energy would destroy the cell. Mitochondria protect the cell by using extra steps so they can store the energy. They store energy in an important molecule with a long name. We will just call it **ATP**. After the mitochondria make energy, ATP can give out the stored energy gently and safely whenever the cell needs it.

shaped like sausages. We sometimes call them the powerhouses of the cell. Heart cells have a lot of mitochondria.

The job of mitochondria is to make energy out of food. God makes them do this job using many complicated steps. Each step is a **chemical reaction**. Chemical reactions change chemicals into different chemicals. The mitochondria change food and oxygen

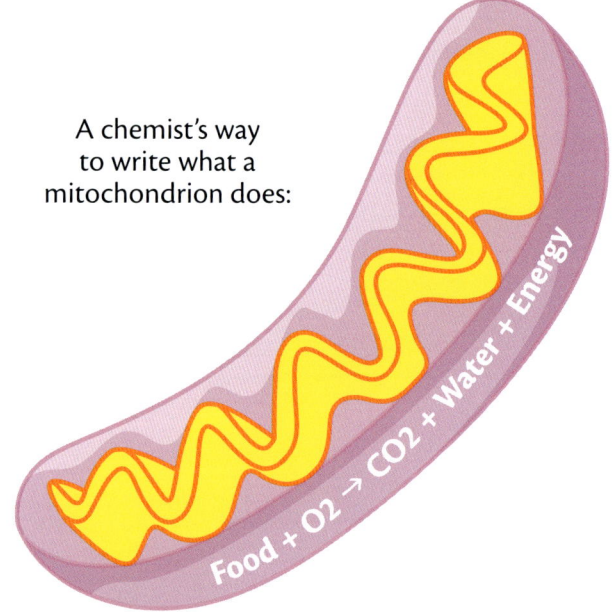

A chemist's way to write what a mitochondrion does:

$Food + O_2 \rightarrow CO_2 + Water + Energy$

 **Time to do Activity 23 in the Activity Book!**

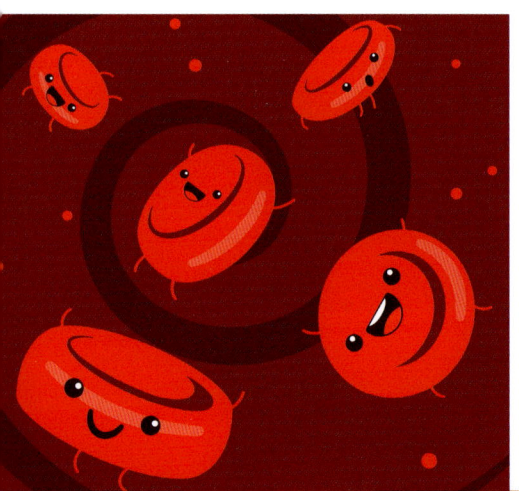

The Path of a Blood Cell

Your heart works hard to pump about 7,000 quarts (liters) of blood a day.

Now let's do something to help us understand God's amazing design for your hard-working heart. Let's take one trip around your body with a red blood cell. We will see the path it takes while it does its job!

84

CHAPTER 8: TAKE A TRIP WITH A BLOOD CELL

 **In the Body** — We'll start with a red blood cell that has just given away its oxygen (O2) somewhere in your body. God gave red blood cells the ability to notice places with low oxygen. This is how your cell knew when it was time to give its oxygen away. To do this, your red blood cell stays in the capillary. But the oxygen jumps from the cell into the plasma around it. The plasma carries the oxygen out through the capillary wall. Then the oxygen goes to a nearby body cell to keep it alive. Back in the capillary, your red blood cell picks up carbon dioxide (CO2). Where has the carbon dioxide come from? The carbon dioxide came through the capillary wall from a body cell that needed to get rid of it. It jumped into the plasma to make the trip.

Gives

Gets

 **Next, the busy red blood cell moves along the capillary.** The capillary joins with other capillaries and becomes a small vein. Small veins join to become larger veins. All this time, your red blood cell and its friends are traveling toward your heart in these veins. Finally, your red blood cell jumps into the last and largest vein before it joins your heart.

 To the Heart — Your heart has just finished a beat. It relaxes. This allows your red blood cell to flow from the large vein into the right atrium of your heart. But it doesn't stay there. Your right atrium gives a little squeeze to push your blood into the relaxed right ventricle below it. The busy red blood cell is pushed along too. Your heart is not a place for the little cell to stop and rest!

 To the Lungs — Next, your right ventricle gives a big squeeze. The red blood cell is pushed out of your heart and into a large artery. That artery divides into two arteries right away. Each artery goes to one of your lungs. In the lungs, your red blood cell gets rid of its carbon dioxide. Then it picks up oxygen. Now it's time for your busy blood cell to leave the lungs. It goes out through a small vein. The small vein joins with other veins from your lungs to make a bigger vein.

Gives

Gets

 To the Heart — Your red blood cell flows from that big vein into the left atrium of your heart. Your left atrium then gives a little squeeze to get your red blood cell down into your left ventricle.

GOD MADE ME

To the Body — Your left ventricle gives a big squeeze. Your red blood cell and its friends squirt out into a big artery (aorta). The aorta divides into smaller arteries. As your red blood cell travels through your body, the smaller arteries divide into even smaller arteries. Soon they have divided into the smallest vessels, the capillaries. Your red blood cell drops off its life-giving oxygen. It then picks up carbon dioxide and starts its journey back to your heart again!

Do you see the pattern?

Heart → Lungs → Heart → Body → Heart → Lungs → Heart → Body

It takes about one minute for one of your red blood cells to take the whole journey God has for it. Then, the next minute, it takes another journey!

Do you see now why God says "the life of the flesh is in the blood"?

The Path of a Red Blood Cell

Right Atrium
Right Ventricle
Left Atrium
Left Ventricle

Can you trace the path of blood through the circulatory system in this picture? To make the picture simple, blood is shown going into one lung and out the other. In real bodies, blood goes into and out of both lungs.

The blood of Jesus Christ His Son cleanses us from all sin. (1 John 1:7)

This verse talks about cleaning. Usually we use water to clean things. Does it seem funny that blood can clean us? It's true, though. That's because the verse isn't talking about physical cleaning like the way you wash your hands. Water cleans away dirt and germs from our hands, but Jesus' blood can cleanse us from the sins inside us.

Do you remember that the Bible says life is in the blood? If we don't have blood, we don't have life. When Jesus lost His life, the Bible says He lost (shed) His blood.

CHAPTER 8: TAKE A TRIP WITH A BLOOD CELL

Jesus died for a reason. He shed His blood to clean away our sins! He did that because He loves us.

Prayer

Jesus, thank You for loving us! You gave Your life and blood to clean away our sins. Thank You for helping us learn about our hearts and blood so we can understand more about You. Amen.

Time to do Activity 24 in the Activity Book!

Caption

UNIT 3
God Gave You a Head

Our memory verse says that all the things God made will praise Him. That's because all God's works show His wisdom! But you can praise God even more than plants, animals, and rocks can. You can use the brain in your head to think about God and His wonderful works. Then you can think of many ways to praise Him!

Our hymn sings about "considering" the things God has made and "thinking" of Jesus taking away our sin. Your brain is where you consider and think these wonderful thoughts!

 ## Memory Verse

All Your works shall praise You, O LORD. (Psalm 145:10)

You can listen to this hymn by searching for "How Great Thou Art" on the internet.

 ## Hymn to Sing: How Great Thou Art*

O Lord my God, when I in awesome wonder
Consider all the worlds Thy hands have made,
I see the stars, I hear the rolling thunder,
Thy power throughout the universe displayed,

Chorus:
Then sings my soul, my Savior God, to Thee:
How great Thou art, how great Thou art!
Then sings my soul, my Savior God, to Thee:
How great Thou art, how great Thou art!

When through the woods and forest glades I wander
And hear the birds sing sweetly in the trees,
When I look down from lofty mountain grandeur
And see the brook, and feel the gentle breeze,
(Chorus)

And when I think that God, His Son not sparing,
Sent Him to die, I scarce can take it in.
That on the cross, my burden gladly bearing,
He bled and died to take away my sin,
(Chorus)

*Rev. 8/30/02/smgWords: Stuart K. Hine. © 1949, 1953 The Stuart Hine Trust CIO. All rights in the USA its territories and possessions, except print rights, administered by Capitol CMG Publishing. USA, North and Central American print rights and all Canadian and South American rights administered by Hope Publishing Company. All other North and Central American rights administered by The Stuart Hine Trust CIO. Rest of the world rights administered by Integrity Music Europe. All rights reserved. Used by permission.

CHAPTER 9
Your Brain Is Amazing

Is your **brain** amazed about your brain? Your brain is amazing! Because of your brain, you can put puzzles together. You can read and write, do math exercises and learn science, play an instrument, and paint a picture. Your brain helps your body balance and move when you play sports. It helps you know where you are so you can find your way home. Because of your brain, you know what to do about things you see, hear, smell, feel, and taste.

Your brain can be aware of many things at once and decide which things to do something about and which things to ignore. And while all these things are going on, your brain pays attention to your whole body and keeps it working inside you. Your brain can think about things like friendship, cooperation, and forgiveness. And best of all, your brain makes it possible to know and love God. Your brain can be amazed by Him!

Your Brain Is in Charge

God put your brain in charge of everything that happens in your body. Because it's in charge, your brain has many jobs, and it does them all well.

Sometimes we say a person is in charge.

Because of your brain, you can figure out how to play games. You can be happy when you win and be a good sport when you lose.

GOD MADE ME

A store owner is in charge of his store. A city has a leader or group of leaders that are in charge. We often say that someone who is in charge is the *head*—the head of a country or the head of a family. If your parents ask you to organize your siblings to pick up toys, you are the head of cleanup. We say things this way because the brain is inside the head, and the brain is in charge of the body.

For so He gives His beloved sleep. (Psalm 127:2)

Your brain also needs sleep! Good sleep makes you better at solving problems. It also makes you more cheerful. Your brain doesn't stop working when you sleep, but it works differently while you're sleeping. Sleep is the time when things you learn during the day become permanent. It's also the time when useless things are forgotten.

About four times during the night, you spend time dreaming. These times are called **REM sleep**. REM stands for Rapid Eye Movement. During this time, your eyes move around quickly under your closed lids. Scientists have discovered certain neurons that are busy during REM sleep. These neurons seem to prevent your dreams from being permanently stored in

Thump's Health Hint

Are you starting to realize that your brain is responsible for a lot of jobs? It's important to remember that your brain needs rest too!

Do you ever feel tired after thinking hard for a long time? Maybe after a long day of difficult schoolwork, you don't feel like playing chess. You probably don't want to make decisions that take a lot of thought. This is because your brain has been releasing a lot of chemicals as it worked. It takes time for the chemicals to return to normal amounts. This could be one of the reasons God wants us to rest from our work and worship Him one day a week. Your brains need rest!

CHAPTER 9: YOUR BRAIN IS AMAZING

Dads' brains need sleep too.

Another interesting thing happens during REM sleep. You become paralyzed! God makes your muscles unable to move so that you won't act out your dreams. This is one way He keeps you safe when you sleep.

your brain. This is why you quickly forget most of your dreams!

Now let's learn about your amazing brain! Let's see what wonderful abilities God gave this perfect "head" that's in charge of your body.

 **Time to do Activity 25 in the Activity Book!**

Let's Look at Your Brain

After you had been growing in your mother's womb about three weeks, God began forming your brain. First, some special cells were gathered into a flat oval "plate." During the next week, this plate folded into what would become your brain. By the time you were born, your brain was almost complete. After birth, your brain would need to keep growing in size and ability. Your brain won't be completely formed until you are about 25 years old. Even after that, you will still be able to learn for the rest of your life.

Your brain is a soft, squishy, delicate organ. It contains billions of nerve cells. God has protected your brain inside a hard **skull**. Your skull is a hollow case made of bone. It's lined with three thin **membranes**.

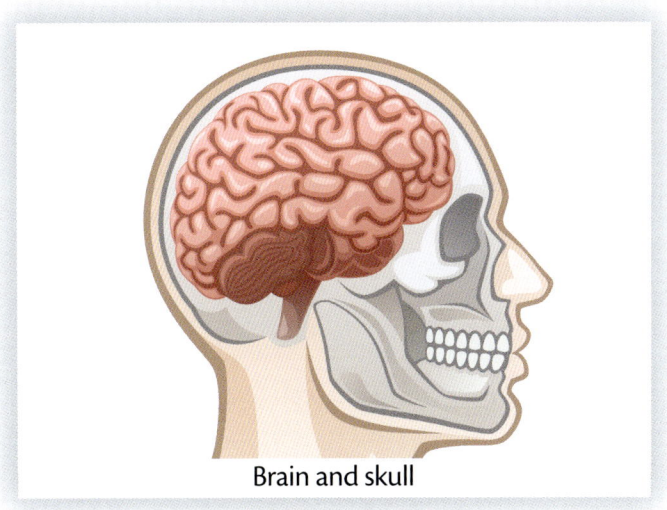
Brain and skull

GOD MADE ME

If you were to fall off your bike and bump your head on the ground, your brain would be protected by your hard skull. But it would also be protected by a special liquid that God has put between two of the membranes. This liquid helps keep your brain from bumping into your skull too hard. Sometimes these protections are not quite enough, and a person can get a **concussion** (kun-kuh-shun) from a hard bump to the head. That's why it's important to wear a helmet while you are riding a bike.

Your brain needs a good supply of oxygen and nutrition—more than the other parts of your body. The membrane closest

to your brain is full of blood vessels that bring oxygen and nutrition to your brain. From the membrane, blood vessels go deep into your brain, delivering good things and taking wastes away.

If your brain doesn't get enough oxygen, you could **faint**. That means you would suddenly fall over and look like you went to sleep. You wouldn't know what was happening around you until you "came to" or woke up. If you fainted, you would probably end up in a lying-down position. Thankfully, that would make your head low enough for blood to quickly reach your brain again and bring more oxygen. That would be good for your brain, but it could

Definitions

A **membrane** is a thin layer of tissue that separates one place from another in the body.

A **tissue** is a group of the same kind of cells that have the same job.

A **concussion** is an injury to the brain caused by a hard bump or sudden shaking as can happen in an accident. A person may feel dizzy or get a headache right away or even a few days later. A doctor can check for concussions and give advice for healing.

Your **senses** are the ways your body receives information from the world around you. You have five main senses: sight, hearing, touch, smell, and taste.

If someone faints, it's good to lift both feet about 12 inches (30 cm) off the ground to help blood flow to the brain.

94

CHAPTER 9: YOUR BRAIN IS AMAZING

be bad if you get hurt falling down.

Now let's look at the parts of your brain! Different structures in your brain are in charge of different things.

Your **cerebrum** (suh-ree-brum) controls all the things you do on purpose. It's where you think, make decisions, plan, and remember. When you talk or figure out math, you are using your cerebrum. Your cerebrum also uses information that comes from your senses. It processes that information along with things your brain already knows. Then it tells your body what to do.

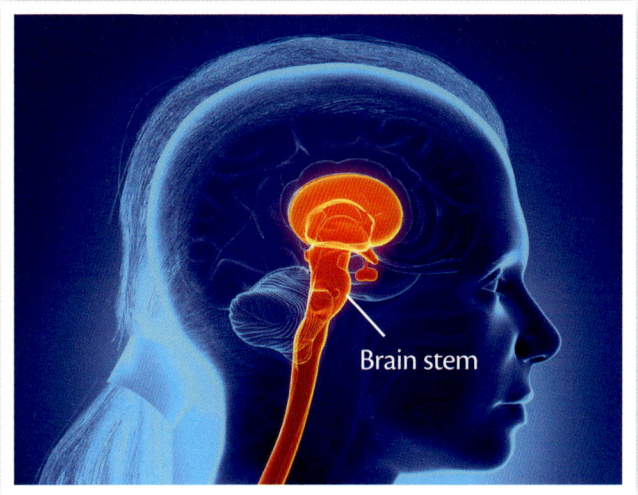

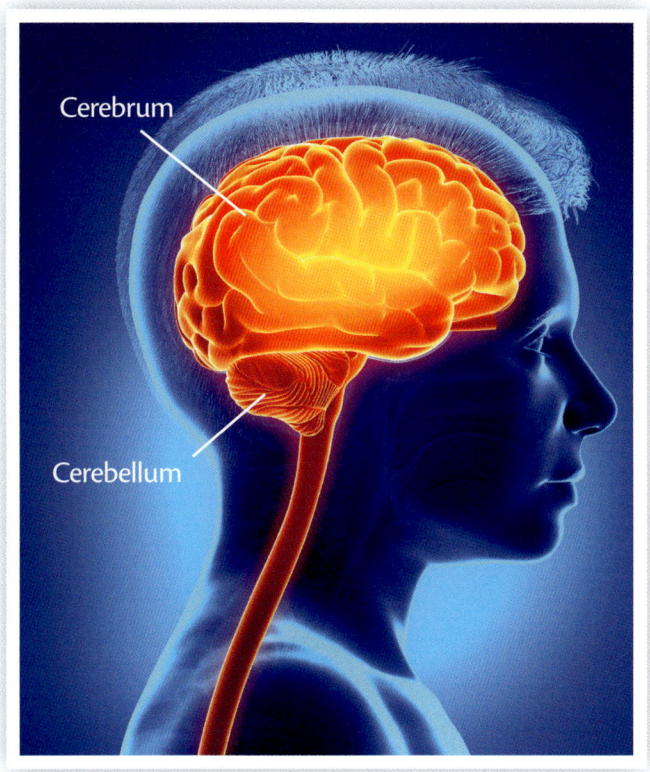

Your **cerebellum** (sarah-bell-um) controls complicated muscle movements like what you do when you play a musical instrument or play sports. When you first learned to walk, your cerebrum carefully thought about what you were doing. But as you practiced, your cerebellum took over. Walking became automatic as you practiced. Now you can walk without thinking about it because your cerebellum controls your muscles.

Deep in your brain are structures that help your body do things automatically. Some of these structures make up the **brain stem**. Your brain stem controls blood pressure, balance, swallowing, hearing, feeling on your face, breathing, and how fast your heart beats.

 Time to do Activity 26 in the Activity Book!

GOD MADE ME

Your Brain Is Too Amazing to Understand

Scientists spend a lot of time learning and studying. They use their brains to understand many things. But scientists' brains don't even understand brains! The more they learn about the human brain, the more they are amazed by it. Only God could make the brain so complicated that it can't even understand itself. And if people can't even understand their own brains, how could we ever understand God's mind?

"For who has known the mind of the LORD?" (Romans 11:34)

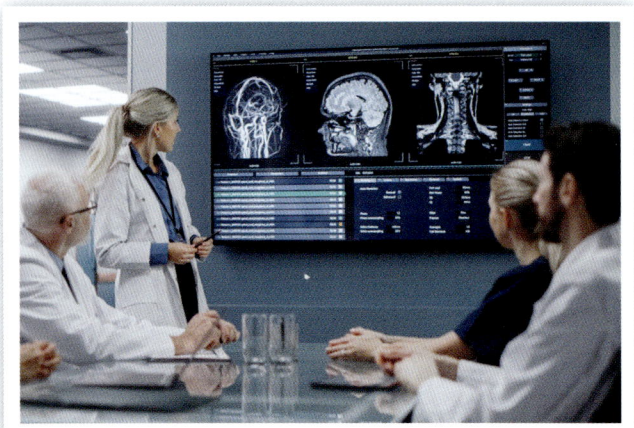

Thump has found some amazing new discoveries about your brain!*

1. A single brain like yours has more places to process information than all the computers on Earth.

2. One brain can have as much memory as the internet.

3. What if there was a robot brain that could work as fast and as well as your brain? The robot brain would need one million times more energy than your brain does. It would need as much electricity as a small hydroelectric power plant could make!

4. We will learn later how your brain and your body communicate with each other using electricity. But scientists think your brain sends messages within itself using light. Your brain needs to use a lot of information from its different parts. It does this all at the same time so it can make decisions and send commands quickly. It uses light because light travels much faster than electricity!

*Tomkins, Jeffrey P. "The Human Brain Is 'Beyond Belief,'" *Acts and Facts*, vol. 46, no. 9, Sep. 2017, p. 10.

CHAPTER 9: YOUR BRAIN IS AMAZING

Small hydroelectric power plants can provide enough electricity for a whole village. A robot brain would use the same amount of electricity as a village if it worked as much as your brain does

Bright messages of light flashing through the brain! That reminds me of this verse. Jesus brought the message of light!

"As the first to rise from the dead, [Jesus] would bring the message of light." (Acts 26:23 NIV)

Prayer

Dear Lord, we can't know what You know. We can't even know our own brain, and that's the place You gave us to know things! Thank You for making our complicated brains. Thank You for giving us the Bible. In it, You tell us many things about Your mind. Thank You that we can know You. Amen.

Time to do Activity 27 in the Activity Book!

97

CHAPTER 10
Your Nervous System

The Parts of Your Nervous System

The rulers of a country are very busy. They don't have time to travel around getting news about their land. Other people bring the news to them. And when something needs to be done, rulers have a system to help them communicate quickly. They send messages to keep their country running smoothly.

Your brain is like a ruler of a country. It must receive and send messages quickly. Your brain is part of a speedy **nervous system** that helps it communicate. This system extends throughout your whole body and keeps it running smoothly.

The three parts of your nervous system are your brain, **spinal cord**, and **nerves**. Your nerves are the first place where things are noticed. They know what's going on inside and outside your body. When something happens, messages speed through your nerves. Nerves in your head send messages directly to your brain. Nerves lower down in your body send messages to your spinal cord. Then your spinal cord takes those messages to your brain. After your brain processes the messages, it sends new messages (commands) through your nerves to your muscles or organs. These commands tell your muscles or organs what to do.

Imagine that you step outside on a cold winter morning. Your foot starts to slip as you put

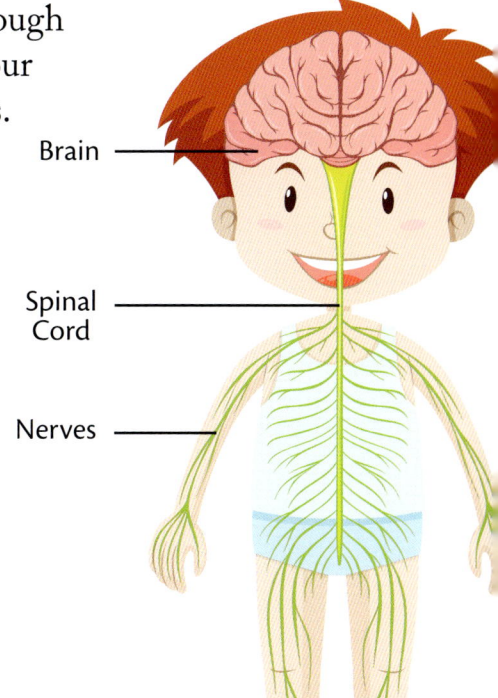

99

GOD MADE ME

Nervous System

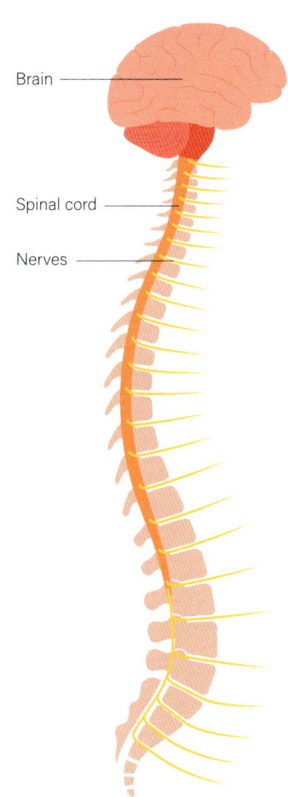

it down. The nerves in your foot feel that something is wrong. They send an "I'm slipping!" message. The message travels through nerves to your spinal cord. Your spinal cord takes the message to your brain.

As your foot slips, you look down and see a sheet of ice on the doorstep. The nerves in your eyes send a picture of the ice to your brain. Now your brain uses its memories of coldness, ice, and slippery things to figure out what to do. It measures how important it is for you to go outside. All this information is coming together from different areas of your brain.

Let's say your brain makes the decision that it would be fun to go outside in spite of the ice. It sends a "Go outside, but be careful!" command to your spinal cord. The message speeds down the spinal cord and travels out through the nerves. Your nerves take the message to your muscles to guide their movements as you carefully walk across the icy doorstep.

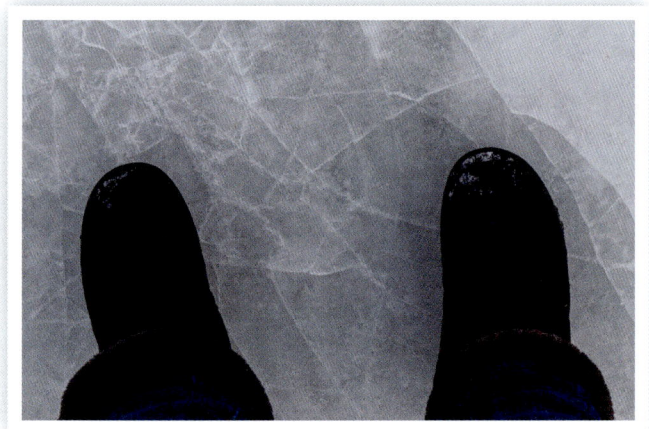

 **Time to do Activity 28 in the Activity Book!**

Like this goat vertebra, your vertebrae have holes inside them where the spinal cord passes through. This is a view from the top.

Your Spinal Cord

Your spinal cord is part of your **spine**. The spinal cord is made of delicate nerve cells. God made it able to bend and twist when your spine does. The special liquid that protects your brain also protects your spinal cord and cushions it.

Your spinal cord is well protected inside a stack of bones. This stack of bones is called your *backbone*. Your backbone is also part of your spine. It starts just below your skull and goes to the end of your back.

100

CHAPTER 10: YOUR NERVOUS SYSTEM

Your spinal cord has two main jobs. One is to carry commands from your brain to your body below your neck. The other is to carry messages from below your neck to your brain.

But God has given your body another ability that uses your spinal cord. He has given you **reflexes**! Reflexes help you react quickly to avoid danger. With reflexes, messages don't have to go all the way to the brain and back. Instead, they go into the spinal cord and commands comes right back out. A command goes to the muscle that takes care of the problem. A separate message goes to your brain so you can learn about the danger and avoid it in the future. But by then, your body has already jerked away from the problem using a spinal cord reflex.

In this illustration, someone's poor finger has touched a cactus spine. But don't worry. The reflexes provided by the spinal cord will take care of things before there's a bad injury. Puff will show us how that works!

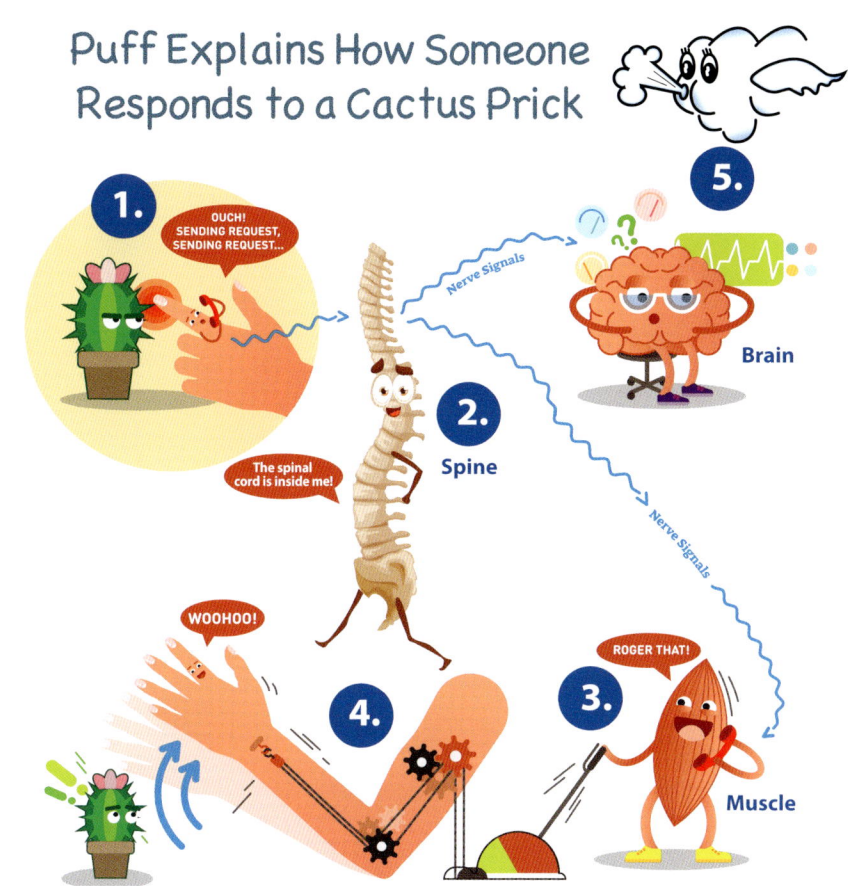

Puff Explains How Someone Responds to a Cactus Prick

1. First, nerves from the finger send a pain message asking for help.

2. The pain message arrives at the spinal cord.

3. The spinal cord immediately sends a problem-fixing message to the arm muscle. This reflex is three times faster than if messages had to go all the way to and from the brain.

4. Your arm muscle is inside your arm and doesn't use gears to do its work. But this picture shows that the muscle moves the arm back and jerks the finger away from the cactus spine. If this doesn't happen quickly, the finger could press the cactus spine in deeper before it's told to stop.

5. The spinal cord will also send a message to the brain about what has happened. The brain can process this information and put it into its memory. It will remember not to allow the fingers to touch cactus spines again!

GOD MADE ME

Learning about the spine

Side view of part of a spine. Each spine bone is called a **vertebra** (vert-uh-bruh). The plural form of vertebra is vertebrae (vert-uh-bray).

 Time to do Activity 29 in the Activity Book!

Your Nerves

Your brain, your spinal cord, and your nerves all contain nerve cells. We will learn about nerve cells in the next chapter. Your nerves each contain several bundles of nerve fibers that belong to nerve cells. Your nerves also contain blood vessels and lymph structures to keep them healthy. A nerve holds many nerve cells where it starts. It divides into smaller branches as it stretches to different parts of your body.

Grouping Nerves by Where They Are

Scientists have come up with several ways of grouping nerves. One way to think of nerves is by where they are. **Cranial** (kray-nee-uhl) **nerves** come out of your brain and go directly to places in your face, head, and neck. One of them also goes to your heart and digestive system. You have 12 pairs of cranial nerves. They help you see,

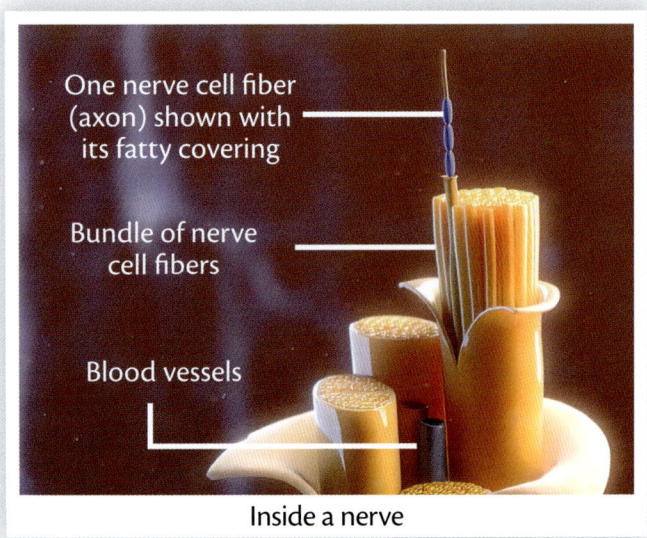

Inside a nerve

102

CHAPTER 10: YOUR NERVOUS SYSTEM

Several of your cranial nerves help you make facial expressions!

taste, smell, hear, and feel things in and on your head. They help you blink your eyes and move your tongue.

The nerves that come out of your spinal cord and go to the rest of your body are **spinal nerves**. You have 31 pairs of spinal nerves. Some spinal nerves carry messages to your spinal cord. These messages tell your spinal cord how your skin, joints, and muscles feel. Other spinal nerves carry commands away from the spinal cord to help you use your muscles. Spinal nerves also control some of your reflexes.

Grouping Nerves by What They Do

Sometimes we talk about nerves in a different way. We talk about nerves according to the jobs the nerve cells do inside them.

Some nerve cells are connected to your senses. They know what is happening somewhere inside or outside your body. These nerve cells send messages to your brain or spinal cord. Nerves that contain these kinds of nerve cells are called **sensory nerves**. Their job is to carry sights, sounds, smells, tastes, and touch feelings to your brain from your senses.

Cranial Nerves

If your tongue could look up at your brain, it would see your brain in this position. (Your nose would be pointing toward the top of the picture.) The lines are pointing to the cranial nerves where they begin in your brain.

Spinal nerves (green)

> Most spinal nerves and cranial nerves can carry messages in both directions—from your body to the spinal cord (or brain) and back the other way.

103

GOD MADE ME

You have different nerve cells that send commands away from your brain or spinal cord to a place in your body. Their job is to tell your body what to do. Nerves that contain these kinds of nerve cells are called **motor nerves**.

Most nerves carry both kinds of cells and are called **mixed nerves**.

Prayer

Lord, our nervous systems do so much communicating! So many messages are being sent at one time. It's complicated, but You have made it work so perfectly and quickly. Thank You for the amazing nervous system and for protecting us with reflexes! Amen.

Time to do Activity 30 in the Activity Book!

CHAPTER 10: YOUR NERVOUS SYSTEM

Can you imagine the many nerve messages that must travel when someone plays a sport? See if you can explain the path of nerve messages while someone waits for and tries to hit a baseball.

CHAPTER 11
Your Neurons

"For as the lightning comes from the east and flashes to the west, so also will the coming of the Son of Man be." (Matthew 24:27)

God can do things quickly. He created quickly—He just spoke, and things instantly came to be. This Bible verse tells us that, when Jesus comes again, He will come as quickly as a flash of lightning.

On the first day of creation, God made light. People have never measured anything that's faster than light. During creation, God made other fast things. He made cheetahs, the fastest runners. He made star-nosed moles, the fastest eaters. God also made nerve cells!

Zoom in to the Cell

Nerve cells are the main kind of cells in your nervous system. Nerve cells are called **neurons** (noo-rons). Neurons are part of your brain, spinal cord, and nerves.

The star-nosed mole can find and eat eight tiny creatures in less than two seconds. The finger-like "star" around its nose is used for finding food, not for catching it. It's a very sensitive tool that can quickly feel things in the darkness underground or underwater. This tool has some of the fastest nerve cells God created. They are as fast as the nerve cells that focus your eyes when you look around at different things.

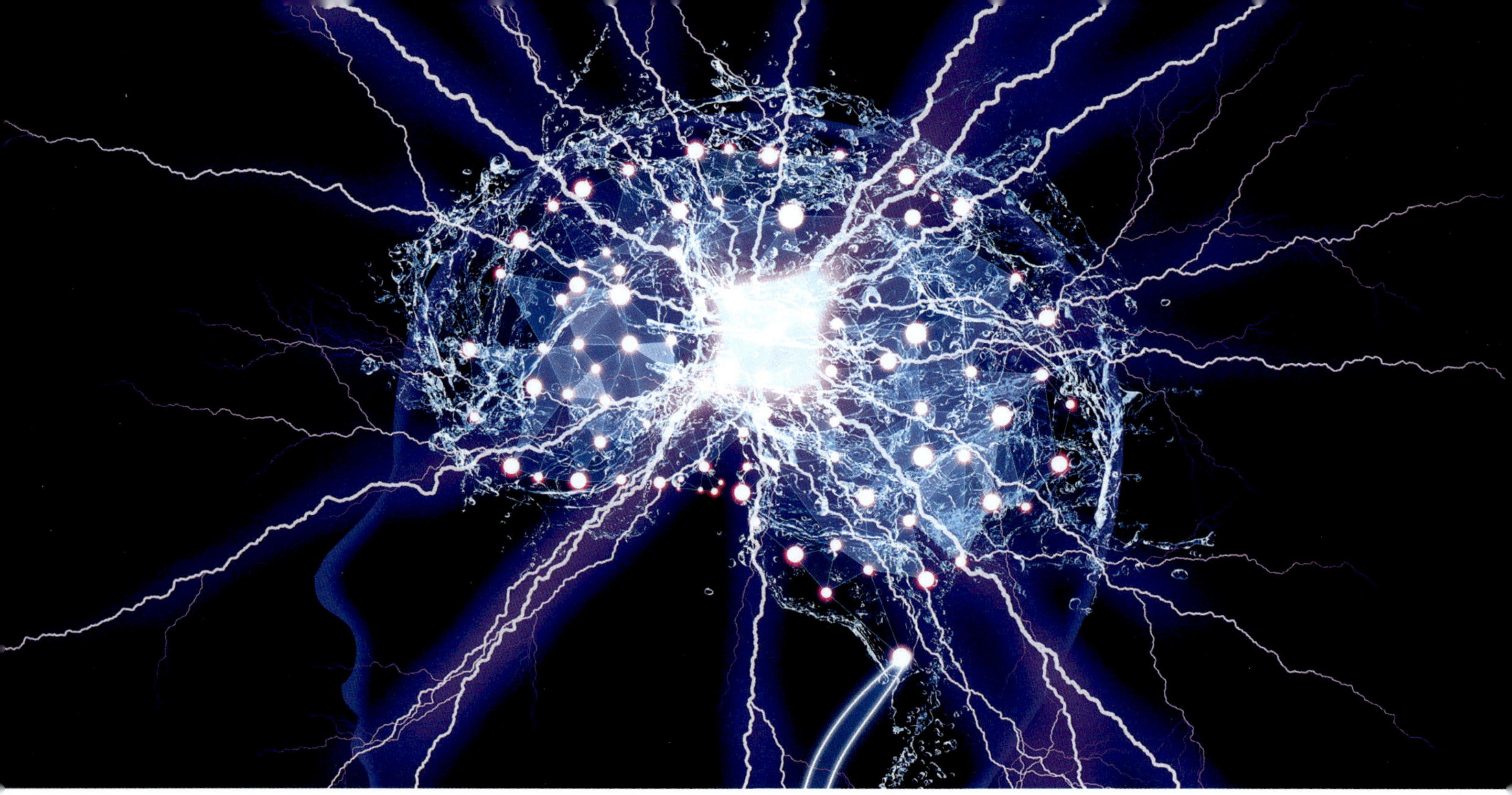

Let's look at the parts of a neuron.

- Neurons have **cell bodies** that contain a nucleus, ribosomes, mitochondria, and other tiny cell parts.
- Neurons also have fibers that come out of their cell bodies. One of these fibers is called an **axon** (ax-on). The axon sends messages out from the cell body. The messages carried by an axon can go to a place in the body (like a muscle) to tell it what to do. Axons don't always give their messages this way. Axons sometimes give messages to other neurons that will pass the information along. An axon is long and is branched at the end so it can send messages to several places. Nerves are bundles of axons belonging to several neurons.
- Axons have a fatty covering called **myelin** (my-uh-lin). Myelin keeps messages from escaping out of the axon. This helps speed them along. Myelin looks like beads on a string. The spaces in myelin help make messages speedier because the messages can quickly jump from one "bead" to another.
- Neurons have other fibers called **dendrites** (den-drites) that receive messages into the cell body. A neuron can have many branched dendrites attached to its cell body, but it can have only one axon.

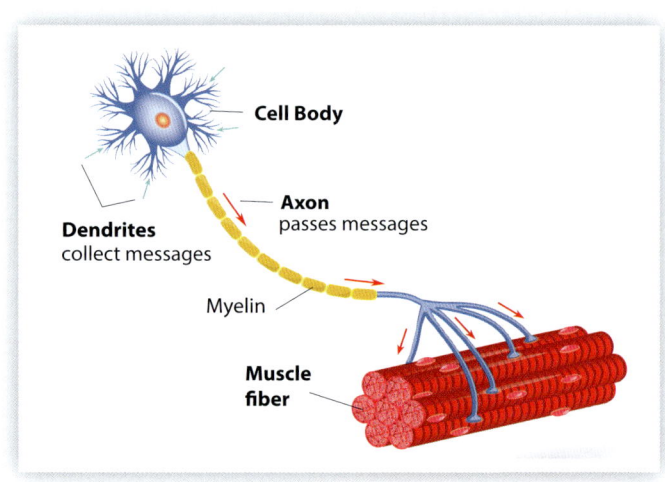

108

CHAPTER 11: YOUR NEURONS

Definitions

Cross section — how something would look if it were cut in half.

Nerve fibers — The axon and dendrites of a neuron.

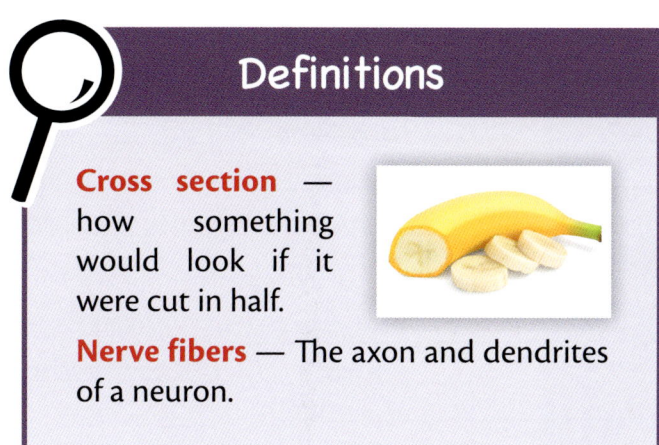

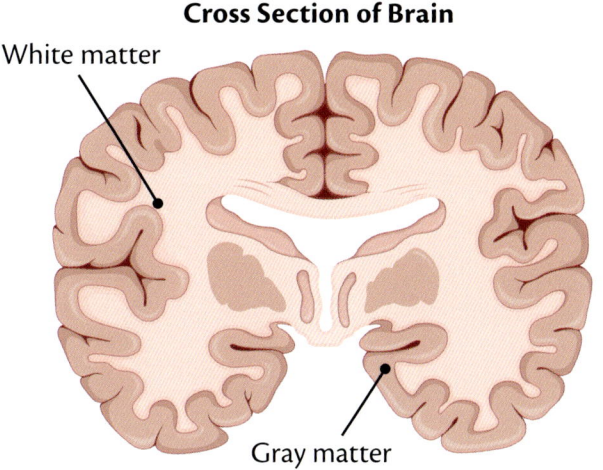

Cross Section of Brain
White matter
Gray matter

Let's look at the neurons in your brain. Your brain has a grayish-pink outer layer made of the cell bodies of neurons. This **gray matter**'s job is to process information. Your brain's inner **white matter** is made of axons. Its job is to send the processed information to other parts of the brain and nervous system. Your white matter is whitish pink because of fatty myelin that covers the axons. In fact, if you look at all the substances your brain is made of, fat makes up more than half of it!

Thump's Health Hint

You can help your brain by eating healthy fats! Your brain needs certain kinds of fat (like omega 3 fat) to help you learn and feel cheerful. Sometimes we don't eat enough of the omega 3 kind of fat. Eating plenty of the foods below will help your brain get this important nutrient!

- Fatty fish like salmon, sardines, and anchovies
- Flax seeds and chia seeds
- Walnuts
- Meat, dairy, and eggs from pasture-raised animals
- Avocados

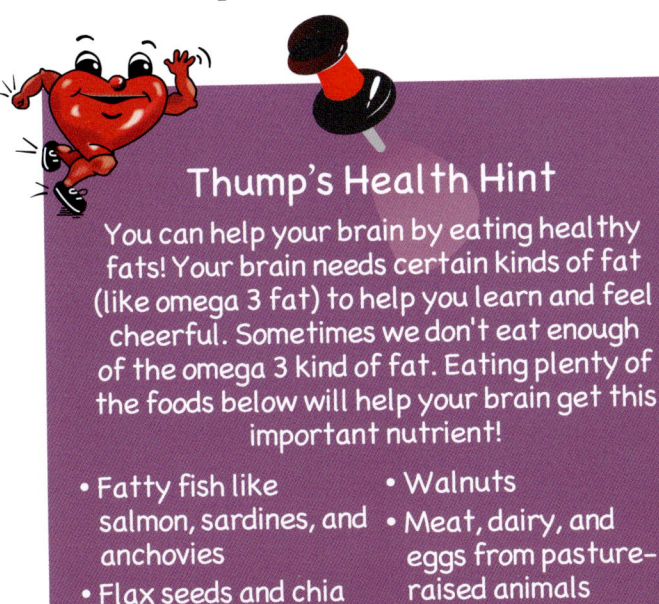

Time to do Activity 31 in the Activity Book!

GOD MADE ME

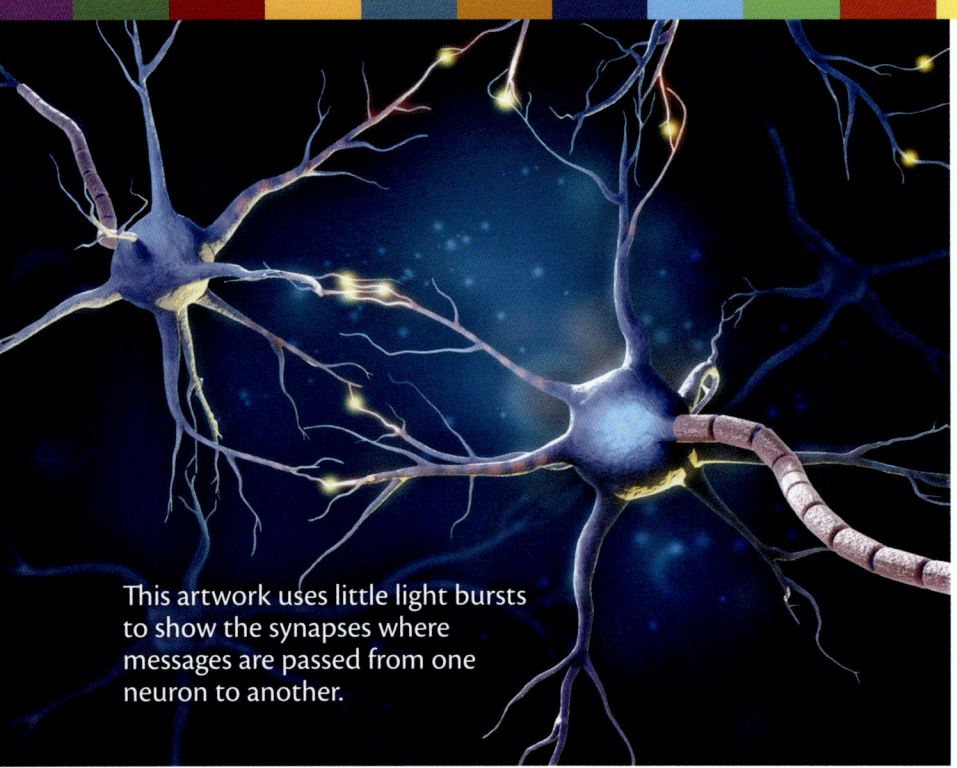

This artwork uses little light bursts to show the synapses where messages are passed from one neuron to another.

Messages Flash between Neurons

Your brain may have about 100,000,000,000 (one hundred billion) neurons. These neurons are very important. They send and receive messages all around your brain. But most neurons do not touch each other to pass their messages. They have spaces between them called **synapses** (sin-apps-sez). Let's learn about the way messages travel through neurons and across synapses to other neurons.

1. A neuron receives a message in some of its dendrites or on its cell body.
2. From the cell body, the message is carried along through the axon, acting like electricity.
3. The message now must cross a synapse to get to another neuron.
4. The branches of the axon release a chemical into the liquid in the synapse. The message has changed from electrical to chemical.
5. Dendrites on a different neuron collect the chemical.
6. Some chemicals can tell the second neuron to go ahead and carry the message. Other chemicals can tell it not to carry the message. Let's call them **"Go" chemicals** and **"No" chemicals**. The dendrites of several neurons can respond to the chemicals of the first neuron's axon.
7. If there are enough "Go" chemicals, the message will become like electricity again. It will pass through the second neuron.
8. If there are enough "No" chemicals, the message will not be allowed to pass through the second neuron.

Synapses are important because they can control whether or not a signal is allowed to travel. They can also turn off a signal when it isn't important anymore. The information our brain receives is important, but it doesn't stay important. Can you remember how nice it is when your brain says that you smell bacon cooking? But what if you kept smelling bacon even when there isn't

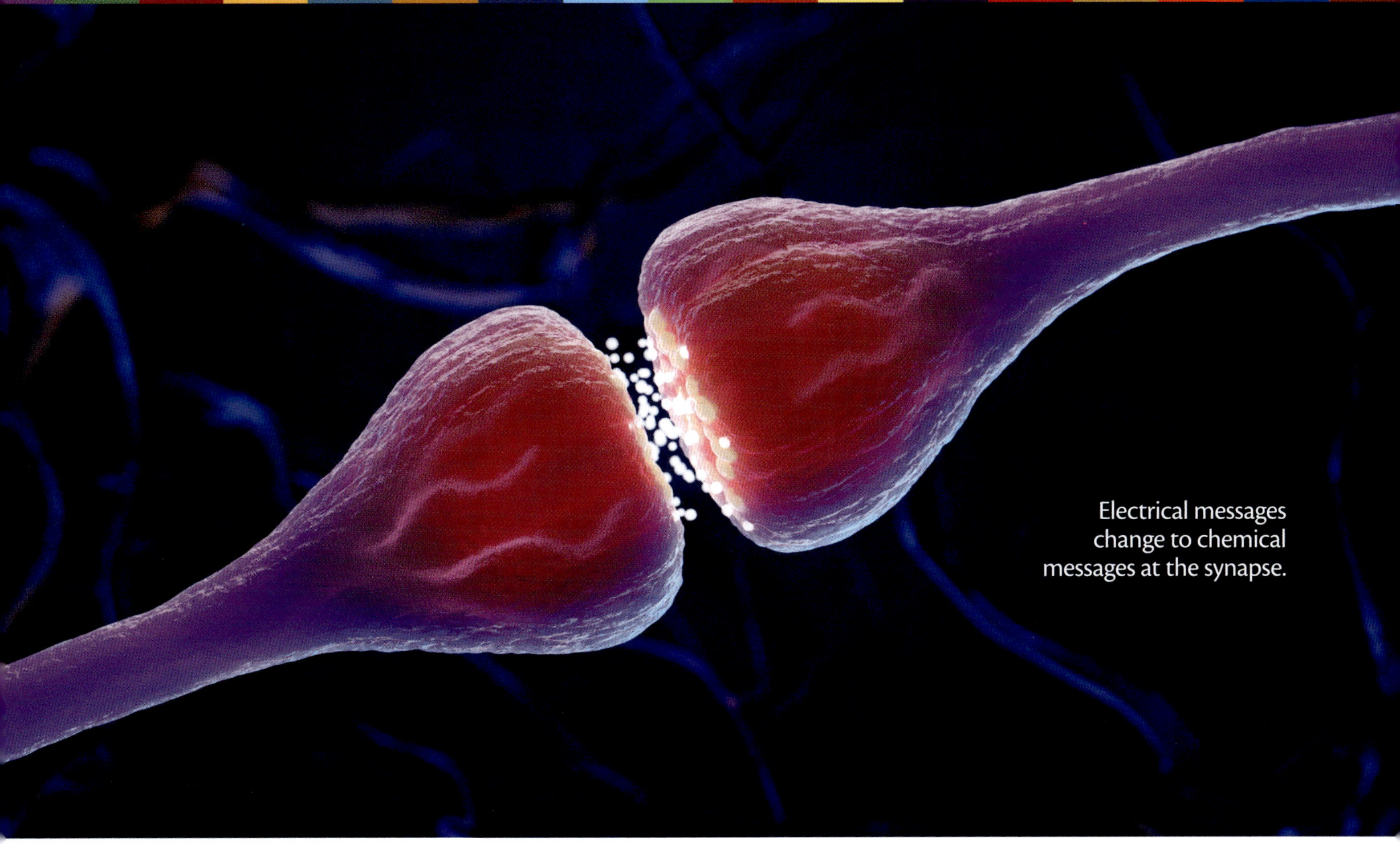

CHAPTER 11: YOUR NEURONS

Electrical messages change to chemical messages at the synapse.

any? Synapses are where messages can be stopped or turned off.

God gave each neuron in your brain about 7,000 connections with other neurons! Each connection has a synapse. Each connection processes a different piece of information or holds a memory.

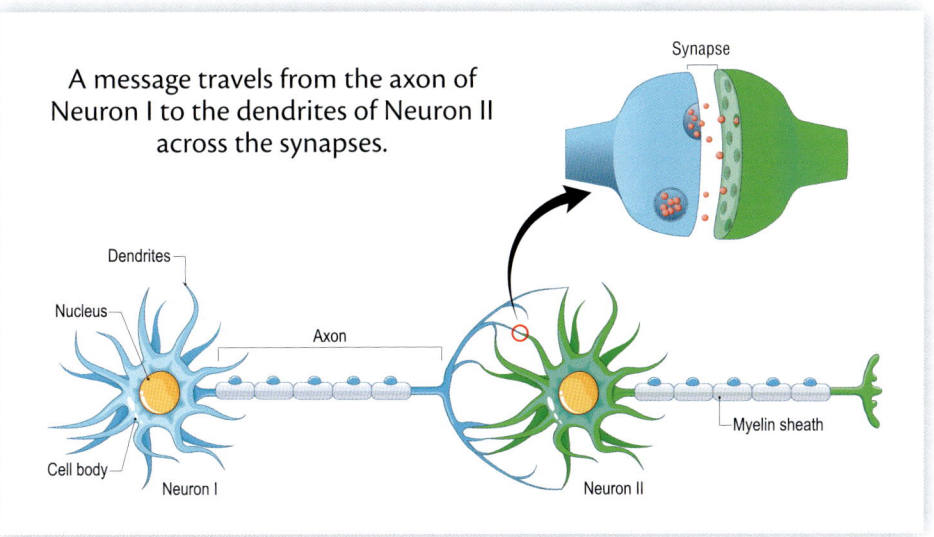

A message travels from the axon of Neuron I to the dendrites of Neuron II across the synapses.

Time to do Activity 32 in the Activity Book!

111

Nerve Pathways

You have three kinds of neurons. Let's see what they are!

1. Your **sensory neurons** receive signals from your senses. Your sensory neurons live in sensory nerves and in mixed nerves.

2. Your **interneurons** receive messages from sensory neurons. They process this information and send commands to motor neurons. Most of your neurons are interneurons. Your interneurons live mostly in the brain and spinal cord.

3. Your **motor neurons** send commands to your muscles and some of your organs, telling them what to do. The axons of your motor neurons live in motor nerves and in mixed nerves. Their cell bodies live outside the nerves.

Your interneurons are amazing! They are the reason your brain is in charge. Interneurons live in your brain and spinal cord. They receive messages from other neurons. They decide which messages are important, and they ignore the others.

Most messages coming to interneurons

Picture A

Sensory neuron | Interneuron | Motor neuron

Sensory Neuron → Interneuron → Motor Neuron

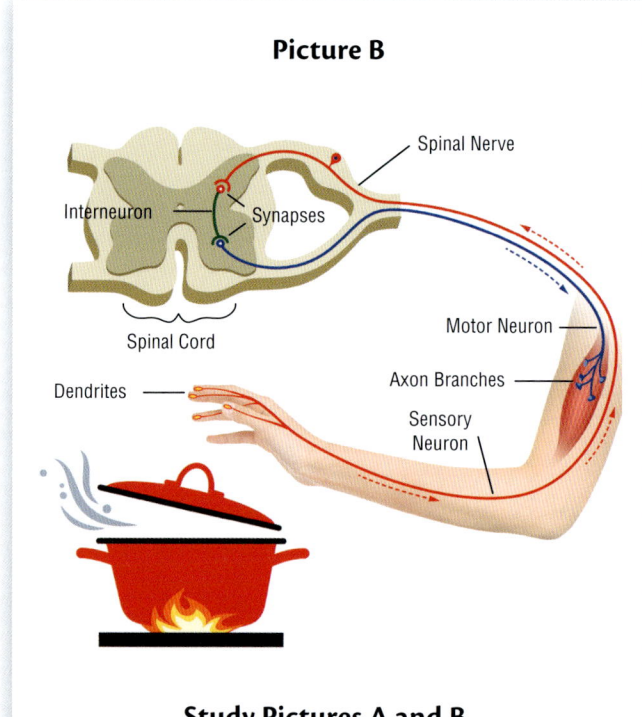

Study Pictures A and B.
In picture B, can you find a sensory neuron (red), its dendrites, its cell body, and its axon? Can you find an interneuron (green), its cell body, and its axon? (Its dendrites don't show in this picture.) Now can you find a motor neuron (blue), its cell body, and its axon? (Its dendrites also don't show in this picture.) Beginning with the burn pain, trace the pathway of messages through the neurons. Say "jump" as you cross synapses.

CHAPTER 11: YOUR NEURONS

are not important. But the messages that are important get special treatment. The interneurons immediately send important messages to the proper place. An important message might be sent to a special area of your brain to tell it something. Or it might be sent somewhere to be put in your memory. It could even be sent to a place used for thinking and problem solving. Your interneurons also give commands to your motor neurons. Your busy interneurons sort out a lot of information and quickly send it along to do its job for you. Your brain talks to itself using its interneurons.

When messages are sent, certain pathways are made between your neurons. As you learn to play a musical instrument, pathways are made throughout your brain. Pathways are also made between your brain and the neurons in your skin, ears, eyes, and muscles. Certain neurons connect with others at the synapses. As you practice music, you play better and better. You are improving because the pathways are becoming more permanent. The axons are making extra chemicals to put in the synapses. Your neurons are making extra dendrites to collect the chemicals. This makes messages travel more quickly and automatically.

Prayer

Dear God, it's amazing how quickly neurons work. Thank You for giving us neurons and synapses that can learn to work even more quickly! Help us to practice the things we need to learn so that we will do them well. When we do things well, help us remember that You are the one who made us able to do them. We give You the glory for our skills! In Jesus' name, Amen.

Time to do Activity 33 in the Activity Book!

CHAPTER 12
Your Nervous System Works Automatically & on Purpose

Do you remember that your nervous system includes your brain, spinal cord, and nerves? We have already divided your nerves according to the jobs they do—your **sensory nerves** gather information from your senses, and your **motor nerves** tell your body what to do. We have also divided them according to where they are—**cranial nerves** are in your head, and **spinal nerves** are in your spine.

We can divide your nerves another way. Some nerves make parts of your body work **on purpose**. They are called **somatic** (so-mat-ick) nerves.

Other nerves make parts of your body work **automatically**. They are called **autonomic** (ah-toe-nah-mick) nerves. We have already learned that you brain controls many automatic jobs in your body. Let's learn about your autonomic nervous system.

Your Nervous System Works Automatically All the Time

Your brain reminds me of this verse. Your brain is awake and in charge even when you are sleeping. It makes sure your whole body works well day and night. God is in charge of helping us all the time. He never sleeps!

My help comes from the LORD, Who made heaven and earth. . .
  Behold, He who keeps [His people]
Shall neither slumber nor sleep.
  (Psalm 121:2, 4)

Your body has many jobs it does automatically, even when you're asleep. You don't even have to think about these jobs. You don't have to tell your body when to start them or when to stop.

115

GOD MADE ME

But how does your body know when to start and stop the jobs that you're not paying attention to? God gave your autonomic nervous system the ability to be in charge of your organs, blood vessels, and glands so that it knows when each part should do its job.

Let's learn a few new words that will help us understand your autonomic nervous system.

Thump's List of Jobs Controlled by Your Autonomic Nerves

Your nervous system automatically controls many things in your body, including:

- Focusing your eyes to see things at different distances
- Changing the size of your pupils for light or darkness
- Releasing your tears
- Making your nose run
- Making your mouth water
- Making you sweat and giving you goosebumps
- Changing the speed of your heartbeat
- Changing your blood pressure
- Telling your body when to fight germs
- Adjusting the amount of air you breathe
- Managing digestion
- Keeping solid waste in your body until you're ready to go to the toilet
- Keeping urine in your body until you're ready to go to the toilet
- Keeping your blood sugar at the right level

Definitions

Hormone — A chemical that's made in one part of your body and causes something to happen in another part. Because of your autonomic nerves, hormones are released automatically when you need them. For example, insulin (in-suh-lin) is a hormone made by your pancreas (pang-kree-us). It keeps your blood sugar at the right level.

Gland — An organ that automatically releases a hormone or other chemical for your body to use. Automatic nerves send messages between your brain and your glands and organs. (Your pancreas is a gland. You also have glands that make tears, saliva, mucus, sweat and many other useful things.)

Time to do Activity 34 in the Activity Book!

116

CHAPTER 12: YOUR NERVOUS SYSTEM WORKS AUTOMATICALLY & ON PURPOSE

Automatic Calming, Automatic Excitement

Your autonomic nerves are divided into two groups. One group contains nerves that cause you to **rest and digest**. The other group contains nerves that make you ready to fight or run away (**fight or flight**).*

This family is watching a movie of happy puppies wrestling with each other outdoors. In the first picture, the family feels that all is well for the puppies. They are eating popcorn and their heartbeats are normal. Autonomic nerves are sending calming messages.

Family using autonomic rest-and-digest nerves

Suddenly, a pack of coyotes appears on the movie, growling and barking at the puppies. The second picture shows the family startled by the coyotes! Their heartbeats have sped up, their breathing has changed, and they are not interested in popcorn. They feel like fighting the coyotes or running away. Different autonomic nerves have taken over and have sent excitement messages to help deal with an emergency.

Family using autonomic fight-or-flight nerves

Fight-or-flight nerves do the opposite of rest-and-digest nerves. Fight-or-flight nerves excite you, and rest-and-digest nerves calm you. But they both work automatically without you having to think about it.

The two branches of your autonomic nervous system are shown on the following page. Notice that your fight-or-flight nerves come from different places than your rest-and-digest nerves. Notice they have opposite jobs. (In this picture, **stimulate** means to encourage or cause more of something. **Inhibit** means to discourage or cause less of something.)

*Scientists use the word *parasympathetic* for rest-and-digest nerves and *sympathetic* for fight-or-flight nerves.

GOD MADE ME

Rest-and-Digest Nerves | Fight-or-Flight Nerves

- Constrict pupils
- Stimulate saliva
- Decrease heart rate
- Constrict airways
- Stimulate digestive activity
- Stimulate activity of intestines
- Contract bladder

- Dilate pupils
- Inhibit saliva
- Increase heart rate
- Relax airways
- Inhibit digestive activity
- Relax bladder

Time to do Activity 35 in the Activity Book!

118

CHAPTER 12: YOUR NERVOUS SYSTEM WORKS AUTOMATICALLY & ON PURPOSE

You Can Purposely Work Your Nervous System

Now let's learn about your somatic (on-purpose) nervous system. Remember, your autonomic nervous system is working all the time. But whenever you are awake, you are also using your somatic nervous system. You use it to move, and you use it to think.

Moving

Often your movements start with a thought. Maybe you wake up, and you immediately want breakfast. Your thoughts tell you to get out of bed. You start to walk to the kitchen, thinking about what you would like to eat. But then your brain remembers that you are supposed to get dressed and make your bed before breakfast. So your brain sends messages to your muscles to make them do those jobs. When you want to move, your brain sends messages to your muscles through your somatic nerves.

But often your movements don't start with only a thought. Often you notice something with your senses. Your sensory nerves send messages to your brain. Maybe you smelled pancakes cooking, and this caused you to get up for breakfast. These sensory messages passed through your somatic nerves. Your brain processed the information and sent messages to your muscles to tell them what to do. These motor messages also traveled through your somatic nerves. Do you remember when we talked about stepping outside on a cold winter morning? When your body reacted to the slippery ice, you were using your somatic nervous system.

You also use your somatic nerves when you are learning to do something with your

Learning to knit uses your somatic nerves.

body. Learning to play soccer, write neatly, play a trumpet, and cut with a sharp knife all use your somatic nerves. Remember, when you practice these skills, your neurons and synapses help you get better at doing them.

Do you remember what happens with your reflexes when you touch a cactus spine or something hot? Your reflex messages make a quick trip from the area of pain to your spine and then right to to the muscles that get you out of trouble. Reflex messages also use your somatic nerves.

verse or math facts. You also use it when you come up with new ideas or invent something. Let's look at the brain and see where thinking happens!

The gray matter covering the outside of your cerebrum has different areas that process different things. These areas are called **lobes**. The different lobes help you do different on-purpose things. If you could see your brain from the top of your head, you would see two halves. The halves have matching lobes.

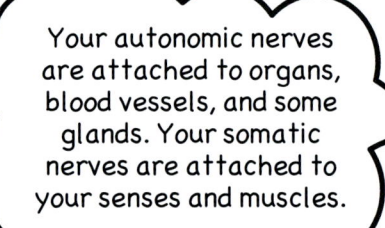

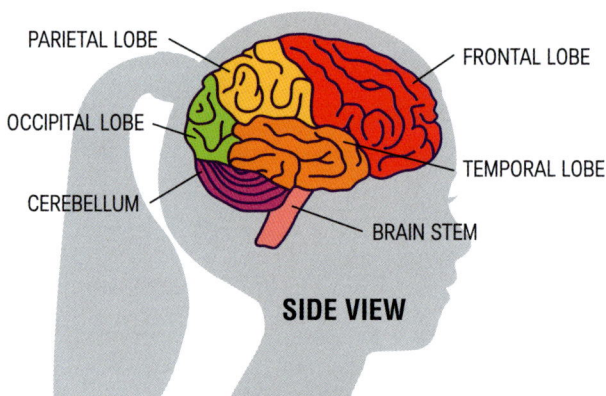

Thinking

What you have learned and received and heard and seen in me—practice these things, and the God of peace will be with you. (Philippians 4:9 ESV)

Thinking does not use your somatic nerves. But it does use the on-purpose parts of your brain. You use your brain when you learn something for the first time. You use it when you work on memorizing a Bible

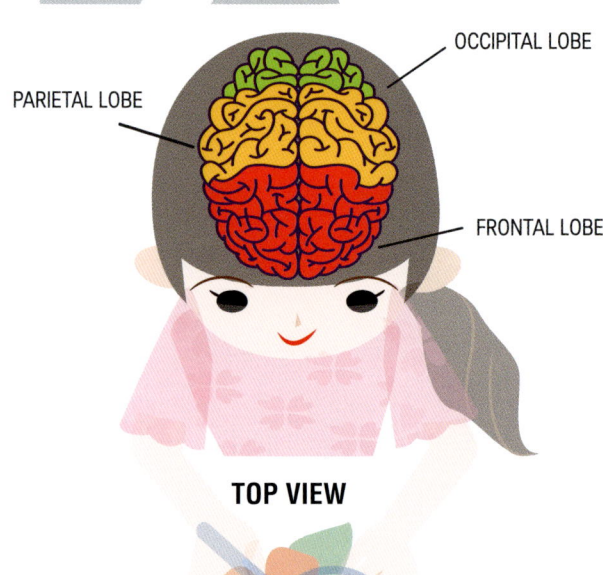

CHAPTER 12: YOUR NERVOUS SYSTEM WORKS AUTOMATICALLY & ON PURPOSE

 ## Puff's List of Brain Lobes and Their Jobs

Frontal lobes are used for:
- Making plans
- Making decisions
- Learning new things
- Remembering
- Intelligence
- Telling muscles to move. (Your right lobe commands muscles on the left side of your body. Your left lobe commands muscles on the right.)
- Our frontal lobes might be where God put our personalities.

You use your frontal lobes when you learn.

Temporal lobes are used for:
- Figuring out messages from the senses
- Remembering complicated things you have seen or heard (for reading, art, or music)

Your temporal lobes help you remember how things look.

Parietal (puh-rye-eh-tuhl) lobes are used for:
- Figuring out touch messages from the body's skin
- Speaking and understanding language
- Knowing your body's position and what's around you

Occipital (ox-sip-eh-tuhl) lobes are used for:
- Receiving messages from the eyes
- Figuring out color, shape, size, movement
- Sending this information to a place where memories help you identify what is seen
- Recognizing faces

You use your parietal lobes when your body moves from one place to another.

Your occipital lobe helps you find someone in a crowd.

121

The things you do on purpose come from several parts of your brain. When you read, your frontal lobe tells your eyes to scan the words on the page from side to side. Your eyes send messages to your occipital lobe. Your occipital lobe recognizes the messages as writing. It works with your temporal lobe to recognize the words. Then your temporal lobe figures out the meaning of the words. At the same time, your parietal lobe knows what the book feels like as your fingers separate a page to turn. When it's time, your frontal lobe tells your muscles to turn the page.

Isn't God amazing to make your brain able to do automatic and on-purpose things all at the same time? Your brain can:

- **Monitor** (be watchful over) your body's insides
- Command your body's inside jobs
- Monitor information from your senses
- Ignore useless information
- Tell your muscles what to do
- Make memories and learn

The brains of animals can also do these things. God gave all animals special gifts that allow them to perfectly live in their neighborhoods. But just think of all the things you can do that an animal cannot do. God made you more complicated than animals.

God has also blessed you more than the animals by giving you a **soul**. We might also call this your *spirit* or *heart*. Your soul is where you know and love God. Because you have a soul, you are responsible to God. You know that He made you. You know right from wrong. You know He expects you to do right and not do wrong.

Your soul doesn't have a place in your body that scientists can label on a drawing. But your soul and body are connected. If you read Philippians 4:8, you'll find a list of good things we are told to meditate on (think about and study). Your body does the reading. Your soul knows that these good things are God's Word to you. Your soul meditates on them. Because your brain is thinking about them for a long time and in different ways, new neuron pathways are formed. This makes it easier to think about things in this good way in the future.

Philippians 4:8 says to meditate on whatever things are true, noble, just, pure, lovely, of good report, virtuous, and praiseworthy.

The Bible says that we can know what kind of person we really are by looking at

CHAPTER 12: YOUR NERVOUS SYSTEM WORKS AUTOMATICALLY & ON PURPOSE

who we are on the inside, in our heart and soul (Proverbs 23:7). Do you meditate on good things? If so, you are building good thinking habits. But if you meditate on bad things, your brain will make bad-thought pathways. Thankfully, Jesus died to free us from the bad things we think and do! You can ask Him to lead your soul and your brain into good pathways.

Prayer

Heavenly Father, You have made our amazing brains! We praise You with our minds and praise You for our minds. Thank You for putting our wonderful brains in charge of our bodies. Help us to meditate on good things so our brains will have good pathways. Thank You for being in charge of us and of everything there is. Amen.

Time to do Activity 36 in the Activity Book!

UNIT 4
God Makes You Aware

In our memory verse, Jesus shows that our senses are blessed when they see (believe) and hear (obey) Him! Because of your senses, you can know God's Word and God's world. Now let's learn about your senses!

God gave you senses so you can be aware of things around you! Our hymn helps us thank God for our senses. Your five main senses are taste, touch, smell, hearing, and sight.

 ## Memory Verse

Blessed are your eyes for they see, and your ears for they hear. (Matthew 13:16)

You can listen to this hymn by searching for "For the Beauty of the Earth for kids" on the internet.

 ## Hymn to Sing: For the Beauty of the Earth

For the beauty of the earth,
For the glory of the skies,
For the love which from our birth
Over and around us lies:

Chorus:
Lord of all, to Thee we raise
This our hymn of grateful praise.

For the wonder of each hour
Of the day and of the night;
Hill and vale, and tree and flow'r,
Sun and moon, and stars of light:

For the joy of ear and eye,
For the heart and mind's delight,
For the mystic harmony
Linking sense to sound and sight:

For the joy of human love,
Brother, sister, parent, child,
Friends on Earth and friends above;
For all gentle thoughts and mild;

For Thyself, best gift divine,
To our world so freely giv'n:
For that great, great love of Thine,
Peace on Earth and joy in Heav'n:

CHAPTER 13
God Gives You Eyesight

Lift up your eyes on high, and see who has created these things. (Isaiah 40:26)

You can use your eyes to look up, down, and all around. Everywhere you look, you will see things God created!

God has created many beautiful and wonderful things for us to see. And He has created our beautiful and wonderful eyes to see with!

God gave your eyes many amazing *structures* that make sight possible. He also gave your eyes *movement* so you can look clearly at the things you want to see. And God gave your eyes wonderful *liquids* to keep your eyes healthy and working properly. We will learn about these eye structures, movements, and liquids in this chapter.

Even with your wonderful eyes, you wouldn't be able to see if God didn't make two other amazing things—light and your brain. Without light and without your brain, your eyes would be useless! Light bounces (reflects) off the world around you and goes into your eyes. Your eyes send information about this light to your brain. Your brain makes sense of it and tells you what you are seeing. Isn't God wise to make things in this world that work so perfectly together? Let's learn about eyesight!

Hi! I'm Beamer, the light beam. I love sending God's light into your eyes to help you see His beautiful creation!

Remember: Light + Eyes + Brain = Sight!

GOD MADE ME

Eye Structures Are Made for Sight

Your eyes are shaped like balls. They live inside safe, hollow places in your bony skull. Between each **eyeball** and its **eye socket** you have a cushion of fat for protection. You also have **eyelids**—skin that can quickly blink to protect your eyes and spread moisture over them.

> ### Definitions
>
> **Eye sockets** — the deep, hollow, bony cups in your skull that form protective shells for your eyes.
>
> **Eyeball** — the ball-shaped part of the eye between your eyelids and the eye socket.
>
> **Focus** — to bend the light coming into your eye so a clear picture hits the inside of your eyeball.

Your eyeballs are about one inch (2.5 cm) wide. Each eye has an **optic nerve** that connects to your brain.

Stand in front of a mirror and try to find these structures of your eyeball:

- **Sclera** (sklare-uh) — the white part of your eye. The sclera is a tough, protective covering that wraps almost all the way around the outside of your eyeball. Underneath the sclera is a layer that contains many blood vessels. Underneath the blood vessels, a third, dark brown layer lines the inside of your eye. This keeps your eye dark inside to help you see.

- **Pupil** (pew-pill) — the black circle you see at the front of your eye. Your pupil is the opening that light passes through. Your pupil looks dark because you are seeing the inside of your eyeball. It opens wider when you are in a darker place. This lets more light inside your eyeball so you can see better. Your pupil gets smaller when you are in a bright place. This protects the inside of your eye from too much light.

- **Iris** (eye-riss) — the ring of color surrounding the pupil. Your iris contains muscles that make the opening of your pupil larger and smaller. The back of your iris blocks light so the only light entering your eye comes through your pupil.

- **Cornea** (corn-ee-uh) — the transparent (clear) covering over your iris and pupil. Your light-blocking sclera

CHAPTER 13: GOD GIVES YOU EYESIGHT

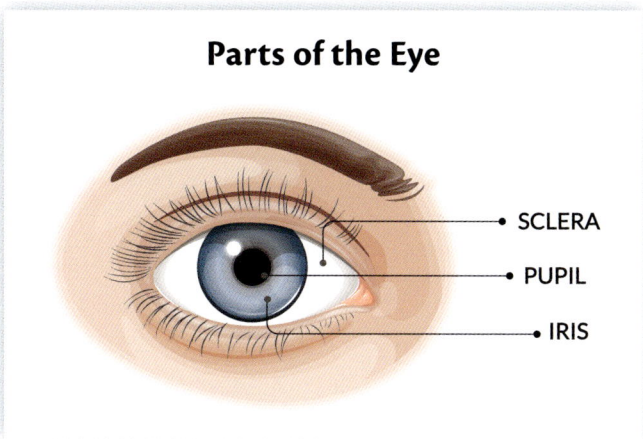

(white) becomes your transparent cornea at this place. The cornea bulges out from your eyeball, causing light to bend toward the pupil. God made your cornea with five layers, all very clear!

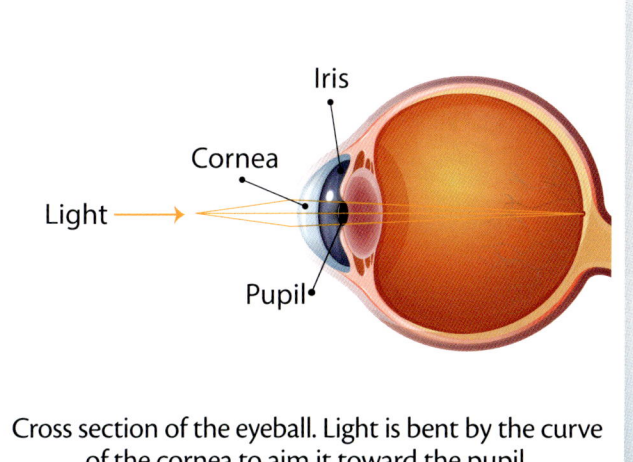

Cross section of the eyeball. Light is bent by the curve of the cornea to aim it toward the pupil.

Notice that light is shining through the transparent cornea but not through the sclera of this beautiful eye. Can you see the pupil, iris, sclera, and cornea?

GOD MADE ME

Now let's learn about the structures inside your eyeballs!

 Lens (lenz) — a wonderful, transparent, rounded structure that focuses light onto the back of your eyeball. Your lenses become more rounded when you look at something up close. They become more flattened when you look far away.

 Retina (ret-in-uh) — the inner lining of your eyeball. Your retina is thin, delicate, and almost transparent. It contains a layer of light-sensitive cells. Your retina also contains a layer of neurons which become the optic nerve.

Optic (op-tick) **nerve** — a nerve that leaves the back of your eye and goes into your brain.

Inside the Eye

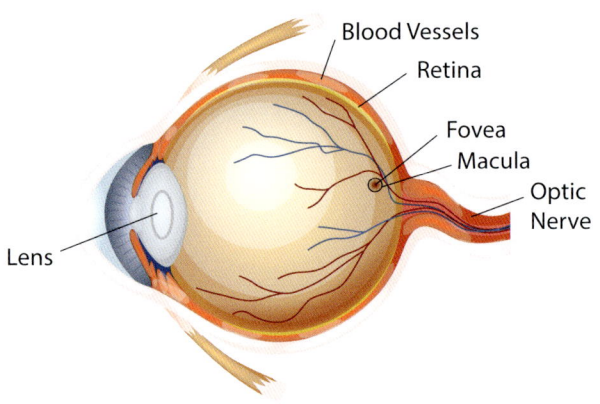

Macula (mack-you-luh) — A yellow spot in the center of your retina. This spot helps you see what's right in front of you, in the center of your vision. Your macula has a dent in its center called the **fovea** (foe-vee-uh). The fovea is the area of your eye used for your sharpest vision.

God gave you something like sunglasses inside your eyes! Your macula has yellow chemicals that protect your light-sensitive cells from blue light. Blue light comes from sunlight and from computer, television, and phone screens. Too much blue light can destroy your yellow chemicals. But God cares for your vision and replaces your yellow chemicals when you eat healthy foods. Yellow chemicals in foods like corn, egg yolks, leafy greens, and pistachios go straight to your macula when you eat them!*

*You can buy supplements to help give your eyes these yellow chemicals. Look for lutein (loo-tee-en) and zeaxanthin (zee-uh-zan-thin).

Your lens sits right behind your pupil.

Time to do Activity 37 in the Activity Book!

130

CHAPTER 13: GOD GIVES YOU EYESIGHT

Eye Movements Are Made for Sight

For the eyes of the LORD run to and fro throughout the whole earth, to show Himself strong on behalf of those whose heart is loyal to Him. (2 Chronicles 16:9)

God gave you eyeballs that move within their sockets and have moving parts inside them. He did this so you can see lots of things. God is a Spirit and doesn't have eyes like we do. But this verse shows that He can see everywhere and help those who love Him!

way? Each eye has six muscles to help with these motions. These muscles are attached to the sclera of your eyeballs.

Why did God give you two eyes? Why did He put your eyes in the front of your head? Why did He put your eyes a certain distance apart and make them move together? One reason is to give you **depth perception**! This means that you can tell how close something is to you. Depth perception helps you take steady steps, catch a ball, and do careful things with your fingers.

Depth perception works because each of your eyes sees a picture that's a little different from the other one. Hold your thumb up at arm's length in front of you. Close only your left eye and look at your thumb. Now close only your right eye. Do this again more quickly, over

First, let's learn about eye movements caused by muscles.

1. **Your eyes use muscles to move in their sockets.** Remember, muscles are groups of cells that can get longer and shorter to make something move. Move your eyes from the right to the left. Move them up and down. Now move your eyes in circles. Did you notice that your eyes move together the same

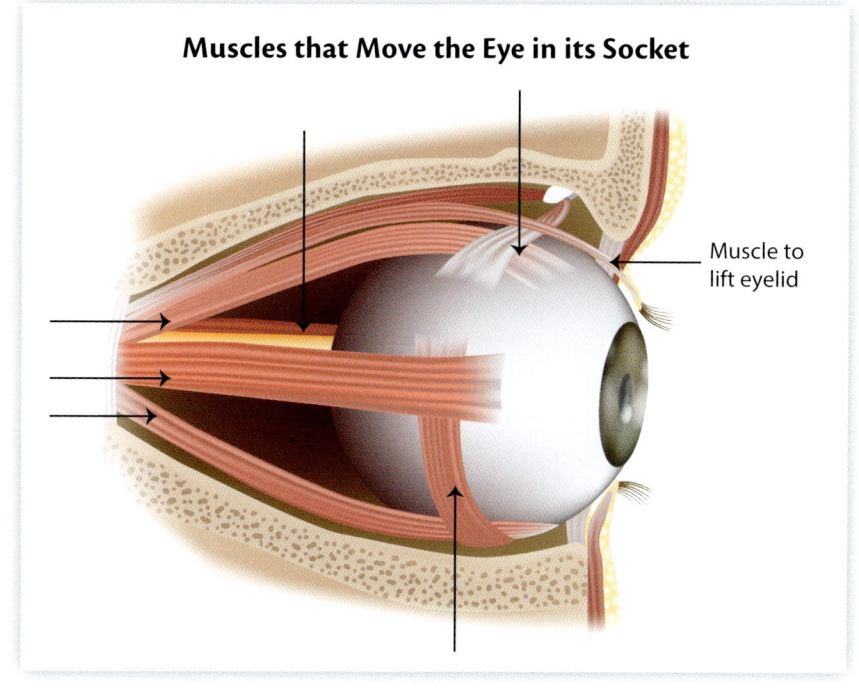

Muscles that Move the Eye in its Socket

Muscle to lift eyelid

131

GOD MADE ME

Aren't you glad you have two eyes that move together to look at the same thing at the same time? Aren't you glad your two eyes are spaced just right on the front of your head? God is very kind to give you so many features that help you see well!

and over again. Does your thumb seem to jump back and forth as your eyes take turns looking? That's because, when you look at something close to you, your eyes are looking from two different angles.

Now look at your thumb with both eyes open. Doesn't that look better? When the two views reach your brain at the same time, you see (and understand) the shapes and distances of objects better.

2. **Your eyes use tiny muscles in your irises to change the size of your pupils.** These muscles move automatically. They are told how much to move by the amount of light hitting your eyes. Your pupils need to get smaller when the light is bright. So a circle of muscles in each iris squeezes and makes itself smaller. This makes your pupil smaller and keeps out some of the light. If the light hitting your eyes becomes dim, different iris muscles pull your pupils open wider. These muscles radiate out from the circle muscles to the outer edge of your irises like spokes on a wheel.

Let's pretend this slice of kiwi fruit is an iris in one of your eyes. The whitish core in the middle would be your pupil. The seeded area would be your iris's circle muscles that squeeze to make your pupil close more. The outer area with whitish "spokes" would be your iris muscles that pull to make your pupil open wider.

3. **Your eyes use muscles to change the shape of your lenses.** When you look at something, the light waves are focused onto your retina by two eye structures—the cornea and the lens. Your cornea's curved shape automatically bends the light toward your pupil. The cornea's shape doesn't change because it has no muscles. It always bends light the same amount. Once light goes through your pupil, it travels to your lens for more focusing.

Your lens shape *can* change. It's

132

squishy. Your lens likes to be plump, but it can be squished flatter. A ring of muscles around each lens helps it squish to change shape. God made your lenses able to change shape so they can bend light different amounts. This helps you focus on things at different distances.

Let's use this orange slice to help us understand your lens muscles. Pretend your lens is the white center of the orange slice. Pretend the orange peel is a ring of muscles around the lens. Now, do you see the membranes separating the orange sections? They are the light lines you see stretched from the center of the orange to its rind. Your eye has rope-like structures that are like the orange membranes. They stretch between the lens and the ring of muscles. In your eye, these "ropes" keep your lens in front of your pupil where it belongs. But they also help change the shape of your lens. Let's learn how that works.

When you look at something that's more than 20 feet (6 m) away, your lens does not need to change its shape to focus. Your cornea helps focus faraway things, and your lens's ring of muscles can rest. This means that the ring is not squeezed into a tight circle. It's relaxed into a big circle like the rind of our orange slice. Because it's a big circle, the "ropes" are pulled tightly across

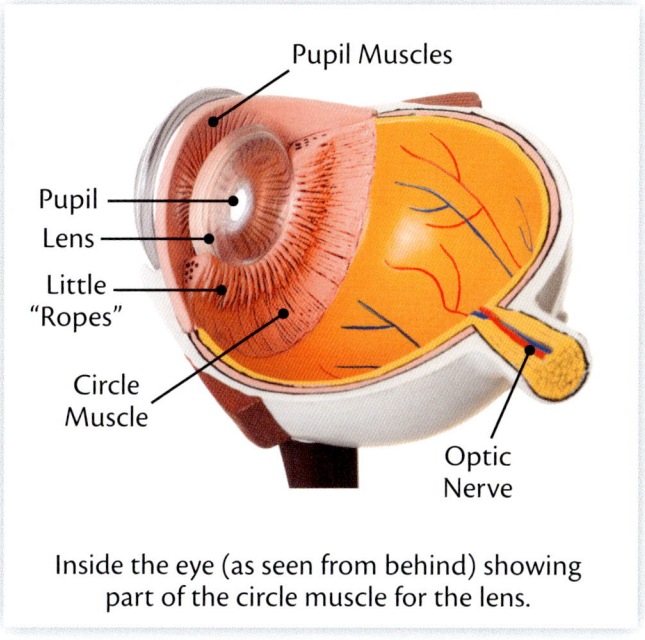

Inside the eye (as seen from behind) showing part of the circle muscle for the lens.

it. These tight ropes are automatically pulling on your lens, making it flatter. Your lenses are flattened when you are looking at something out the window.

Now, what if you squeezed the juice out of the orange slice and somehow made the rind a smaller circle? What would happen to the membranes between the orange sections? They would become loose! The same thing happens in your eye when you look at things closer than 20 feet. The circle of muscles around your lens squeezes to make itself smaller.

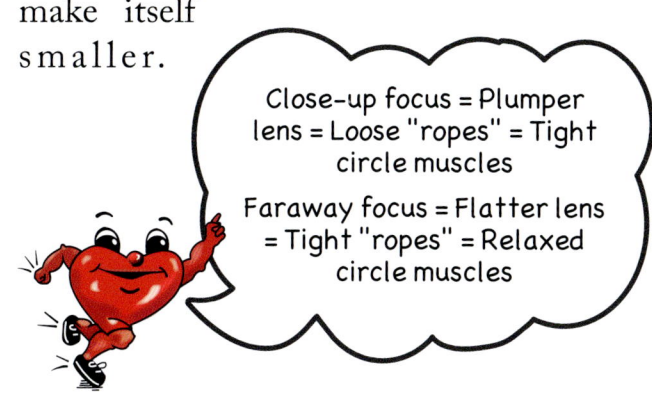

Close-up focus = Plumper lens = Loose "ropes" = Tight circle muscles

Faraway focus = Flatter lens = Tight "ropes" = Relaxed circle muscles

GOD MADE ME

Puff's Health Hint

I hope you understand God's amazing gift of focusing. It's complicated, but it shows us another thing to praise God for!

Now you know why your eyes get tired after reading too long. The muscle circles around your lenses have been squeezing a long time as they helped you focus on close-up words.

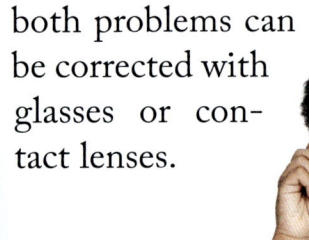

Sometimes people who do a lot of reading and close-up work lose their ability to focus on faraway objects. The next time you read, try looking at something outside the room every time you turn a page. This will relax your focus and may protect your distance vision. Eye doctors might suggest wearing special glasses for close-up work. These glasses may also help protect distance vision.

A smaller circle makes the "ropes" looser. Loose "ropes" don't pull on your lens to flatten it. It becomes plumper. Your lenses are plumper when you are reading or writing. The closer you look at something, the plumper your lenses become.

Now let's learn about the movement of light through your eyes!

As light comes into your eye, the first structure it goes through is the cornea. Remember, the cornea bends light to aim it through your pupil. Your iris has opened the pupil to allow just the right amount of light to get through. After it passes through your pupil, the light goes through your lens. The lens muscles try to bend the light so that, when it passes through your eyeball, it focuses exactly on your retina.

For some people, the light focuses too near, before hitting the retina. These people are **nearsighted**. For other people, the light focuses too far. These people are **farsighted**. Thankfully, both problems can be corrected with glasses or contact lenses.

Here's a way to rest and protect your eyes when you use the computer or read: Every 20 minutes, look at least 20 feet away for 20 seconds.

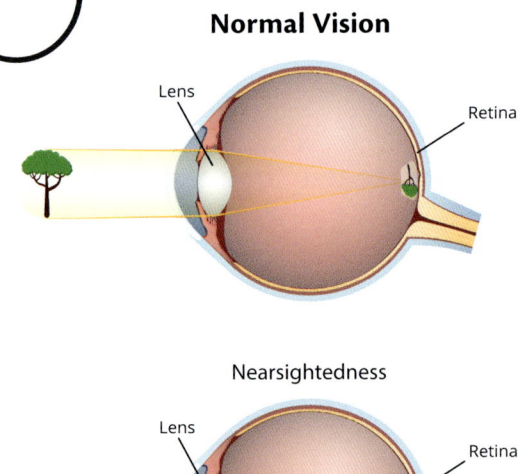

Normal Vision

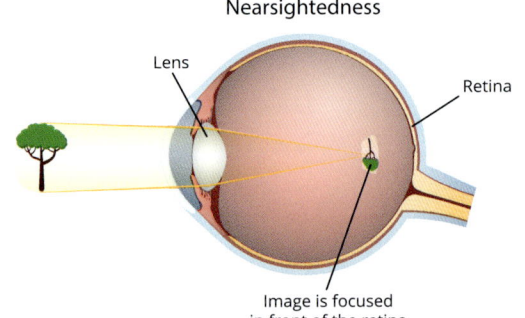

Nearsightedness

Image is focused in front of the retina

134

CHAPTER 13: GOD GIVES YOU EYESIGHT

Zoom In to the Cell

God gave each of your senses special cells that notice the thing they are supposed to notice. These cells are called **receptors**. Sometimes receptors are neurons. Sometimes there are different cells that give their messages to neurons.

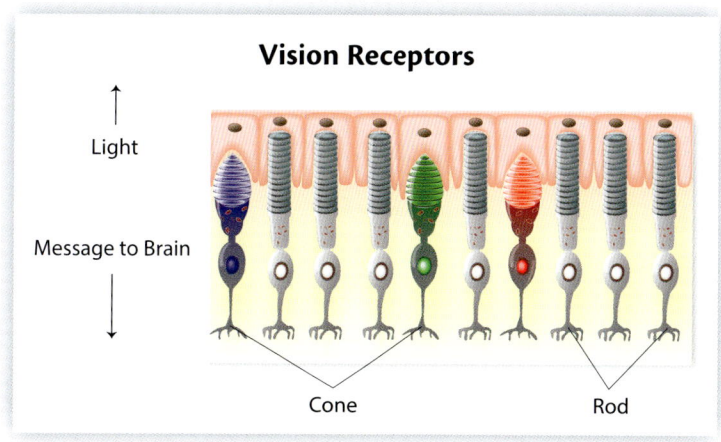

In your eyes, your vision receptors are neurons. They live in your retina.

As light hits your retina, it first meets some neurons that are not your vision receptors. These neurons sense light and tell your irises how much to open or close your pupils. They also help your brain know when daylight is ending so you will get sleepy. Then, just past these neurons, the light hits your vision receptors!

The smallest piece of light is called a **photon**. It only takes one photon to cause some receptors to send a message to your brain. That message tells your brain one tiny piece of what you see. Your retina has millions of receptors. They all send their tiny pieces of information through your optic nerve to your brain. Your brain puts all the tiny messages together to make a whole picture.

Your eyes' receptor cells are wonderful inventions of God! Your retina has two kinds of receptor cells: **rods** and **cones**.

Thump Tells Us the Differences Between Rods and Cones

Rods are shaped like a **R O D.**

Cones are shaped like a **CONE.**

Rods sense only **BLACK** and **WHITE**.

Cones sense only **COLOR**. There are three kinds of cones for sensing three colors—**RED**, **BLUE**, and **GREEN**. All the other colors you see come from combinations of these three colors.

Rods help you see **BRIGHTNESS** and *MOTION*.

Cones help you see DETAIL.

135

GOD MADE ME

> **Each eye has about 100,000,000 rods. Each eye has about 6,000 cones.**

Rods are more plentiful farther away from the macula. Cones are more plentiful in and near the macula. The fovea has only cones.

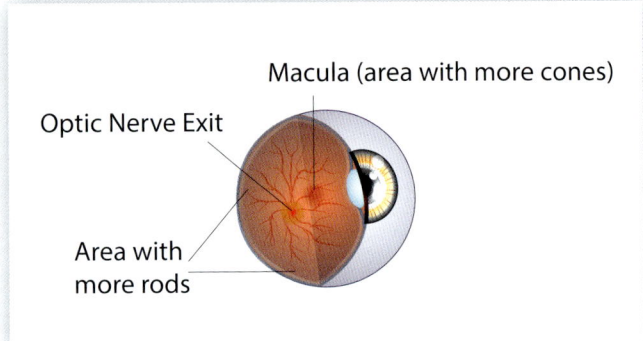

You have a place in your retina where the optic nerve leaves your eye. Blood vessels also pass through this area. This optic nerve exit has no rods and no cones. Any light that hits this **blind spot** will not be seen by your eye. Is this a problem? No! God works it out so that you don't even notice it. Your blind spots are in different places in your two eyes, so what one eye misses will be seen by the other one. Also, your eyes move around fast enough that you have already seen what the blind spot covers, and your brain fills in any missing information.

Your retina has other kinds of neurons that live among the ones we have learned about. These neurons help organize and combine the messages as they are passed along.

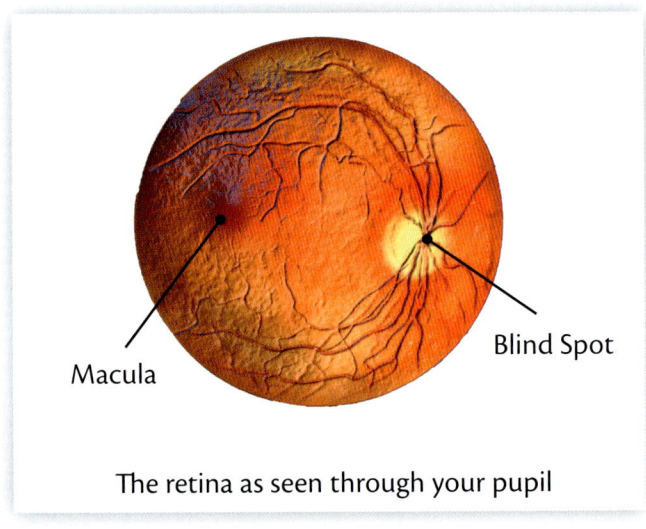

The retina as seen through your pupil

Rods are 100 times more sensitive to light than cones are. They help you see at night. Cones help you see in the day.

Time to do Activity 38 in the Activity Book!

Eye Liquids Are Made for Sight

Thanks to God's wisdom, your eyes make some very important liquids for themselves. First, let's look at the liquids that cover the outside of your eyeball.

Outer-Eye Liquids

Do you like the way your eyes always glisten? They glisten and sparkle because of **tears**. Tears keep the eye from drying up. Without tears, your eyes could become damaged enough to harm your vision. Most of your tear liquid is made continually in a **lacrimal** (lack-rim-uhl) **gland**. You have one for each eye. The lacrimal gland is above your eye's outer corner, just below your eyebrow.

Tears contain three liquids. These liquids stay in three layers on your eyes:

- **Bottom layer: Mucus** — a slimy liquid that helps tears spread evenly over the eyeball. Eye mucus is made by a clear tissue that covers the whites of your eyes and the inside of your eyelids.[1]
- **Middle layer: Water** — to keep the eye wet and clean. This salty water is made by the lacrimal gland.
- **Top layer: Oil** - to keep the tears from evaporating. Glands found near your eyelashes make this complicated mixture of oils, waxes, and other chemicals.

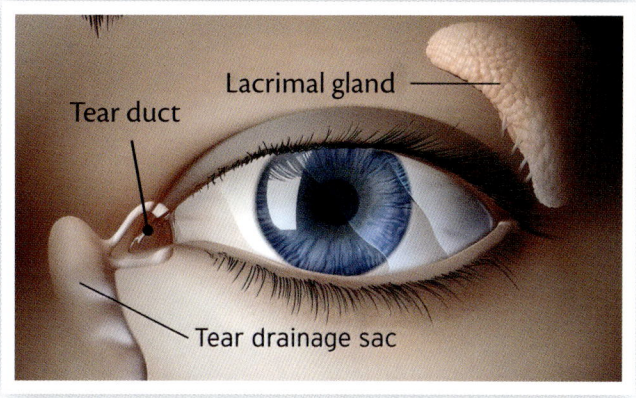

Tears deliver oxygen and nutrition to your outer eye since there are no blood vessels there. Tears also contain chemicals and microscopic creatures that protect your eyes from germs.

Humans are the only creatures that cry because of emotions. Our loving God gives us a comforting chemical in our emotional tears. The tissues around our eyes absorb this chemical from the tears. The chemical then travels to the brain to give a feeling of comfort.[2]

Extra tears leave your eye through tiny holes in your upper and lower eyelids. These drain holes are in the inner corner of your eye, close to your nose. In fact, the holes lead to little tubes (**tear ducts**) that join and carry the liquid to your nose through another duct. This is the reason you get a runny nose when you cry.

Inner-Eye Liquids

You have two different liquids inside your

---

[1] This covering is the *conjunctiva*. Conjunctivitis (pinkeye) is an infection of this covering.

[2] Randy J. Guliuzza, M.D., "Made in His Image: Tiny Parts Are Big Players in Human Vision," Acts and Facts, vol. 44, no. 6, June 2015, p. 18.

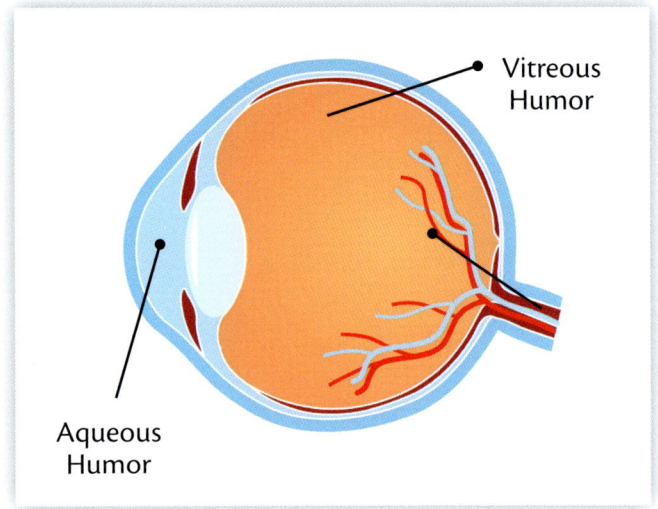

When you get something in your eye, don't rub it. This may press the painful object into the surface of your eye and make it harder for tears to wash it out.

eyeballs. They are transparent of course!

Aqueous (ack-we-us) **humor** fills the space between your cornea and the front of your lens. Your cornea and lens don't have capillaries to supply them with good things. Instead, oxygen and nutrition are filtered out of your blood and placed into your watery aqueous humor. The lens and cornea take what they need from this liquid. The aqueous humor also helps focus light!

Vitreous (vit-ree-us) **humor** fills the space between the back of your lens and your retina. This liquid is gel-like and helps your eyeball keep its shape. Eyeballs need to keep their shape so light will focus at the right place on the retina. The consistency of vitreous humor also helps focus vision—your eyes would have to be two times bigger to focus correctly if they were filled with air instead of this liquid.

Our eyes see many things! We can't see God, but we can see His Word. His Word helps us know Him.

Open my eyes, that I may see Wondrous things from Your law. (Psalm 119:18)

CHAPTER 13: GOD GIVES YOU EYESIGHT

Prayer

Lord, You made such wonderful, beautiful, and complicated things when You made eyes! Our eyes have so many parts that have to work together at the same time for us to see. And You make them work with our brains and the light You created. We praise You for eyesight! Thank You that we can use our eyes to read Your Bible and see so many colorful and interesting things in Your world. Amen.

Time to do Activity 39 in the Activity Book!

139

CHAPTER 14
God Gives You the Sense of Touch

> That which was from the beginning, which we have heard, which we have seen with our eyes, which we have looked upon, and our hands have handled concerning the Word of life . . . we declare to you. (1 John 1:1-3)

Your Covering of Skin

Most of your sense of touch is felt on your skin. Scientists don't know everything about how we feel things, but God does. Let's learn about your skin!

Your skin is one of your body's largest organs. If you weigh 60 pounds, your skin will be about five pounds of that weight. Your skin is not just a tight bag you wear to hold your insides together. Your skin is God's wonderful creation of layers and structures that protect your whole body. Let's look at some of the things skin does!

Jesus' disciples were blessed to be with Him. With their senses, they saw, heard, and handled (touched) the things He taught them about eternal life. They were very excited to "declare" to others the things they knew about Jesus.

We were not with Jesus when He was on Earth. But we can know the things declared about Him in God's Word. Then, if we believe, we will be blessed too (John 20:29)!

GOD MADE ME

Puff's List of Skin's Protective Jobs

- 👉 Your skin is a barrier between your insides and the outside world. Many germs can't grow on healthy skin. Germs need moisture to grow, and skin is not very moist. Skin also coats itself with chemicals that keep germs from growing.
- 👉 Your skin's sense of touch gives your brain valuable information about air temperature and the position of things around you.
- 👉 Your skin can alert you to danger when it feels pain, pressure, heat, and cold.
- 👉 Your skin is waterproof. It keeps your insides from drying out.
- 👉 Your skin contains a pigment called **melanin** (mell-uh-ninn) which protects you from the sun's harmful rays.
- 👉 On a hot day, your skin's blood vessels can make you cooler. The vessels open up to let more blood come to your body's surface where it gets rid of heat.
- 👉 Your skin can also make you cooler by releasing **sweat** (swett). The water in sweat evaporates, which makes your skin and body cooler.
- 👉 On a cold day, your skin's blood vessels can preserve your body's heat. The vessels close partway so less blood will come to the surface where it would lose heat.
- 👉 Your skin also preserves heat with its bottom layer of fat. Cold takes a lot longer to pass through fat than through muscles and other body tissues.
- 👉 Your skin's fat also protects the tissues under it when you bump into something.
- 👉 Your skin uses sunshine to make Vitamin D. Vitamin D is good for your bones and helps your body fight germs.

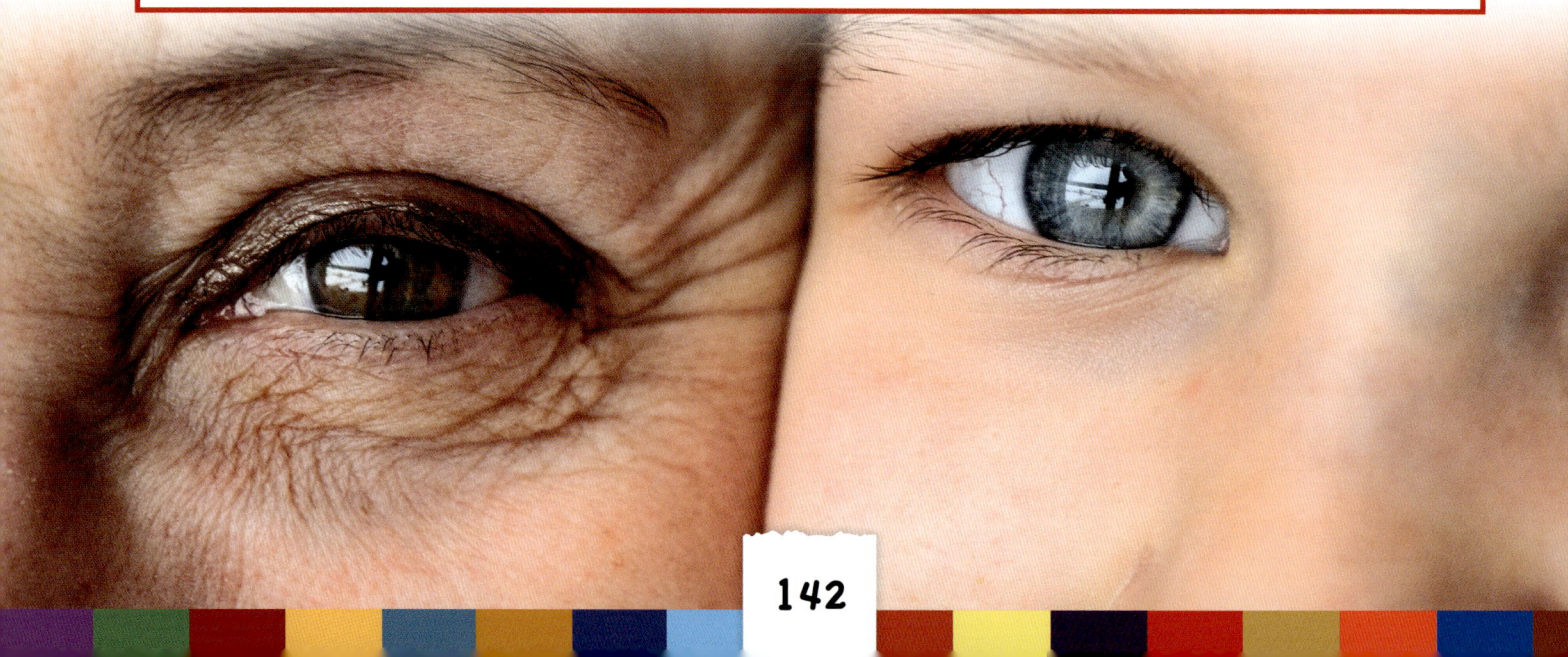

CHAPTER 14: GOD GIVES YOU THE SENSE OF TOUCH

Your skin is made of layers. The top layer, the **epidermis** (epp-ih-derm-iss), is made of dead skin cells. These cells are continually falling off—about half a million every day! But don't worry. Your skin keeps making new cells and pushing them up from underneath.

Your epidermis is where the special cells that make melanin live. The more melanin-making cells you have, the darker your skin will be. When you spend time outdoors, these cells make even more melanin and give you a suntan. Melanin is the pigment that makes the inside of your eyes dark. It also gives brown-eyed people their iris color.

Below the epidermis is the **dermis** (derm-iss) layer. The dermis is where new skin cells are made and pushed up into the epidermis. Hair grows from the dermis.

The bottom layer of your skin is the **hypodermis** (hy-poe-derm-iss). It's made of fat and **connective tissue**. Half of your body's fat lives in your skin. Besides protecting you from cold and injury, the fat layer can give you energy if you don't get enough energy from your food. Connective tissue in the body connects things. The connective tissue in your hypodermis connects your skin to the muscles underneath it.

These children have different amounts of melanin in their skin and eyes.

Skin Layers

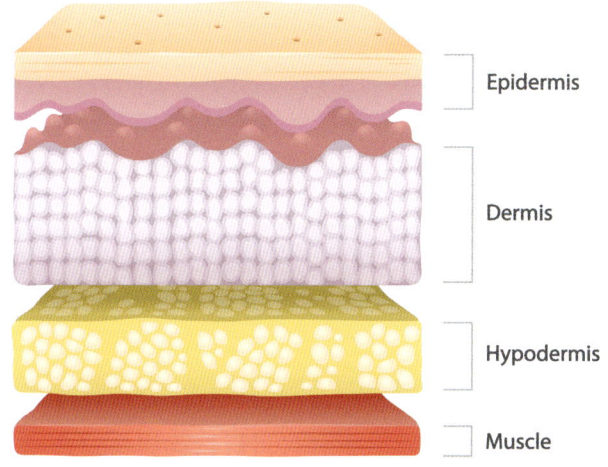

Time to do Activity 40 in the Activity Book!

Touch Receptors in Your Skin

Do you remember that your eyes' receptors (rods and cones) are unusual neurons made to sense light? Your skin doesn't use whole neurons as receptors for sensing things. Instead, your skin's neurons have unusual dendrites that are your skin's receptors.

143

GOD MADE ME

Some of these dendrites have interesting structures on their ends. The different structures help you feel different things. Some, near your skin's surface, sense gentle touch; deeper ones sense hard pressure or pinching of the skin. Some receptors sense when your skin is stretching. Some help you feel texture as you gently rub against something. Your skin receptors sense these feelings because they notice when there is movement of the skin—either a lot or a little.

You have other dendrite receptors in your skin that don't have special structures. They are called **free nerve endings**. Free nerve endings don't sense movement like the dendrite structures do. Instead, they sense temperature changes and pain.

Scientists don't understand exactly how skin receptors sense temperature changes. They do know that the receptors for sensing cold do not send very many messages to the brain when your skin is a normal temperature. They send more messages when things get chilly. Then, when things are very cold, they stop sending messages. This is why your feet get numb when you wade in a very cold stream.

The skin receptors that sense heat also do not send very many messages to your brain when your skin is normal temperature.

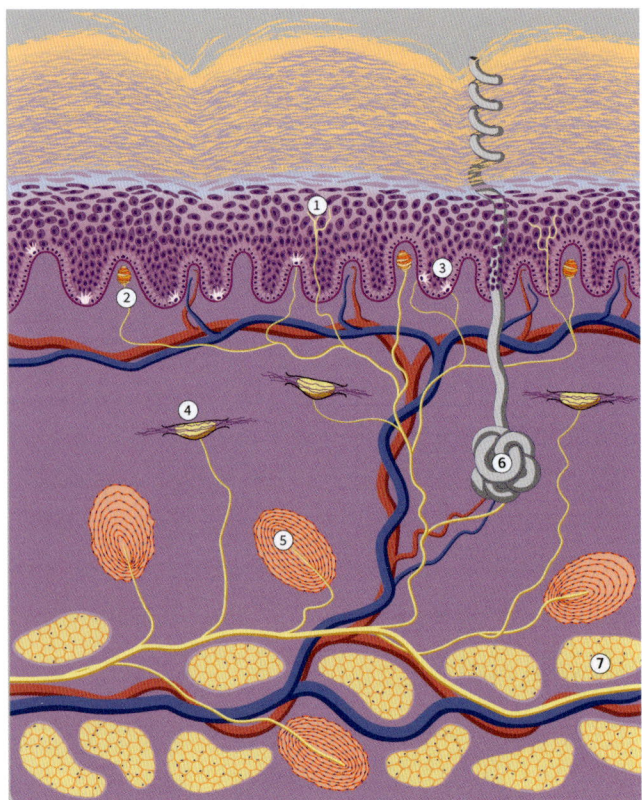

Receptors in the Skin
#1: Free nerve endings (dendrites).
#2–5: Special structures on dendrites
#6: Sweat gland
#7: Fat

CHAPTER 14: GOD GIVES YOU THE SENSE OF TOUCH

They send more messages with higher heat. Then, if the temperatures are burning, pain messages are sent!

Pain receptors in your skin sense something different than movement or temperature. Scientists think pain receptors sense chemicals. These chemicals come out of cells when they get damaged. The cells might break if they are pinched, cut, or burned. Damage causes the liquid inside the cell to leak out, where it's noticed by your pain receptors. Then the pain receptors send an alarm to your brain!

Pressure receptors on your feet help you keep your balance.

Your body has more than three million pain receptors. They live in your skin, muscles, bones, blood vessels, and some organs. They help keep your body safe from harm by sending early warning messages to your brain.

A bee sting will cause a sudden pain message to take a fast pathway to your brain. A longer-lasting ache will cause a pain message to take a slower pathway to your brain.

Thump's Health Hint

When a bee stings, she leaves behind her stinger and a little pouch of venom attached to it. Bee venom is a chemical that gives you more pain than just the stinger alone would. If you get stung, you should remove the stinger and pouch as soon as possible. Muscles in the pouch continue to squeeze out venom even though they are no longer connected to the bee! It's best to do this by sliding a knife under the pouch and scraping it out. If you use tweezers, you could squeeze more of the painful venom into your skin!

Light-touch receptors help your skin feel crawling bugs.

 **Time to do Activity 41 in the Activity Book!**

145

GOD MADE ME

Things That Come Out of Your Skin

Isn't your stretchy, bendable skin amazing? Imagine stretching a sock over one foot and pretending it's your skin stretching over your bones and muscles. But the sock can't feel anything. It can't help you balance. It can't feel if an ant is crawling on your foot. If your sock is too close to the fireplace or buried in snow, it won't tell you to move your foot to safety. There are other things your sock won't do. Your sock won't grow hair or toenails. And even though it might be a sweaty sock, it didn't make the sweat. But your real foot can do all these things. It does them because of its amazing skin! Now let's look at the things God brings out of your skin.

Your Skin Grows Nails

Your toenails protect the ends of your toes when you stub them. Your fingernails protect the ends of your fingers. Fingernails also give you built-in tools like tweezers, scrapers, and scratchers. Let's take a closer look at nails!

Your fingernails and toenails are made of a wonderful, tough protein called **keratin** (carrot-in). Keratin is strong and light. Keratin is also found in your top skin cells. It isn't as hard in your skin as it is in your nails, and it can bend more easily.

Mammal hooves and horns are made of keratin. So are bird feathers and beaks.

CHAPTER 14: GOD GIVES YOU THE SENSE OF TOUCH

Let's Have Thump Tell Us About the Structure of Your Nails!

- The main part of the nail you see is called the **nail plate**. When you clip, cut, or file your nails, you are shortening the nail plates.
- The nail plate grows from your **nail root**. As it grows, your nail is pushed from the nail root toward the end of your finger or toe.
- The **lunula** (loon-yuh-luh) is the part of your nail root that you can see through your nails. Lunula means "little moon." Can you see why?
- The **nail bed** is underneath your nail plate. The nail bed contains blood vessels and nerves for your nail. As the nail root grows, the nail bed adds keratin to the underside of the nail plate to make it thicker.
- Your **cuticle** makes up one edge of the waterproof seal between your nail and your skin.

Definitions

Scalp — the skin that covers the top of your head. The borders of your scalp are your face and the back and sides of your neck.

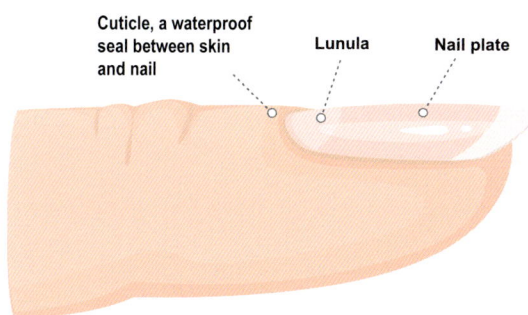

Side View of Finger and Nail

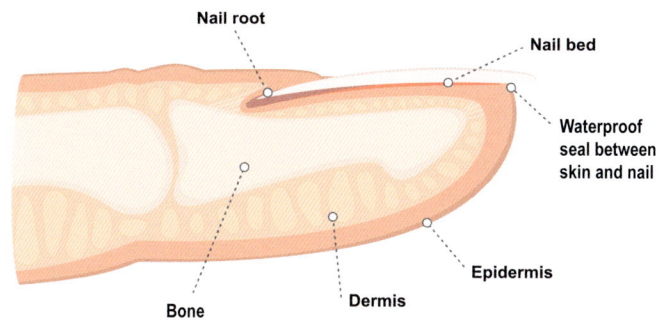

Your Skin Grows Hair

Hair is a very interesting thing that grows out of your skin! Can you imagine how hard it would be to tell people apart if none of them had hair on their heads? The color and style of your hair is part of what makes you look unique.

A single hair will grow for a while. Then it rests but stays attached. Later, a new hair starts to grow from the same place. The new hair pushes the old hair along until it falls out. Your **scalp** has a lot of thick hairs that grow long. Scalp hairs usually need to be cut before they fall out. Your eyebrows and eyelashes are thick hairs that stay short. Many parts of your body grow short, fine hair. The palms of your hands and soles of your feet grow no hair at all!

GOD MADE ME

Each hair has special receptors under your skin that help you sense when your hair moves. This helps you know if something is touching your skin or hair. These nerves might also help your scalp feel nice when someone arranges your hair. Or they tell you when your hair is stuck on something and about to be painfully pulled. Of course, if hair is pulled too hard, pain receptors take over!

Now Let's Have Puff Tell Us About Hair!

- Hair grows out of hair follicles. **Hair follicles** are deep pits that live mostly in the skin's dermis.
- Hair is made from special cells that live in the bottom of the hair follicle.
- Melanin is also made by these cells to give hair its color.
- As these cells divide and grow, they are pushed up the narrow follicle. The shape of a follicle gives the hair its shape. If the cross section of your follicles are round, the cross section of your hairs will be round. You will have straight hair. If your follicles are oval, the cross section of your hairs will be oval, and you will have wavy or curly hair.

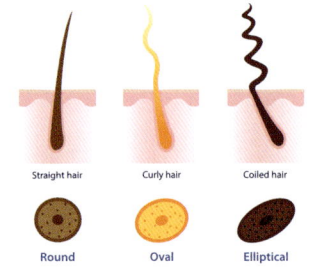

- As the hair cells are pushed up the follicle, they begin making keratin, and they die. By the time you see a hair outside your skin, it's mostly keratin, and it's dead.
- Goosebumps are caused by tiny muscles attached to hair follicles. When the muscles shorten, the hairs stand up and goosebumps appear where the hairs come out of the skin.

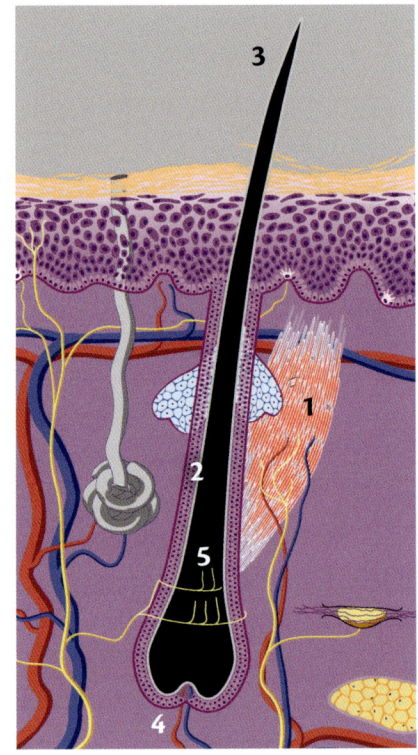

#1 — Muscle to make hair stand up
#2 — Hair follicle
#3 — Hair
#4 — Blood vessels to feed growing hair cells
#5 — Free nerve endings to feel hair movement

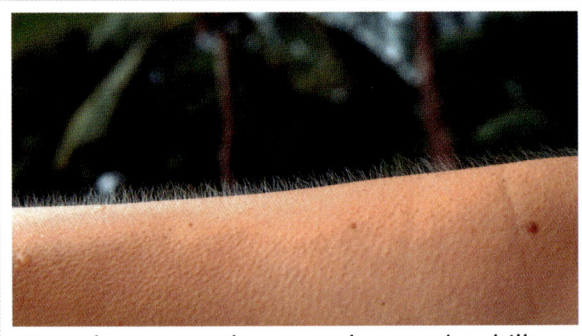

Goosebumps can happen when you're chilly or feel certain emotions.

CHAPTER 14: GOD GIVES YOU THE SENSE OF TOUCH

Your Skin Oozes Sweat

Do you remember that glands can be a special group of cells that make something for your body? Your skin has a lot of glands that make sweat or **perspiration** (purr-spur-ay-shun). Sweat glands live in almost all areas of your skin. The palms of your hands and the soles of your feet have the most sweat glands. This sweat helps you grab things with your fingers and hands. It helps your bare feet grab the ground so you don't slip and slide.

Each sweat gland is made of a tiny tube. The tube is coiled into a ball shape at its deeper end. Sweat is made in the coiled ball. Above the coil, the tube straightens out so the sweat can be carried to the surface of your skin.

Your skin has two kinds of sweat glands. One kind does its job on hot days and when you exercise. To cool you off, it releases sweat directly to the surface of your skin through a tiny hole called a pore. The other kind of sweat gland releases sweat into the hair follicle when a person has certain emotions.

Sweat is mostly water. Sweat also has small amounts of salts and wastes. Sweat sometimes smells badly because of germs that grow easily in moisture.

Your Skin Oozes Oil

We have learned about many tiny structures in your skin. God has given

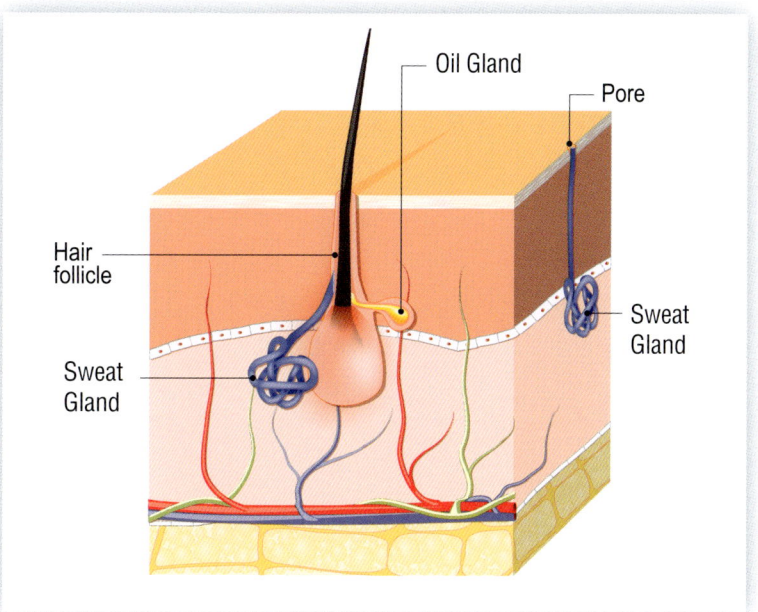

Skin oil is important! It keeps your skin and hair soft. It keeps your skin flexible and waterproof. Skin oil contains chemicals that kill germs on your skin.

When you sweat a lot, your blood loses water. Your brain notices this and makes you feel thirsty. At the same time, your brain tells your body to take less water out of your blood as it makes urine.

149

GOD MADE ME

them all important jobs. We have one more important skin structure to talk about: oil glands! Oil glands are found all over your skin except for the palms of your hands and the soles of your feet. Oil glands are usually found next to a hair follicle where they release oil onto hairs. But you do have a few oil glands in the corners of your mouth that empty directly onto your skin.

**You clothed me with skin and flesh,
and knit me together with bones and sinews.
You have granted me life and steadfast love,
and your care has preserved my spirit.
(Job 10:11-12 ESV)**

Prayer

Father, as this verse says, You are the One who gave us our covering of skin. You clothed us with an amazing organ. Our skin is one of the many ways You give us life and show us Your steadfast love. Thank You for preserving us by giving our skin millions of tiny parts. They do such important jobs! Amen.

Time to do Activity 42 in the Activity Book!

CHAPTER 14: GOD GIVES YOU THE SENSE OF TOUCH

CHAPTER 15
God Gives You Hearing

[Jesus said] "If anyone has ears to hear, let him hear." And He said to them, "Pay attention to what you hear." (Mark 4:23-24 ESV)

Jesus tells us to use our ears to purposely listen to His important teachings. Let's pay attention and not treat what He says as background noise!

Let's try to be still for a minute and pay attention to the sounds our ears are hearing. (Listen for one minute.) When you listened just now, did you notice sounds that you usually don't pay attention to? Often these "background noises" are quiet sounds that go on and on. The humming of a refrigerator or a breeze whispering in the trees might be background noises. Your ears get used to hearing constant sounds. You may not think about these sounds unless they suddenly stop.

Other sounds are louder and more sudden like a bird call, a knock at the door, or the voice of someone speaking. These important sounds interrupt our thoughts and cause us to notice them.

God Made Sound Waves

All the sounds you hear come into your ears as **sound waves**. Sound waves are not like ocean waves, traveling up and down. Sound waves travel forward in bursts. Waves

When you are daydreaming, you may not hear the sounds around you.

Sometimes you hear a noise that makes you alert.

GOD MADE ME

Hello-oh-oh-oh. My name is Song. I'm a sound vibration. I'd like to take you on a ride through your ear. But first, we need to learn about sound so we can be amazed at how God made sound and ears work so perfectly together!

of sound start when something happens that moves the surrounding air. If you snap your fingers, a little blast of air bursts away from your finger as it hits your palm.

To help you understand sound, imagine that you were a big orange ball. You and a lot of other colored balls were walking around in a room. As you walked past a closed door, you thought, "Hmm, I wonder what's in there." You stopped and put your ear to the door, listening for clues. Suddenly, the door burst open and hit you hard! You rolled forwards. You bumped into a yellow ball. The yellow ball was pushed and bumped into a green ball. The green ball bumped into a blue ball. The blue ball bumped into a purple ball, and the purple ball was pushed through the back door. You (the orange ball) did not get pushed all the way across the room. But the *energy* from the door hitting you did travel across the room. Let's call this energy a *wave*.

Sound works the same way! The snap of your fingers is like the sudden opening of the door. The air molecules are like the colored balls. Your ear is like the back door. Air molecules get pushed, and they bump into other air molecules. The air molecules next to the snapping sound do not travel all the way to your ear. But the energy does get passed from one air molecule to the next when they bump. That energy reaches your ear, and you hear the snap!

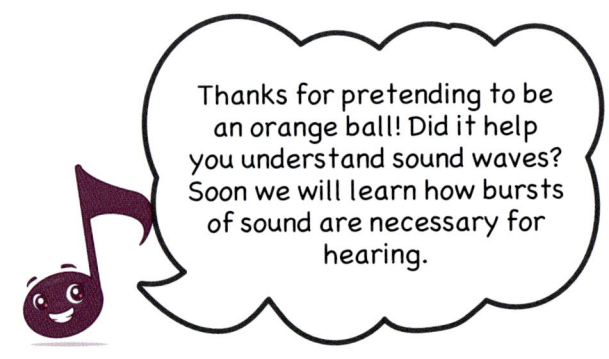

Thanks for pretending to be an orange ball! Did it help you understand sound waves? Soon we will learn how bursts of sound are necessary for hearing.

CHAPTER 15: GOD GIVES YOU HEARING

Definitions

Sound waves are a kind of energy that's released when something vibrates. As it vibrates, it pushes air into your ears.

Vibrate means to move back and forth very quickly.

A quick finger snap makes several waves, but the long whistle of a train makes even more!

Sound spreads in every direction—forwards, sideways, up, down, and around corners. We can think of each color in this sphere as a sound wave spreading in all directions.

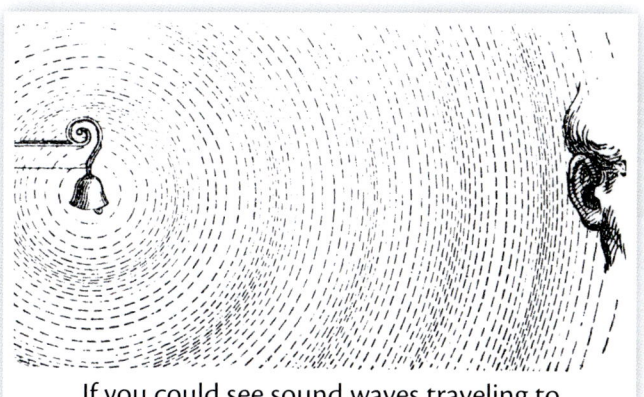

If you could see sound waves traveling to someone's ear, they might look like this.

In the roomful of colored balls, we only talked about the balls that would go in a straight line toward the back door. But when the balls were bumped, they would have also bumped into several other balls. This means the wave would have spread in other directions too. Sound spreads in all directions. This is why your whole family can hear, "Dinner's ready!" from anywhere in the house.

 **Time to do Activity 43 in the Activity Book!**

The Path of Sound Through Your Outer and Middle Ear

Your ears are made of wonderful structures! God has given each part of your ear an important job. Each part does something special with sound wave vibrations.

155

For you to hear, vibrations must pass through three parts of your ear—the outer ear, the middle ear, and the inner ear. Your outer and middle ear sections are filled with air. Let's follow Song as she travels through these two areas in one of your ears!

Through the Outer Ear

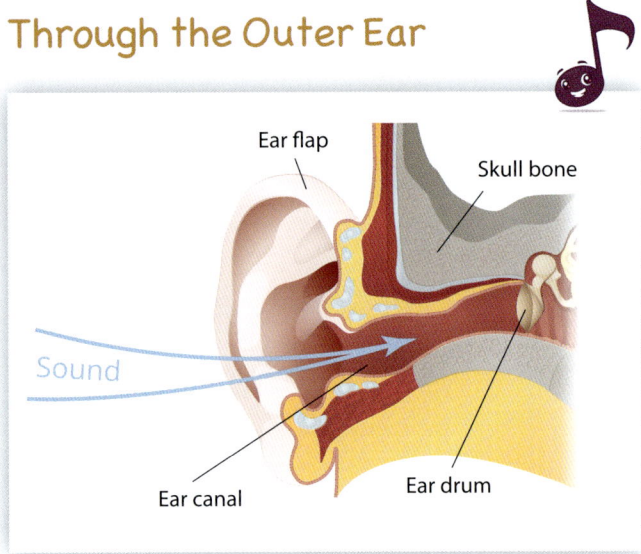

First, Song travels through your **ear flap**, the first part of your **outer ear**. The ear flap is the only part of your ear that you can see when you look in a mirror. The ear flap gathers sound vibrations in the air and directs them inside your ear. The ear flap is made of strong, bendy cartilage covered with skin. Its interesting ridges move sound into your ear in a way that helps you know which direction the sound is coming from. Your ear flap ridges also help sort out sounds to especially bring in sounds from human speech. Isn't God kind to help us hear what others have to say?

Next, Song passes through your **ear canal**. The ear canal takes Song through a hole in your skull. The ear canal is about one inch (2.5 cm) long. It's lined with skin that makes oil and other chemicals. These products trap dust and dead skin cells, and they mix with them to become **earwax**. Earwax feels waxy, but it isn't wax. Earwax cleans your ears, protects them from germs, and keeps them from drying out.

Now it's time for Song to bump into your **eardrum**! The eardrum is a cone-shaped membrane that's stretched tightly across the end of your ear canal. Song's vibration energy makes the eardrum vibrate too.

Through the Middle Ear

Just past your eardrum, Song moves into your **middle ear**. Your middle ear is an air-filled space surrounded by bone. Air from the outside does not travel past your eardrum. Your middle ear is filled with different air.

Do you ever feel like your ears are plugged? Sometimes this happens when you're on a car trip in the mountains or flying in a plane. It happens because the air pressure outside your ear has changed. It makes you want to yawn or swallow so you can "pop" your ears and feel normal again. Your ears feel plugged when the air pressure inside your middle ear is different from the air pressure on the outside.

To solve this problem God has given you a **Eustachian** (you-stay-she-uhn) **tube** that goes from your middle ear to your throat. Your throat is open to the outside air and has the same pressure as the outside air. When your Eustachian tube "pops"

CHAPTER 15: GOD GIVES YOU HEARING

open, the air in your middle ear becomes the same pressure as the outside air. Doesn't that feel better?

Inside your middle ear, you have three tiny bones that could all fit inside an orange seed! These bones are joined together like a bridge between your eardrum and another membrane farther in. Now Song will travel from the ear drum to the first bone—the malleus or **hammer**. The next bone in the bridge is the incus or **anvil**. The last bone is the stapes or **stirrup**. These tiny bones are attached to each other and to the walls of the middle ear. Their most important job is to make sound louder! They take the vibrations from the eardrum's big membrane and squish them onto a small membrane that the stirrup is attached to. By doing this,

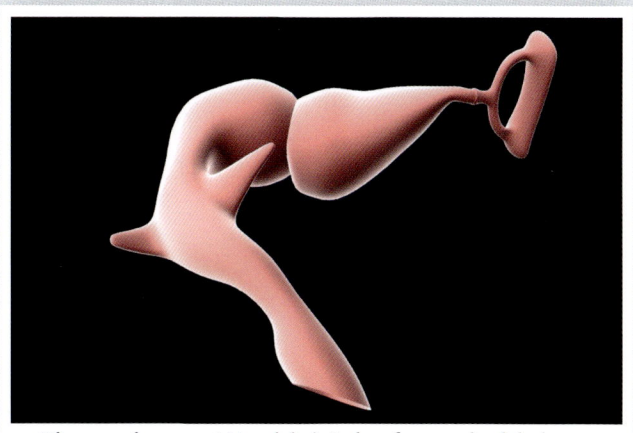

The ear bones. Wouldn't it be fun to hold these tiny bones in your hand and look at them with a magnifying glass?

the bones make sound 20 times louder! You can hear very quiet sounds because God gave you your ear bones.

God also made a way for your middle ear to protect your hearing. Normally, vibrations make your ear bones move as they pass the sound along and make it louder. But when you hear a loud noise, the muscles attached to these bones freeze up. This makes the bones stop! Now harmful vibrations are not passed along as loudly!

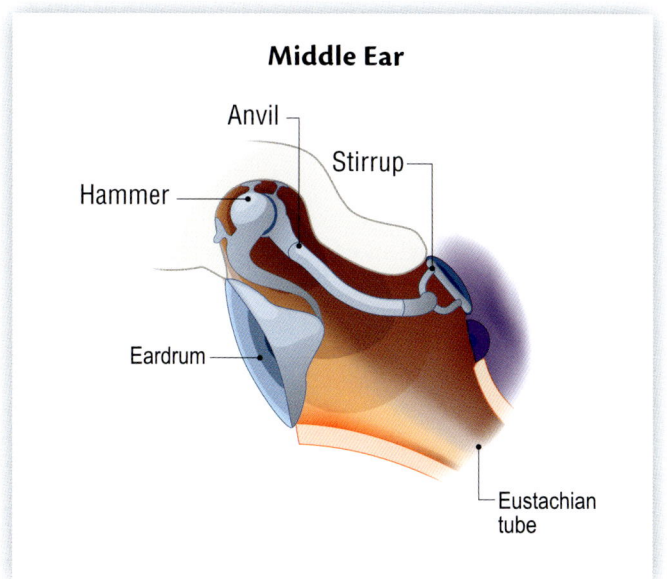

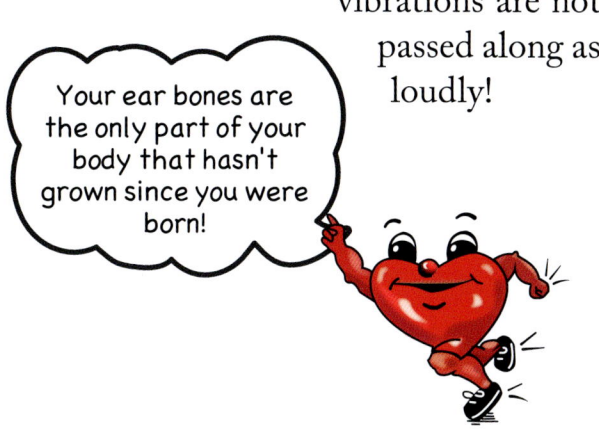

Your ear bones are the only part of your body that hasn't grown since you were born!

Time to do Activity 44 in the Activity Book!

GOD MADE ME

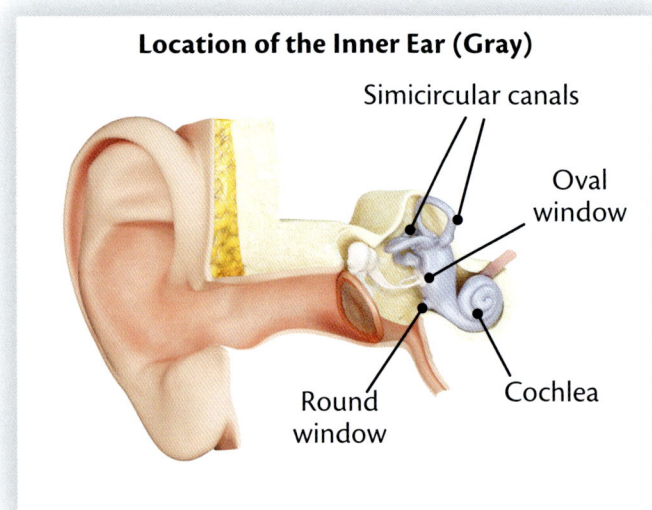

Location of the Inner Ear (Gray) — Simicircular canals, Oval window, Cochlea, Round window

The Path of Sound in the Inner Ear

Your amazing **inner ear** is made of bony structures inside your skull bone. These structures are filled with liquid instead of air. Have you ever noticed that it's harder to wade in water than to walk on land? This is because it's harder for your legs to move through water than through air. It would also be hard for sound vibrations to move through the liquid of your inner ear if your ear bones didn't make the waves stronger.

Now Song is going to travel into your inner ear. As Song makes your stirrup bone move, her vibration is passed to the membrane of your **oval window**. This is where the air vibration of your middle ear becomes the liquid vibration of your inner ear.

The oval window is a small opening in your ear's **cochlea** (cock-lee-uh). The cochlea is a bony structure shaped like a snail's shell. Inside your cochlea, three tubes stretch along its spiral from the beginning to the end. Two of the tubes are bony, and one is made of flexible membranes. Song will travel from the oval window through the cochlea, spiraling along in one of the bony tubes. At the tip of the spiral, she will jump into the other bony tube and spiral back to the beginning. There, she will bump into the membrane that covers the **round window**.

Let's see if we can understand why God put a round window in your cochlea. Imagine blowing up a balloon. Your breath is able go into the balloon because the balloon stretches to

Someome could climb up this spiraling staircase and slide down the spiraling slide next to it. In the same way, Song spirals up one tube in your cochlea and down another.

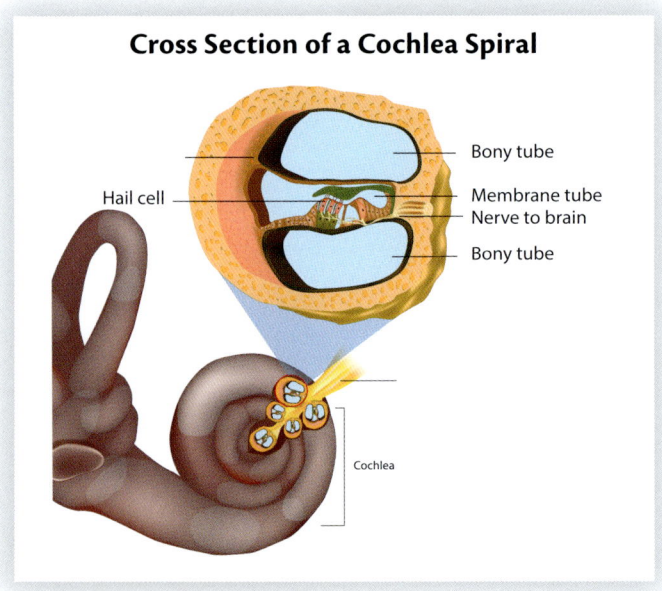

Cross Section of a Cochlea Spiral — Hail cell, Bony tube, Membrane tube, Nerve to brain, Bony tube, Cochlea

CHAPTER 15: GOD GIVES YOU HEARING

make a place for the air to go. Now imagine trying to blow up a glass bottle. You can't blow any more air into the bottle because the bottle won't stretch.

It's the same idea with vibrations that need to enter your cochlea. Your cochlea is hard like a bottle. Vibrations couldn't go in if God didn't invent a stretchy place for the vibrations to bump into after they come out of the cochlea. That stretchy place is the membrane covering your round window. Now, when a new vibration begins at the oval window, its bursts have a stretchy place to end up at!

You can understand this better by puffing up one of your cheeks with air. Now, keeping your mouth and throat tightly closed, press on the puffy cheek. What happens to the air? It must go into the other cheek. This is what happens between the oval window and the round window as the sound vibrations push. When the stirrup pushes the oval window inward, the round window is pushed outward. When the oval window is pulled outward, the round window is pulled in. Sound waves arrive one after another very quickly. The stirrup moves quickly and so do the oval and round windows.

Song has an important job to do as she travels through your cochlea. She does something to the membrane tube that sits between the two bony tubes. Her sound

Bird chirps are high pitched. A lion's growl is low pitched. Aren't you glad God made so many different sounds to hear?

has its own special vibration. Her vibration causes a certain spot in the membrane tube to react. That spot sends a message to your brain about Song's sound. It's as though Song is traveling next to a long piano, and she reaches out to press the key that makes her certain sound. If Song's vibration affects a place where she enters your cochlea, you will hear a high-pitched sound. If her vibration affects a place farther in, you will hear a low-pitched sound.

So far we have been learning about the path of sound vibrations through your ear. Now let's learn how those vibrations become messages to your brain!

The receptors for hearing live inside your cochlea. Hearing receptors are called **hair cells**. Each hair cell has 50–100 "hairs" which are not the same as the hair that grows out of your skin. When a certain sound vibration reaches a certain place in the cochlea, it causes the hairs on those hair cells to bend. They bend because that vibration pushes them up against a membrane above them. When the hairs bend, an electrical message is sent to your brain.

To understand this, imagine that you are sitting on the upper mattress of a bunk bed. Someone is lying on the bed below,

GOD MADE ME

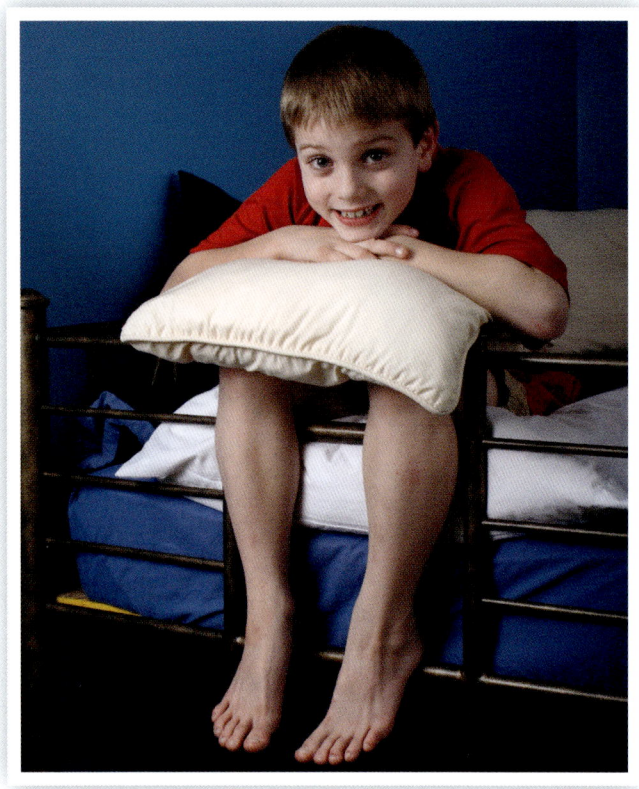

> Now we know why God made sound wave vibrations travel in bursts! Each burst makes a certain hair cell move. Then that certain sound message gets sent to your brain!

also not sleeping. He decides to surprise you by suddenly pushing your mattress up with his feet. He pushes so hard that your head touches the ceiling! You say, "Hey!"

On the bunk bed, you were like a hair cell. The person below was like the sound vibration that pushed you. Your "Hey" was like a message that headed to the brain.

Song has done her job! Her sound vibration has been changed to an electrical message by the hair cell.

In this chapter, you may have noticed some bony loops in pictures of the inner ear. These interesting structures don't have anything to do with hearing. We will learn about these inner-ear structures later.

"Hairs" at the top of a frog ear's hair cell

Thump's Health Hint

Lawnmowers and fireworks can make very loud noises. Noises this loud can bend the hairs of your hair cells so much that they could break off. This would permanently damage your hearing. Be sure to avoid loud noises or wear protection on the ears God gave you!

CHAPTER 15: GOD GIVES YOU HEARING

So then faith comes by hearing, and hearing by the word of God. (Romans 10:17)

People and mammals have an extra kind of hair cell that other animals don't have. These cells help you notice very small differences in sound—especially in the complicated sounds of speech and music!

Prayer

Lord, You have given us the amazing gift of ears. We want to use this gift to listen to the wonderful sounds You made. Thank You for bringing birdsong, music, thunder, and loving voices to our ears with sound waves. Most of all, as this verse says, thank You that we can learn faith by hearing Your Word! Help us to purposely listen at church, during family Bible reading, and anytime we can learn about You. Amen.

Time to do Activity 45 in the Activity Book!

CHAPTER 16
God Gives You Taste, Smell, and Other Senses

Do you remember that your sense receptors for vision are switched on by light? And that the sense receptors in your skin are switched on by temperature or pressure? You've learned that the sense receptors for hearing are switched on by sound waves when they hit the "hairs" of the receptor cells. In this chapter, we are going to learn how the receptors for smell and taste are switched on by chemicals. God gave your senses these different receptors and switches. They work perfectly to make you aware of the things around you!

No matter what kind of receptor we look at, its job is to change your sense's message into an electrical message. All the electrical messages that travel from your senses to your brain are alike. The sensory nerves they travel through are all alike. But the parts of your brain where the messages end up are different! In fact, if we could take the nerve that goes from your nose and attach it to the hearing part of your brain, the smell of pizza might be changed into the sound of crunching leaves! But don't worry. God did not create us to have this confusing problem.

For God is not the author of confusion. (1 Corinthians 14:33)

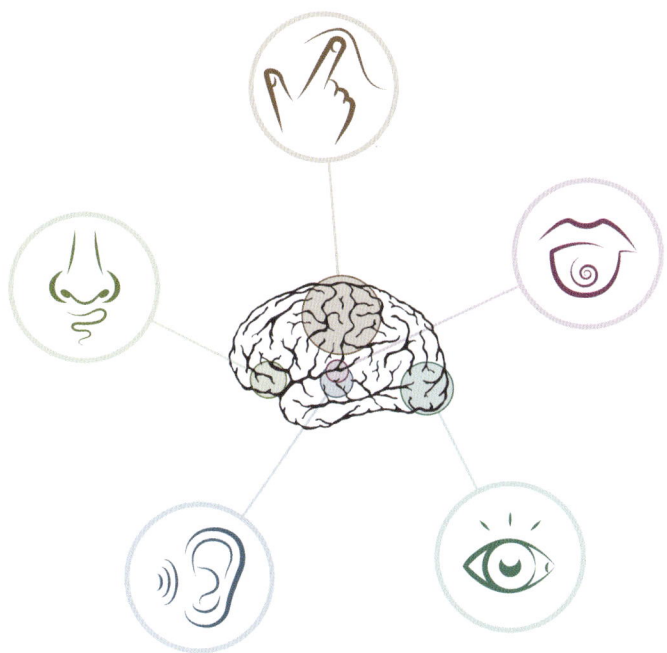

GOD MADE ME

God Gives You the Sense of Smell

**Now thanks be to God who . . . through us diffuses the fragrance of His knowledge in every place.
(2 Corinthians 2:14)**

You have about six million smell receptors, but a dog has about 300 million. And the smelling area of a dog's brain is 40 times larger than yours. God gave dogs the sense of smell as their most important sense!

Isn't it pleasant when the smell of baking cookies is diffused (or spread) throughout the whole house? It's a good, sweet smell—a fragrance. This verse thanks God because He diffuses His knowledge through us. It goes everywhere, as a fragrance does.

Your sense of smell is not as sensitive as a dog's is. But your nose is still very amazing as it notices molecules floating around in the air. Scientists think that people can smell one trillion different smells!

One reason God created our sense of smell is to warn us of danger. Our sense of smell might alert us to a fire when we first smell a tiny bit of smoke. We can use our sense of smell to know if food is rotten or if a dangerous chemical has spilled.

God also gives us the sense of smell because He loves us and wants us to enjoy His creation. Freesia flowers, forests, and fur have certain smells. So do coconuts, caves and cows. You may like some smells and dislike others. Often a smell can bring back a memory. Maybe the smell of roast beef reminds you of Christmas dinner at your grandparents' house. God's kindness

CHAPTER 16: GOD GIVES YOU TASTE, SMELL, AND OTHER SENSES

Freesia

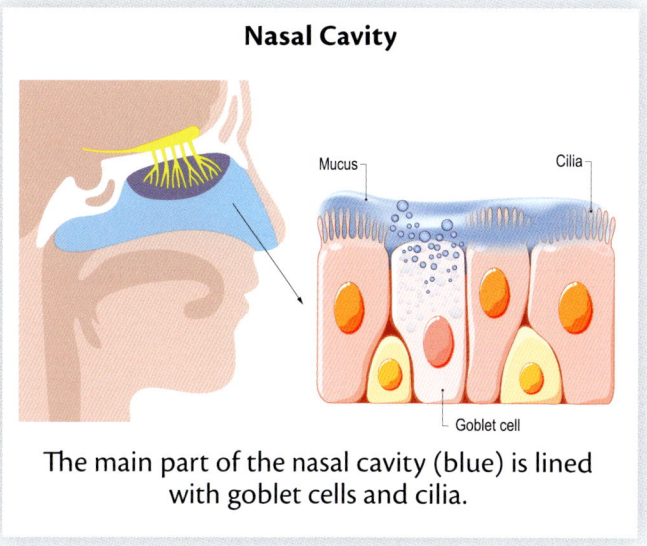

The main part of the nasal cavity (blue) is lined with goblet cells and cilia.

makes our lives rich with the sense of smell!

Let's learn about your nose!

Your **nasal cavity** (nose space) is made of the bones, tissues, and blood vessels inside your nose. It's divided in half by your **septum**. God gave your nasal cavity three important jobs.

1. **Your nasal cavity moistens the air you breathe.** Moist air is friendlier to your lungs than dry air. Air is moistened when it passes over the liquid mucus that lines your nasal cavity. Mucus is produced by **goblet cells** in the main part of your nasal cavity.

2. **Your nasal cavity warms the air for your lungs.** Your nasal cavity has many blood vessels very close to the surface. This allows warmth to transfer from your blood to the air you are inhaling. But it's also easy for these vessels to bleed if your nose is dry or gets bumped.

3. **Your nasal cavity keeps harmful germs and dust from entering your body.** Special cells in your nasal cavity have **cilia** that can hold liquid mucus where it should be. When germs, pollen, or other dusty things get trapped in the mucus, the cilia can move in a way that sweeps them out of your nose.

The nose septum starts between the nostrils. It continues up inside the middle of the nasal cavity.

It's a good thing that God gave these children nasal cavities with cilia "brooms" to clean out dust!

Definitions

Cilia are fingerlike things that stick out from some cells. God gave them the ability to move so they can do jobs. Cilia is plural. One of them is called a cilium.

An **odor** is any smell, pleasant or unpleasant.

Now let's learn how smelling works.

The word **olfactory** (ole-factory) is often used for things dealing with smell. Your sense of smell is your olfactory system. On the "ceiling" of each side of your nasal cavity, you have a special area of **olfactory tissue**. Each olfactory area is about one square inch (6.5 square cm).

The receptors for your sense of smell live in these two patches of olfactory tissue. Smell receptors are a kind of neuron. The dendrites of these **smell receptor neurons** are the only dendrites in your body that live outside of you. They stick out of your olfactory tissue into the mucus that covers that area. God put them in the open so they can grab the molecules of smells that come into your nasal cavity.

Everything is made of chemicals. Many things release some of their chemicals into the air as molecules. Some things, like steel, don't release any molecules into the air, so we don't smell them. Other things, like gas for your car, do release molecules, but we don't have smell receptors that can smell them. Since gas is dangerous, gas companies add bad-smelling chemicals to their gas. This helps people know if gas is leaking somewhere.

Smell receptor neurons are the only neurons in your body that are continually replaced. With fresh neurons, your olfactory system keeps working well!

166

CHAPTER 16: GOD GIVES YOU TASTE, SMELL, AND OTHER SENSES

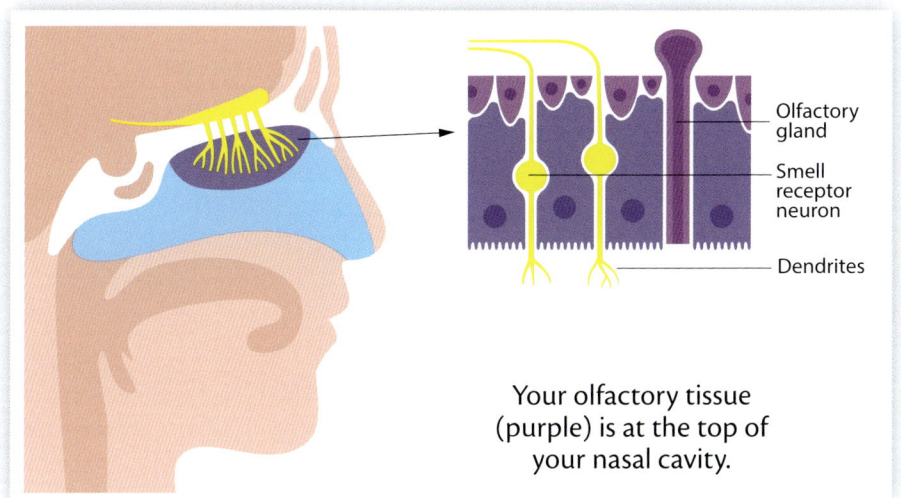

Your olfactory tissue (purple) is at the top of your nasal cavity.

If your nasal cavity were dry, your smell receptors wouldn't be able to sense molecules in the air. **Odor** molecules must be dissolved in liquid before your sense of smell can recognize them. Mucus does the job of providing this liquid! In the olfactory area of your nose, mucus is made by **olfactory glands** instead of goblet cells. Once odor molecules hit the mucus of your olfactory tissue, they can be grabbed by the dendrites of your smell receptors.

You have several kinds of smell receptor neurons. Each kind can grab only certain kinds of odor molecules. Scientists aren't sure how smell receptors recognize the molecules meant for them. But when their dendrites do grab their kind of molecule, the chemical message becomes an electrical message. Then the electrical message travels through its neuron.

Above your olfactory tissues, you have a bony plate that's part of your skull. How do smell messages get through the bony plate and to your brain? Before a neuron reaches the bony plate, it's joined by other neurons from the same kind of smell receptors. Joined together, they are called *nerves*. These nerves pass through holes that God has put in your bony plate.

Once the nerves go through your bony plate, they enter your **olfactory bulbs** (you have one on each side of your brain). The olfactory bulbs are the part of your brain where smell messages are combined. Odors are complicated. The odor for chocolate might be a combination of messages from several different kinds of smell receptors. These different smell messages are sent to different places on the olfactory bulbs. From there, they are sent to other parts of your brain where the smells are figured out. Scientists don't know exactly how this happens.

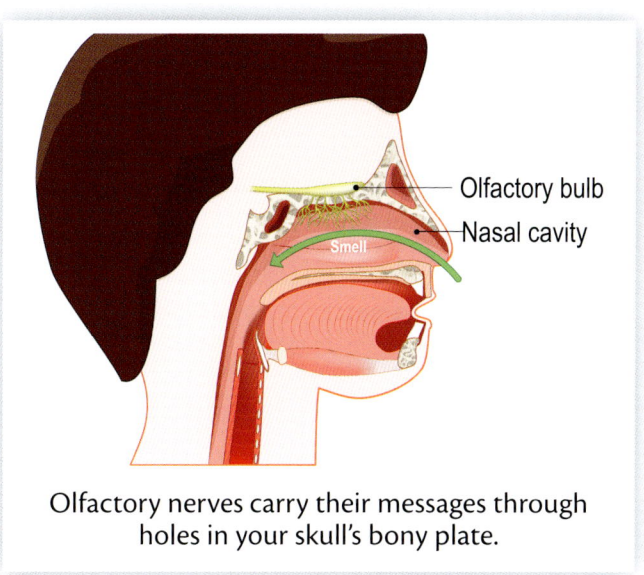

Olfactory nerves carry their messages through holes in your skull's bony plate.

167

GOD MADE ME

Usually, when we sense something, we soon forget it. Of all the things you have ever heard, seen, read, and dreamed, you have only remembered a small amount. Your brain decided that some of those things are not important enough to remember a long time. It put them into **short-term memory**. But, once your brain figures out a smell, it puts it into your **long-term memory**. Certain smells can bring back your memories of places and things that happened long ago.

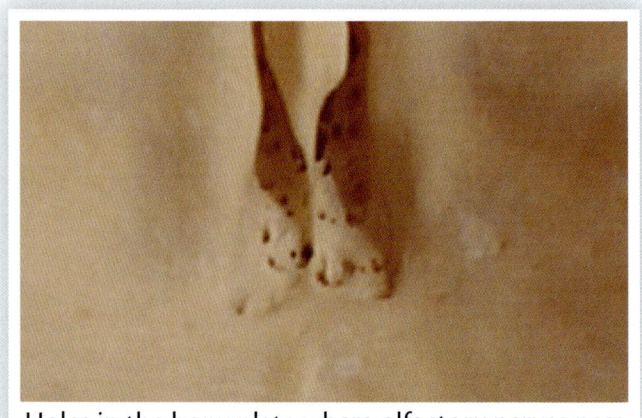
Holes in the bony plate where olfactory nerves pass through

Time to do Activity 46 in the Activity Book!

God Gives You the Sense of Taste

Oh, taste and see that the LORD is good. (Psalm 34:8)

I like this verse! It reminds us to think of the goodness of God when we use our senses!

Think of one of your favorite foods. What do you like about it? Do you like how sweet or salty it is? Do you like how cold or warm it feels in your mouth? Does it look beautiful? Maybe you like its crunchy sound. Or maybe you like its smooth texture? Does it have an interesting smell inside your head? When you eat it, do you feel contented and thankful? All these things are a part of taste! But we usually think of our **tongue** as the place we taste.

Your tongue is a wonderful, wiggly thing inside your mouth. Your tongue has eight muscles inside it and eight muscles that attach it to you. Because of these muscles, you can wiggle your tongue to make all the words you speak. These muscles also help you hold food in your mouth, move food between your teeth when you chew, and push food back so you can swallow it.

CHAPTER 16: GOD GIVES YOU TASTE, SMELL, AND OTHER SENSES

Your tongue is also wonderful because it's wet. If your tongue were dry, you wouldn't be able to taste things. Just like odor molecules, taste molecules need to be dissolved in moisture for you to sense them. Your mouth makes watery **saliva** to mix with food and help you taste it. Saliva also contains chemicals that begin to break down food so your body will be able to use it.

Another wonderful thing about your tongue is its covering of funny bumps. Your sense of taste happens in **taste buds** on your tongue. But did you know that the bumps you see on your tongue are not your taste buds? The bumps you can see are called **papillae** (puh-pill-ee) and some of them *contain* your taste buds.

Some papillae look like pointed fingers. These finger-like papillae don't have taste buds, but they do make your tongue rough enough to grab your food. They help your tongue move food around.

You have three more kinds of papillae on your tongue. These papillae are rounded and look like mushrooms and other funny bumps. Each of these papillae is the home of several taste buds. Taste buds are tiny, barrel-shaped structures that live just under the surface of your papillae.

Your tongue can sense five tastes: **sweet**, **sour**, **salty**, **bitter**, and **savory** (like the taste

God gave your mouth six **salivary glands**. Salivary glands make saliva and deliver it to the inside of your mouth through tubes called **ducts**.

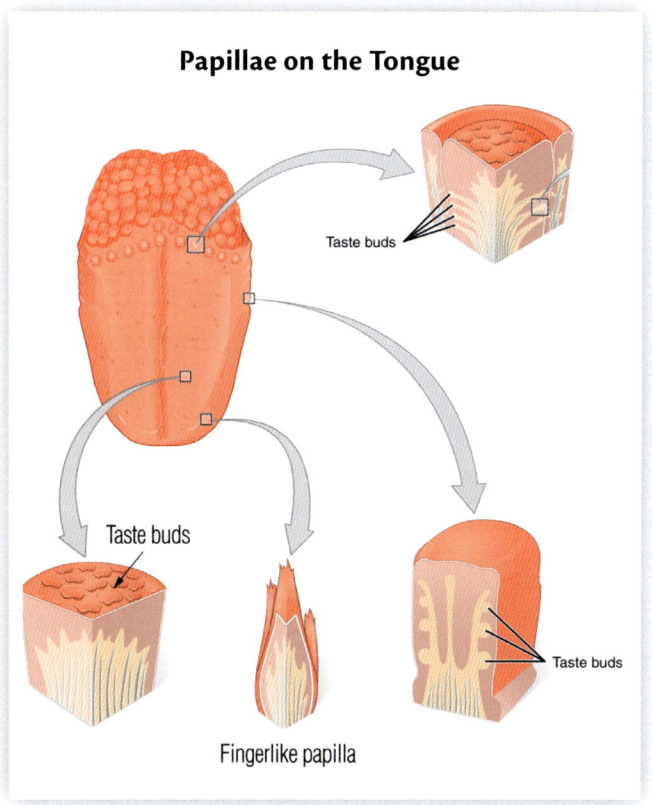

Papillae on the Tongue

Taste buds

Taste buds

Fingerlike papilla

Taste buds

169

GOD MADE ME

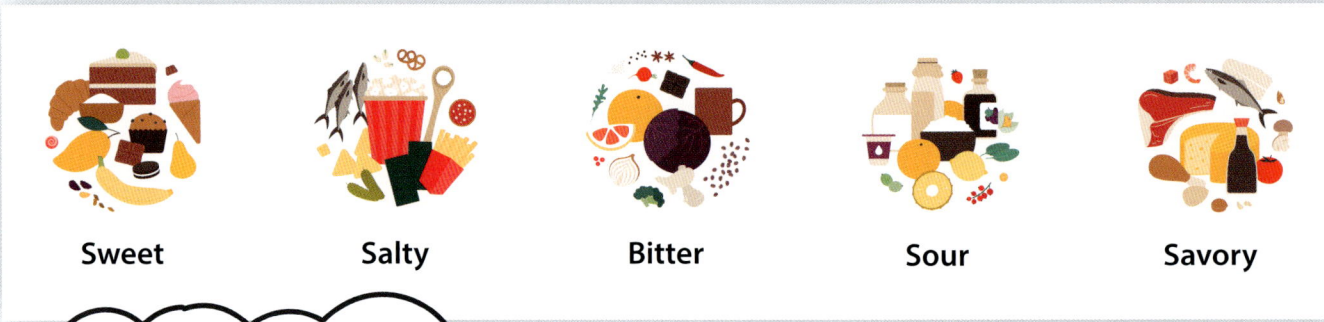

Sweet   Salty   Bitter   Sour   Savory

Most of your tastebuds are on your tongue's papillae. But you also have a few taste buds on the roof of your mouth and in your throat!

of browned meat). Any part of your tongue can taste any of these tastes, but some parts of your tongue are more sensitive to certain tastes.

Inside each barrel-shaped taste bud, you have many **taste receptor cells**. Each taste bud contains 30-100 of these taste receptors. Taste receptor cells can sense chemicals in saliva. Certain taste receptors sense certain tastes. If a taste receptor cell for "sweet" senses a sugar molecule, an electrical message travels through its cell.

Taste receptor cells are not neurons. But they pass their messages to neurons. Several neurons join to become a nerve that takes an electrical message to the sweet-sensing part of your brain.

A taste receptor cell lives only about two weeks. When it dies, God provides a new one to take its place!

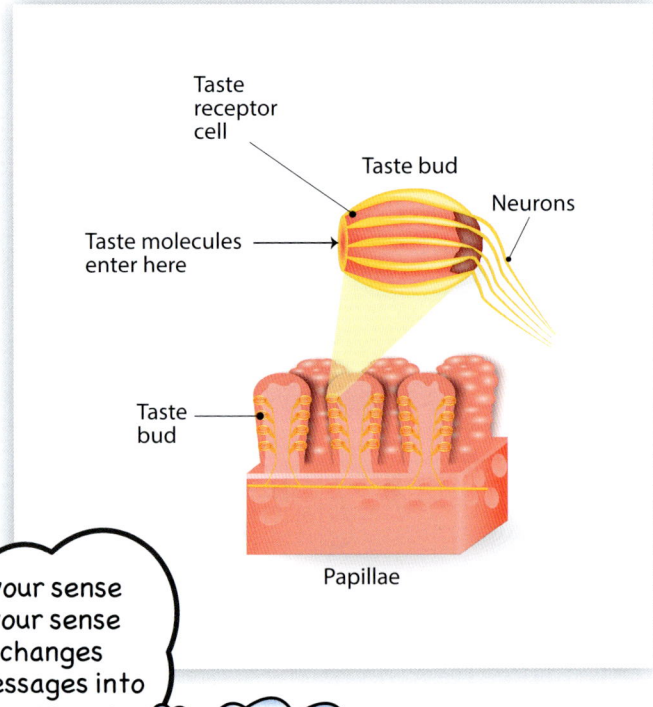

Just like your sense of smell, your sense of taste changes chemical messages into electrical messages.

170

CHAPTER 16: GOD GIVES YOU TASTE, SMELL, AND OTHER SENSES

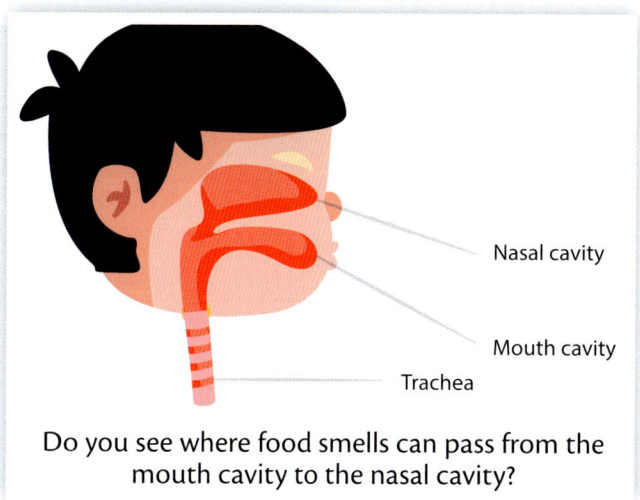

Do you see where food smells can pass from the mouth cavity to the nasal cavity?

With all the different foods God made, doesn't it seem like you can taste more than five tastes? It seems like there are thousands of different tastes! The reason for this is that God adds to your sense of taste by using your sense of smell. As you eat, odor molecules from food travel inside your head from the back of your mouth into the back of your nose. So, instead of only tasting the sweetness of a strawberry ice cream cone on your tongue, you can also smell strawberries, cream, and the toasty cone. God wants you to enjoy eating!

Sometimes when your nose is stuffy, food doesn't have much taste. Your nose mucus has become so thick that it covers your smell neurons too deeply. The food's odor molecules can't reach your smell neurons' dendrites.

 **Time to do Activity 47 in the Activity Book!**

God Gives You Automatic Senses

Our bodies do many things because of what our senses tell us. If a mosquito lands on your dad's arm, he sees it with his eyes and feels it on his skin. But he also has another sense that automatically knows where his swatting hand is. It helps Dad's brain know just how to aim that hand for

Proprioception helps you know where your body is in space. It can keep you from walking too close to the edge of a cliff.

GOD MADE ME

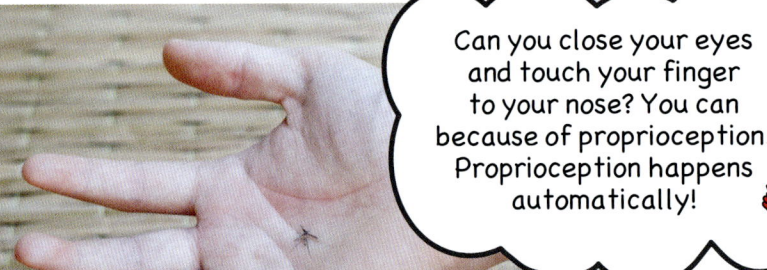

Can you close your eyes and touch your finger to your nose? You can because of proprioception. Proprioception happens automatically!

a good splat. He can concentrate on the mosquito and not have to plan how to move his hand. This sense is called **proprioception** (pro-pree-oh-sep-shun). It helps you automatically know where your arms, legs, and other body parts are.

Proprioception uses receptors inside your body near places that move. You have receptors near your muscles and in the joints between your bones. These receptors sense your movements by noticing how your muscles and joints stretch as you move. They send messages to your brain so it can guide your movements.

God has also given you a sense of **balance**. Your sense of balance is not just a single sense like eyesight. It's a sense that receives messages from all over your body. And it's always automatically aware!

Try to stand on only one foot. Does the bottom of that foot feel pressure as gravity presses it against the floor? This sense of touch helps you know which way is up and which way is down so you can stay balanced. Close your eyes while you stand on one foot. Is it harder to balance? Your sense of sight helps you balance too.

You probably can't stand perfectly still on that one foot. As you wave your arms and tilt your body to stay balanced, you are depending on your sense of proprioception to tell you when and where to move.

God has also given you some amazing structures in your inner ear that help you balance. Do you remember the cochlea? It's the bony, coiled-up tube in your inner ear that's used for hearing. We learned

172

CHAPTER 16: GOD GIVES YOU TASTE, SMELL, AND OTHER SENSES

to the side, as when you lean one ear down to listen to a little child.

> Because He has inclined his ear to me, Therefore I will call upon Him as long as I live. (Psalm 116:2)

Isn't God kind to listen to your prayers? It's as though He leans down to listen to you. You can always "call upon Him!"

that your cochlea has "hairs" to sense the motion caused by sound waves. But you also have bony structures in your inner ear that use "hairs" to sense motion. Instead of sensing sound wave movement, these hairs sense their own movement as you move your head and body. They help you balance. Let's learn how this happens.

You have two different kinds of structures to help you sense motion—your **semicircular** (sem-eye-sirk-you-luhr) **canals** and your **otolith** (oh-tuh-lith) **organs**.

Semicircular Canals

You have three semicircular canals in each inner ear. They are anchored in your skull as the cochlea is. The semicircular canals are amazing bony tubes, each shaped like the letter C. God put them in different positions so that each one helps you sense a different motion of your head. The canal in one position notices when you turn your head side to side like when you say "no." Another canal notices when you tilt your head up and down as in saying "yes." The last canal notices when you tilt your head

Each semicircular canal has a bulge near one end. A set of bendy "hairs" is rooted inside each bulge. As your head moves, the semicircular canals of course move with it. The canals are filled with liquid. As the canals tip, their liquid tries to stay level. The liquid acts the same way as it would in a cup of water. You can tip the cup, but the surface of the water stays level.

As your head moves, the loose ends of the hairs try to stay with the unmoving liquid. This means the hairs get pulled

173

GOD MADE ME

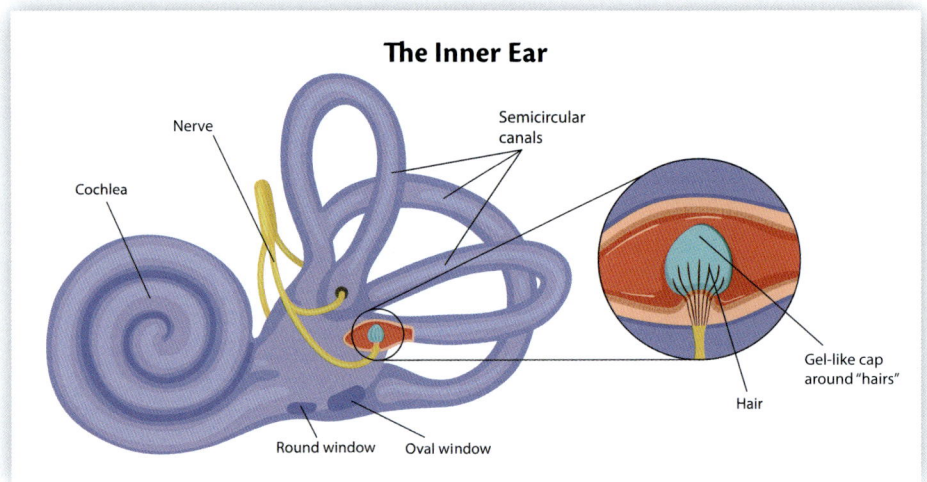

sideways. The hairs notice they are being pulled, and they send messages to your brain. You can picture this by thinking of what happens to your hair when you swing on a swing. When you swoop back and forth, your hairs try to stay behind in the still air. But since the hairs are attached to you, they get bent!

Otolith Organs

Your otolith organs live inside the bony area between the cochlea and the semicircular canals. In each ear, you have one otolith organ that lies horizontally (flat) and one that is vertical (up and down). The horizontal one helps you notice when your body moves forwards, backwards, or sideways, as when you ride in a car. The other otolith organ helps you notice when your body moves up or down, as when you ride in an elevator.

God has given "hairs" to your otolith organs so you will notice these motions. These motions are not usually as strong as the head-turning motions noticed by your semicircular canals. The otolith hairs must be able to sense the light force of gravity pulling down. They must also notice the forces when you gently move forwards, backward, and sideways. God has given your otolith organs wonderful, tiny structures to help you notice these gentle forces. He has made little "stones" that cause light forces to push harder. These stones or otoliths are inside a blanket-like membrane that covers the hairs. This "heavy blanket" helps push on the hairs harder so they will bend and send their messages to your brain!

Did you know that your ears help you see? The amazing structures that God has put in your inner ear are always sending messages to your brain about the position of your head. Then your brain can adjust the movement of your eyes to match. You can look back and forth at different things even if your head changes positions. You can walk or run without your eyes seeing a bouncy picture.

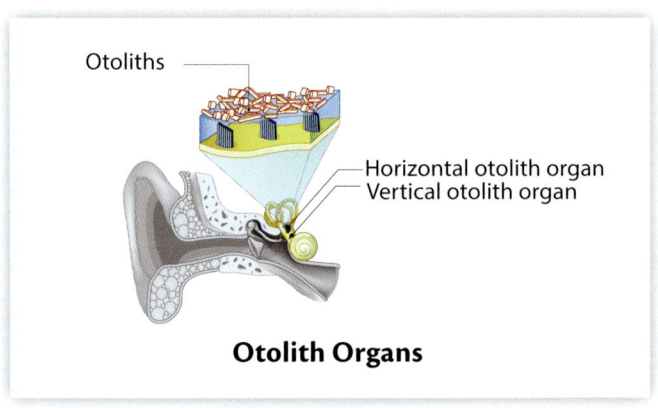

Otolith Organs

CHAPTER 16: GOD GIVES YOU TASTE, SMELL, AND OTHER SENSES

Puff's Health Hint

Your automatic senses help you pay attention. If you have been exercising, you will be more alert when you sit down to do your schoolwork. If you have been sitting all day, you will feel sluggish and have a hard time thinking. So get up! Use your proprioception and balance to make your brain alert and active again!

And you can do all this without getting dizzy and without your sight getting blurry. How amazing it is that God takes care of all this automatically, without us even having to know it's happening!

Prayer

Lord, our senses are so complicated! You have made many wonderful receptors so our bodies can gather information. We gather light to see with, sound waves to hear with, temperature and pressure to feel with, chemicals to smell and taste with, and movement to notice what is happening around us. You made these receptors able to change their information into electrical messages. You made perfect pathways to take the messages to the exact parts of our brains so we can understand them. Thank You for making our receptors and the world around us work so perfectly together! Help us use our wonderful bodies to give You glory. In Jesus' name, Amen.

Time to do Activity 48 in the Activity Book!

All good gifts around us are sent from heav'n above.
So thank the LORD, O thank the LORD, for all His love.

UNIT 5
God Feeds You

Our memory verse tells us that God provides food for every living thing. He satisfies the hunger of people and animals all around the world all year long!

As we sing this hymn, our voices thank God for the way He provides food. He uses water and warmth to grow it. Then He uses food to give us health and life. Let's learn how God made our bodies able to use His gift of food!

 **Memory Verse**

The eyes of all look expectantly to You,
And You give them their food in due season.
You open Your hand
And satisfy the desire of every living thing.
(Psalm 145:15-16)

You can listen to this hymn by searching for "We Plough the Fields and Scatter" on the internet.

 **Hymn to Sing: We Plow the Fields and Scatter**

We plow the fields and scatter
the good seed on the land,
But it is fed and watered by
the LORD's almighty hand.
He sends the snow in winter,
the warmth to swell the grain,
The breezes, and the sunshine,
and soft refreshing rain.

Chorus:
All good gifts around us are sent
from heav'n above.
So thank the LORD, O thank the LORD,
for all His love.

He only is the Maker of all things near and far.
He paints the wayside flower,
He lights the evening star.
The winds and waves obey Him,
by Him the birds are fed;
Much more to us, His children,
He gives our daily bread. (Chorus)

We thank you, then, O Father,
for all things bright and good,
The seedtime, and the harvest,
our life, our health, our food.
The only gift we offer for all your love imparts,
Is what you most would welcome:
our humble, thankful hearts. (Chorus)

CHAPTER 17
What Is Food?

God Feeds All Living Things

We can talk about food in a very big way: **Food** is something a living thing uses to give it energy. People use food to get energy. So do animals, cells, and even plants.

How do plants get food? God has put sunlight, oxygen, and water all over the earth. Plants can live almost everywhere on the earth because they use these three things to make their own food. This food is made within their cells and helps them live and grow. The food-making process in plants is called **photosynthesis** (foe-toe-sin-thuh-siss).

God created plants to make extra food, which they store in their cells. This extra food becomes food for animals and people when they eat the plants. Besides plants, people and animals can also eat meat, eggs, and dairy products. These things all come from animals, but the animals would not exist if they didn't have plants to eat.

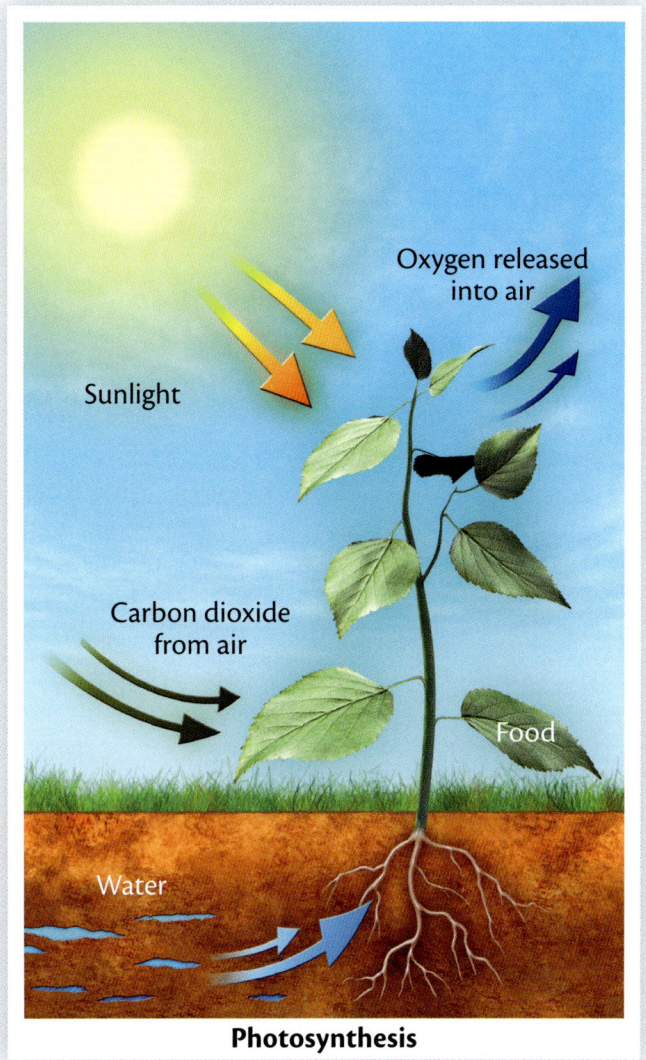

Photosynthesis

GOD MADE ME

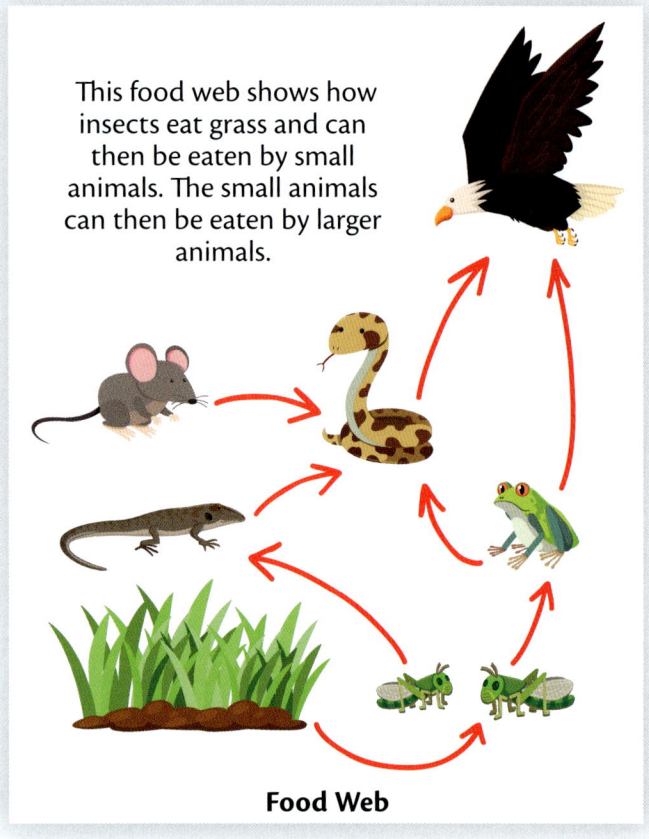

This food web shows how insects eat grass and can then be eaten by small animals. The small animals can then be eaten by larger animals.

Food Web

Definitions

When food is **digested**, it's broken down into pieces small enough to be absorbed and used by the body. All the structures, organs, and glands that work to break down and absorb food are called your **digestive system** (**digestive tract**).

A **nutrient** (new-tree-unt) is something that people and animals eat to get energy and to help them grow and live. Plants get their nutrients from the soil. An **essential nutrient** is a nutrient that is necessary for life but can't be made by a person's body.

Some creatures are so tiny that they only have one cell. We call them *one-celled* creatures. These creatures can absorb the smallest bits of food through their skin-like coverings (cell membranes). Your body cells can absorb the smallest bits of food this way too. Some of your body cells can also bring in larger bits of food a different way. They wrap their cell membranes around the food. Once the food is inside the cell, it can be **digested** (die-jest-ud).

Now let's talk about *your* food. Your food is anything you eat that gives your body the energy it needs for everything you do. Food contains nutrition to help you live, grow, heal, think, move, and even sleep.

The amount of energy a food can

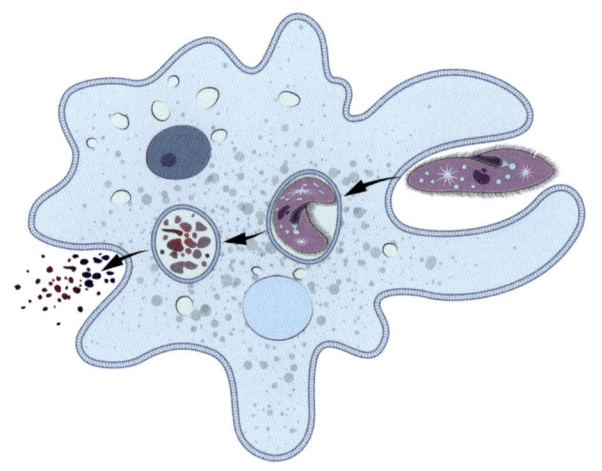

Some one-celled creatures like this amoeba (uh-mee-buh) eat other one-celled creatures by surrounding them with their cell membrane (another word for surrounding them is **engulfing** them). The amoeba engulfs its food, the food is sealed off inside, chemicals digest the food, and waste is released. Inside your body, you have cells that can kill germs by engulfing them in the same way.

CHAPTER 17: WHAT IS FOOD?

Calorimeter

After a day of hiking, your body needs a lot of food to replace the energy you used.

produce is measured in **calories** (kal-oh-reez). Scientists measure the calories in food by burning it in a **calorimeter** (kal-oh-rim-uh-tur). The amount of heat produced helps them know how much energy the food contains for you.

Your body constantly uses calories. (When a person's body uses calories, we say the body *burns* calories.) Your body burns calories all day and all night long.

Even while sitting still, your body still needs energy for all its automatic jobs. But you use a lot more calories when you exercise. Do you notice yourself feeling hot when you've been running or riding a bike? That's because God makes your body burn more calories when you need more energy!

 **Time to do Activity 49 in the Activity Book!**

What Do You Need From Your Food?

When something is necessary to keep things working as they should, we say it's **essential**. You breathe air so you can get the oxygen that's essential for your body. You drink water, which is also essential. Food provides the other essential things your body needs to live. Essential nutrients are things we must eat because our body can't make them for itself.

181

Thump's List of Things You Need from Food

- **Carbohydrates** (car-boe-hide-rates) give you quick energy. **Sugars** and **starches** are carbohydrates. Your body uses up carbohydrates quickly, but it saves a small amount in case you need it later. **Fiber** is a kind of carbohydrate that your body doesn't break down for energy. Instead, fiber passes through your digestive system to scrub it out.
- **Fat** is important for your brain. You also use fat from your food to store energy, keep warm, and protect your organs. Fat helps control your growth and fights germs.
- **Protein** helps your body grow and repair itself. It makes muscles, sends neuron messages, and gives red blood cells the ability to carry oxygen. Protein also makes chemicals that run the jobs in all your cells.
- **Vitamins** are essential molecules that help your body do its chemical reactions.
- **Minerals** in food are chemicals that plants take from the soil and pass along to us. Your bones are made of several minerals. Your blood has the mineral iron to help it carry oxygen.
- **Calories** give energy to all your cells so they can do their jobs. When your cells do their jobs, your body can do its jobs. You can get calories from carbohydrates, protein, and fat. Your body gets calories quickly from carbohydrates and more slowly from the protein and fat you eat. You don't get calories from water, minerals, or vitamins, but you must have them. They are essential for helping your cells do their jobs. Without vitamins, minerals, and water, you couldn't use the calories from carbohydrates, fats, or proteins.

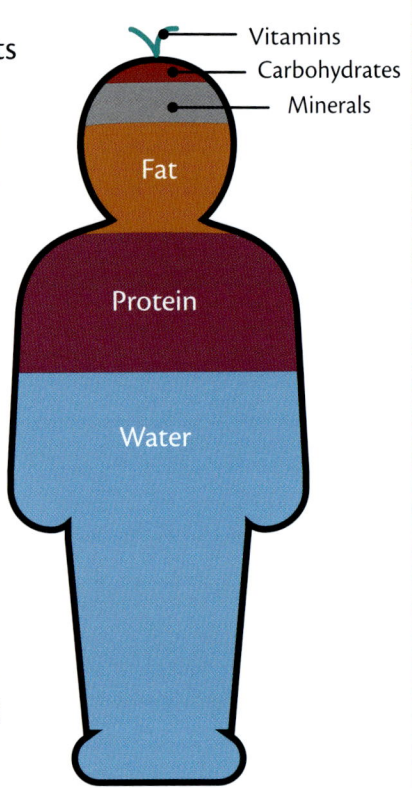

If your body could separate the essential things it's made of, it might look like this. You are more than half water!

Time to do Activity 50 in the Activity Book!

182

CHAPTER 17: WHAT IS FOOD?

God Puts Essential Nutrients All Over the Earth

Since our bodies need essential nutrients, God provides them in many different foods all over the earth. Wherever it's possible for people to live, He provides water, plants, and animal products for us. Some parts of the earth would not be friendly places for people, but God places special animals there and gives them their essential nutrients too.

Most foods contain several essential nutrients. Usually, a food has a lot of one or more nutrients and only a little of others. When a food has a lot of one nutrient, we say it is *high* in that nutrient. Let's look at the essential nutrients and where we can find them:

Carbohydrates

God gives us carbohydrates in many plants all over the earth. Fruits are high in sugar carbohydrates. Your body uses energy from sugar quickly.

Antonovka apples can grow in the cold areas of Russia.

Grains and root vegetables like potatoes and sweet potatoes are high in starch carbohydrates. You use energy from starch more slowly than energy from sugar because your body must break down starches into sugars. It takes time to break down the starch into sugar.

Pineapples can grow in hot tropical areas.

God feeds penguins even though He created them to live where people cannot.

GOD MADE ME

Rice is a grain that feeds people in warm, wet areas like Thailand.

Oats are a grain that feeds people in cooler areas like North America.

Vegetables like broccoli and lettuce are low in energy-giving carbohydrates but high in fiber carbohydrates.

Animal products usually have no carbohydrates. But God has put a special sugar carbohydrate in milk. It's called *lactose*. Can you guess why God put special things in milk? He put all the essential nutrients in milk so babies and baby mammals will get everything they need from their mothers.

Definitions

Grains are the seeds of grasses. Wheat, rice, oats, and corn are grains.

Legumes are plants in the bean and pea family. We eat their pods and their seeds.

Fats

People can easily get the fat they need by eating animal products. God gave animals the ability to live in both cold and warm places around the world. He gave them ways to keep warm and ways to keep cool.

Cassava has starchy roots that feed people in warm places like Rwanda.

White potatoes are starchy tubers that feed people in cool areas like the Netherlands.

CHAPTER 17: WHAT IS FOOD?

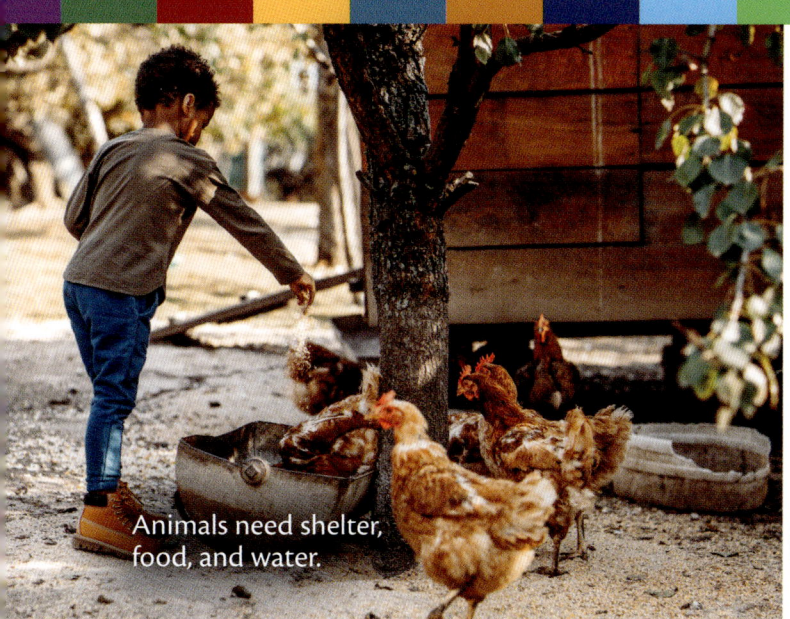
Animals need shelter, food, and water.

Avocados and coconuts are high-fat foods that grow in warm places.

With people to care for them, animals can be raised for their food products anywhere people can live.

Many plant foods are also high in fat. Nuts, seeds, and some vegetables can provide the fat people need around the world.

Carbohydrates give you quick energy to run a fast race. But fat gives you energy for slower activities. Your body receives energy from fats slowly. If you take a long walk, you will be using energy from fat. Your body stores fat so it's available for energy when you need it gradually.

Protein

Protein is found in animal products everywhere. Protein is made of smaller parts called **amino** (uh-mean-oh) **acids**. Animal products contain all the amino acids your body needs. We say they are *complete proteins*.

God put different combinations of amino acids in different plants. Most vegetables, fruits, and grains do not contain all the amino acids necessary to make complete proteins. A food might have some amino acids but not others. If you ate only

Walnut trees need to grow where there are warm summers and cold winters with temperatures below freezing.

Brazil nuts must grow wild in hot, moist jungles.

185

GOD MADE ME

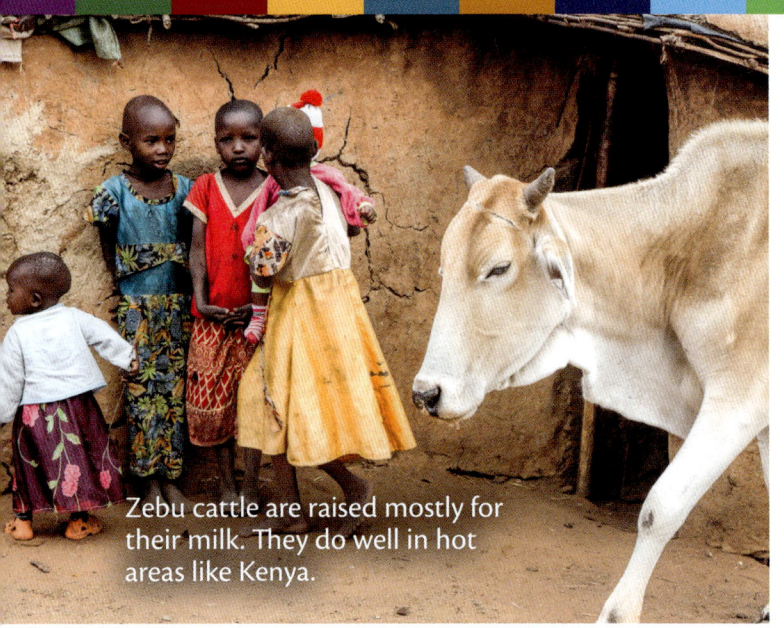

Zebu cattle are raised mostly for their milk. They do well in hot areas like Kenya.

Yaks are raised for their milk and meat in cold areas like Mongolia.

God put protein and fat in nuts and seeds all over the world.

and get complete protein. This means that when you eat refried beans on a tortilla or peanut butter on bread, you are getting complete proteins!

Vitamins

Essential vitamins can also be found all over the earth. God put each vitamin in many foods and in many places. He is so kind to give people around the world the vitamins they need.

Minerals

Minerals can be found in soil all over the earth. God gave plants the ability to absorb minerals with their roots. When people or

rice, or only apples, your body couldn't make all the kinds of protein you need. Certain amino acids would be missing or very low. But you could get complete proteins if you combined certain foods. For example, you could eat **grains** and **legumes** together

CHAPTER 17: WHAT IS FOOD?

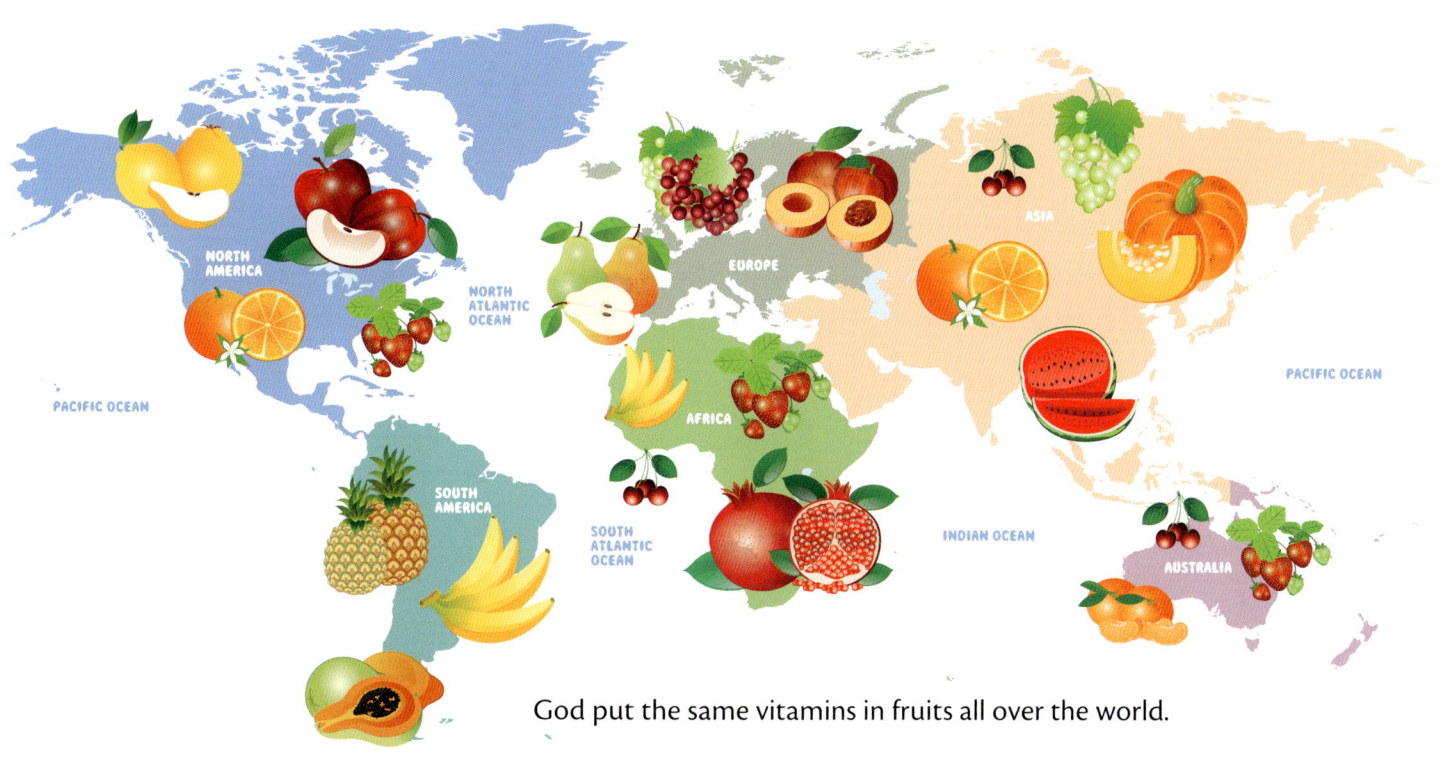

God put the same vitamins in fruits all over the world.

animals eat the plants, they are eating the minerals too. The minerals you eat in animal products were once part of the animal's food, which grew from the soil.

> **Ask the LORD for rain in the springtime; it is the LORD who sends the thunderstorms. He gives showers of rain to all people, and plants of the field to everyone. (Zechariah 10:1 NIV)**

Isn't God kind to provide water and food all over the earth?

These foods are high in protein

Cranberries grow in cold climates and are high in vitamin C.

GOD MADE ME

This picture shows different types of food that are high in carbohydrates, protein, or fat. The overlapping areas show foods that are high in two of these nutrients.

Prayer

Lord, You are so kind to provide exactly what our bodies need! You have put essential nutrients in food all over Your world. We can see that You are wise in planning how our bodies work and making the world work perfectly to be a part of that. And we praise You for loving us so much that You make eating fun and tasty. We are amazed and thankful! Amen.

 Time to do Activity 51 in the Activity Book!

CHAPTER 17: WHAT IS FOOD?

Thump's Health Hint

God made it easy to find the nutrients we need. It's up to you to eat the healthy foods He created. Most desserts and packaged snacks are missing important nutrients. They fill you up when you eat them, but they don't build a healthy body and a happy mind. With thanks, eat the healthy foods God gives. And when you eat a treat now and then, thank Him for that special experience too!

Oranges are high in vitamin C. They will only grow in warm climates.

CHAPTER 18
To Your Stomach

When you eat, food begins its trip through your body, starting at your mouth. You're probably very aware of food while it's in your mouth because of your sense of taste. You are also aware of food as you decide when to swallow each mouthful. If the food is really cold or really hot, you might feel it as it goes down your throat after you swallow. Once you have eaten enough, you are probably aware that your stomach feels full of food. After awhile, though, you are no longer aware of the food inside you. That's because it's being digested and is moving past your stomach.

Let's learn about the first part of digestion by following the path of food from your mouth to your stomach.

Digestion Starts in Your Mouth

You've already learned that digestion starts by breaking down food into smaller pieces that your body can absorb. There are two ways our bodies break down food. One way is to take big pieces and mechanically (meck-an-ick-lee) make them smaller. When you use a knife in your kitchen or at your table, you are **mechanically** breaking the food down. It's as though you are a mechanical machine.

As you chew your food, your mouth is like a machine. It's designed to use your

GOD MADE ME

teeth to mechanically tear and mash your food into smaller pieces. Your cheek muscles and tongue keep the food between your teeth as you chew.

Primary (Baby) Teeth

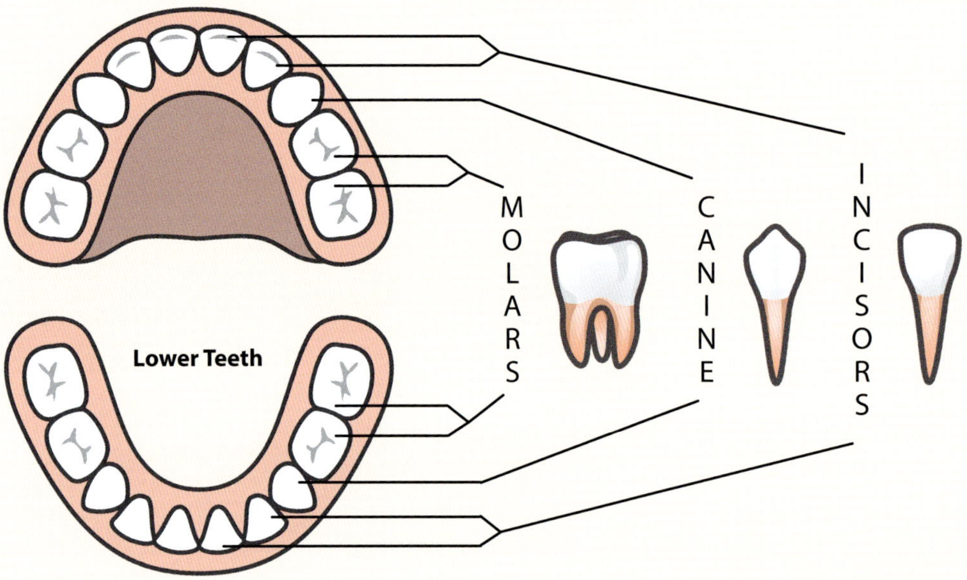

Upper Teeth

Lower Teeth

MOLARS   CANINE   INCISORS

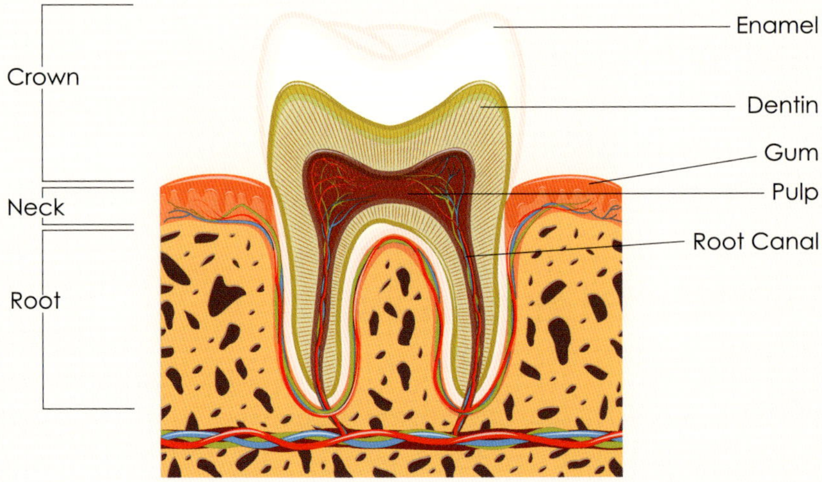

Crown

Neck

Root

Enamel
Dentin
Gum
Pulp
Root Canal

192

CHAPTER 18: TO YOUR STOMACH

From the time you were around three years old until you were around six, your mouth probably had about 20 teeth. The 20 teeth in the picture on the previous page are the first teeth to grow in children. They are the baby teeth or **primary teeth**. Primary teeth fall out one by one and are replaced with **permanent teeth**. God created your permanent teeth to last a long time. There will be no new teeth to replace them. Maybe you have a mixture of primary teeth and permanent teeth in your mouth right now. Since age six, some of your primary teeth have probably fallen out and been replaced. Other primary teeth have not fallen out yet.

At about age six, your mouth grew four new molars behind the ones you see in this picture. These *six-year molars* did not replace any teeth. They just grew in the space God saved for them. Your six-year molars are permanent. They won't fall out, and there are none to take their place.

Another way our bodies break down food is to break it into smaller pieces **chemically** (kem-ick-lee). This means our bodies produce chemicals that break large food molecules into smaller molecules that our bodies can use. Remember, molecules are the smallest piece of something that still is that thing.

You have already learned about your salivary glands that produce saliva. You learned that saliva moistens your food to help you taste it. Saliva has another job. Saliva is the first part of your chemical digestion.

Starches are long chains of sugar molecules stuck together. Saliva contains a chemical that breaks starch molecules into smaller pieces. These pieces are made of two or three sugar molecules stuck together. A later process in digestion will break these pieces into single sugar molecules.

The starch-breaking chemical in your saliva is a digestive **enzyme** (en-zime). Enzymes are amazing chemicals God made to speed up jobs in your body. If it weren't for digestive enzymes, digestion would happen too slowly for you to use your food. God made enzymes perfectly. They do their job, but they don't get used up. Your body can use them over and over again!

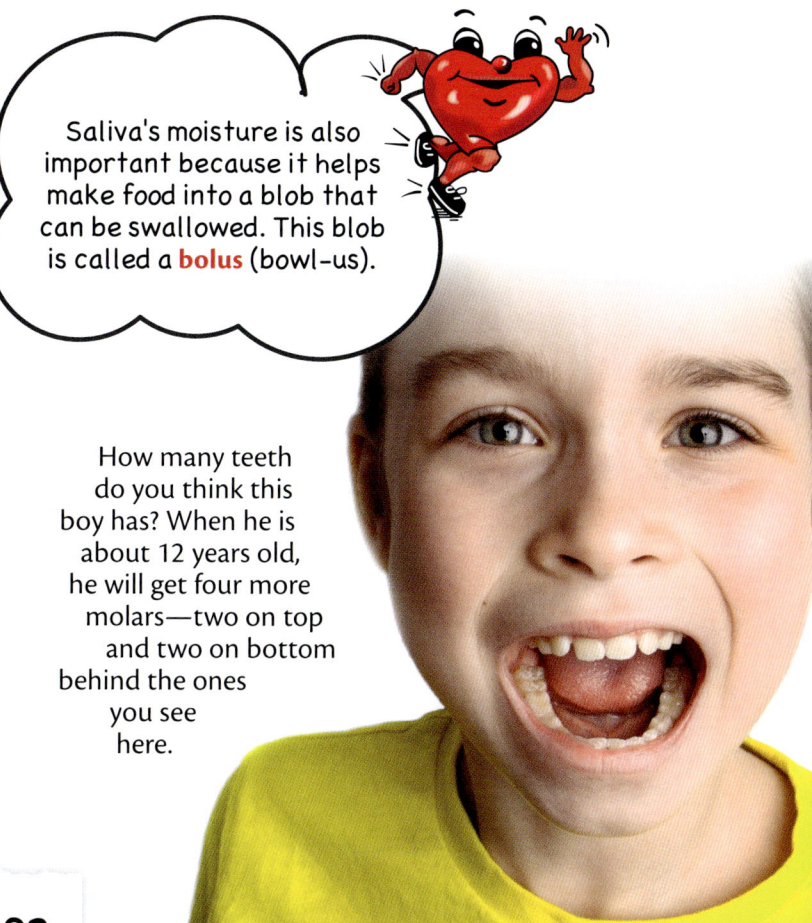

Saliva's moisture is also important because it helps make food into a blob that can be swallowed. This blob is called a **bolus** (bowl-us).

How many teeth do you think this boy has? When he is about 12 years old, he will get four more molars—two on top and two on bottom behind the ones you see here.

193

GOD MADE ME

Your mouth makes another chemical for digestion. This chemical is also an enzyme. It's made under the papillae on your tongue. It helps break down fats into **fatty acids** and other things. When you swallow, this fat-breaking enzyme moves along through your digestive system. It will be joined by other fat-breaking enzymes later. The fat-digesting enzyme made by your tongue breaks down about one third of the fat you eat.

Ice cream contains a lot of fat, but your tongue quickly helps digest it.

Time to do Activity 52 in the Activity Book!

The First Tube in Your Tract

Your digestive system takes food from top to bottom through hollow tubes and hollow organs. Let's learn about your **esophagus** (iss-off-uh-gus), your first hollow tube.

When you eat, you purposely tell your mouth to do its mechanical jobs. These on-purpose jobs do the chewing and swallowing. But once you swallow, your esophagus takes over and does its job automatically. Your brain tells it to begin pushing food toward your stomach.

God gave you an epiglottis to protect the breath He gave you. Remember, your epiglottis closes off your trachea as you swallow. That way, food goes through your esophagus instead of into your lungs!

Your esophagus doesn't have stiff rings of cartilage holding it open like your trachea does. Your esophagus is normally pressed flat by the other things inside your chest. But it's still possible for food to move through it. Your esophagus has muscles!

As soon as you swallow, food moves into your esophagus. Muscles behind the bolus squeeze and push the food along. At the same time, the muscles in front of the bolus relax to let the food through. This

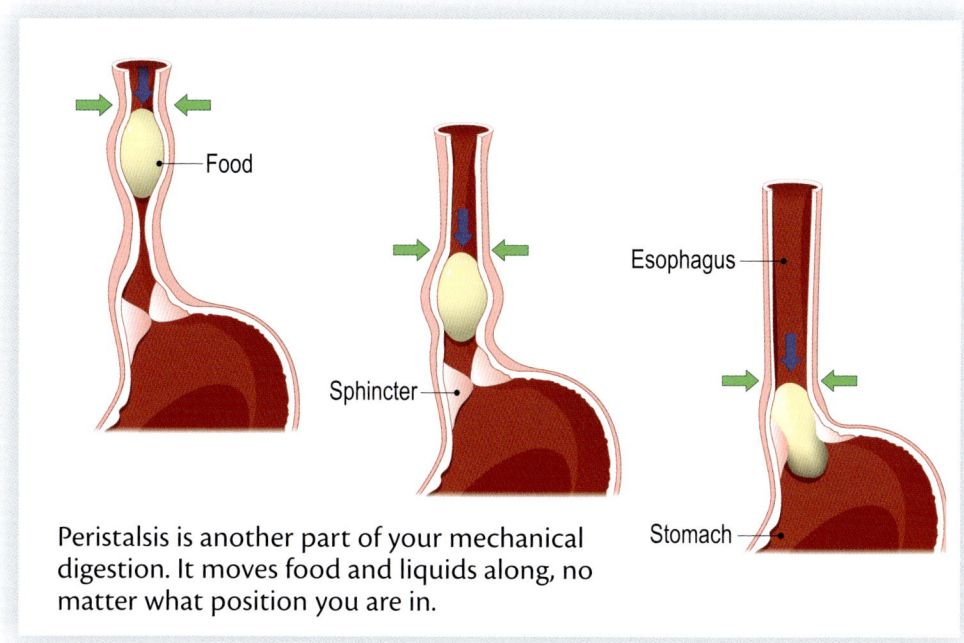

Peristalsis is another part of your mechanical digestion. It moves food and liquids along, no matter what position you are in.

muscle movement along the esophagus is automatic. You don't have to think about it. It's a moving squeeze called **peristalsis** (pare-ih-stoll-siss).

If you swallow a glassful of water and then do a somersault, what happens to the water? Does it run out your mouth? No, it stays inside you. God gave you one special ring of muscles near the beginning of your esophagus and another ring of muscles at the end. These **sphincters** (sfink-turz) are almost always tightly squeezed to keep your esophagus closed. This keeps food and liquid inside your digestive tract when you lie down or bend over.

At the right time, your sphincters automatically open to let food pass through. When you swallow, your upper sphincter is told to open and let the bolus through. When the bolus reaches your lower sphincter, the sphincter automatically opens to let the bolus into your stomach.

Sometimes air gets into your stomach as you eat and drink. When this happens, your lower sphincter opens to let the air out of your stomach and back into your esophagus. Then the air travels up your esophagus until it reaches your upper sphincter. As air gets pushed past that sphincter, it can make a burping sound as the sphincter tissues vibrate. As children

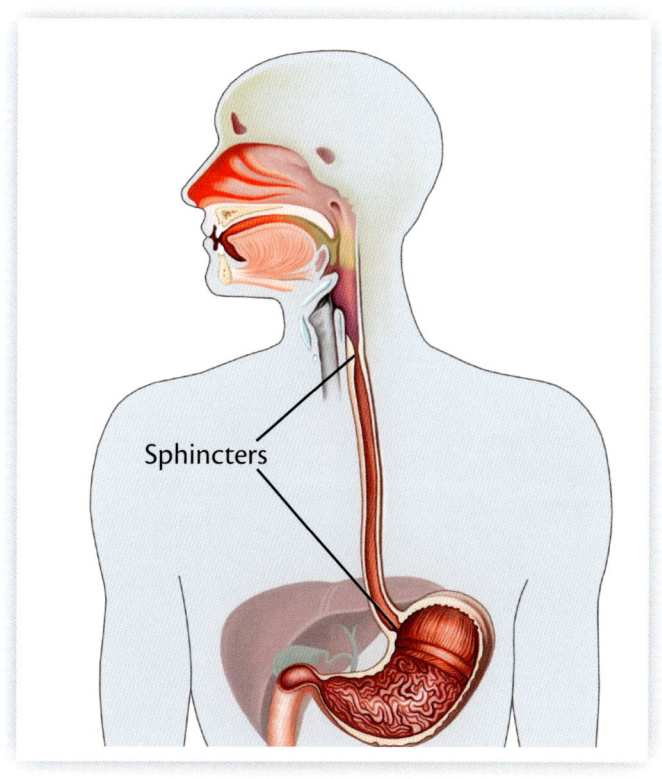

grow up, they learn to control when they burp, and they learn to do it more quietly.

Your esophagus uses a moving-squeeze action. Think of it as the way these girls are squeezing and sliding their hands down the icing bags to push the icing out.

 **Time to do Activity 53 in the Activity Book!**

Your Hungry Stomach

Do you ever feel so hungry that your stomach growls? It might even hurt. These signs of hunger are caused when the stomach shrinks because it's empty. The shrinking can cause the stomach's muscles to painfully squeeze and cramp. Then your stomach will relax for a half hour or more before cramping again. But usually, before you have more hunger pains, you've asked for some food. Your parents like to give you the food you need because they love you.

[God] gives food to the hungry. (Psalm 146:7)

Mechanical Digestion in Your Stomach

Your stomach uses both mechanical and chemical digestion. Its mechanical digestion works by using muscles in the stomach wall to move food around inside it. These are the same muscles that shrink

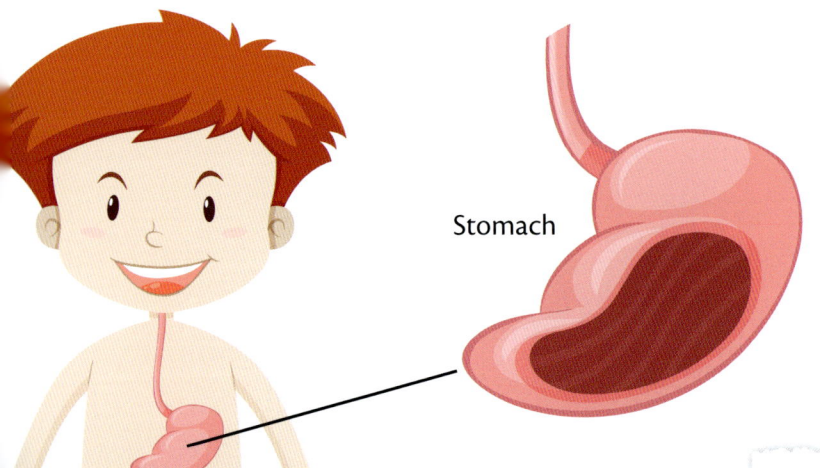
Stomach

CHAPTER 18: TO YOUR STOMACH

God is our heavenly Father. He loves us and feeds us!

You have a nerve that goes between your digestive tract and the place in your brain where you feel emotions. When you are happy, your digestive system works better. You will digest well if your family eats together lovingly and thankfully!

when you're hungry. After you eat, the muscles first squeeze at one place and then at another. Think about what you would do if you were making trail mix in a plastic bag. You would pour in the nuts, then the raisins, then the chocolate pieces. Next you would seal the bag and keep squishing it with your hands until everything was mixed. In the same way, your stomach keeps changing its shape to mix your food. When the food has been worked on in your stomach for about three hours, it's a pasty mixture called **chyme** (kime).

Chemical Digestion in Your Stomach

Your stomach makes three kinds of chemicals. Two of the chemicals are strong chemicals—strong enough to digest food. The third chemical is thick mucus. God designed this mucus to coat the inside of your stomach so your stomach won't digest itself with its strong chemicals.

One of the strong chemicals made by your stomach is **stomach acid**. Acids taste sour. The more **acidic** something is, the more sour it tastes. Water is not at all acidic. Tomatoes are acidic. Lemon juice is twice as acidic as tomatoes. Stomach acid is twice as acidic as lemon juice. Stomach acid is so acidic that, over time, it could dissolve bones and teeth!

God gave your stomach acid two jobs:
- To kill germs you may have swallowed.
- To help break down proteins into amino acids.

What is the other strong chemical made by

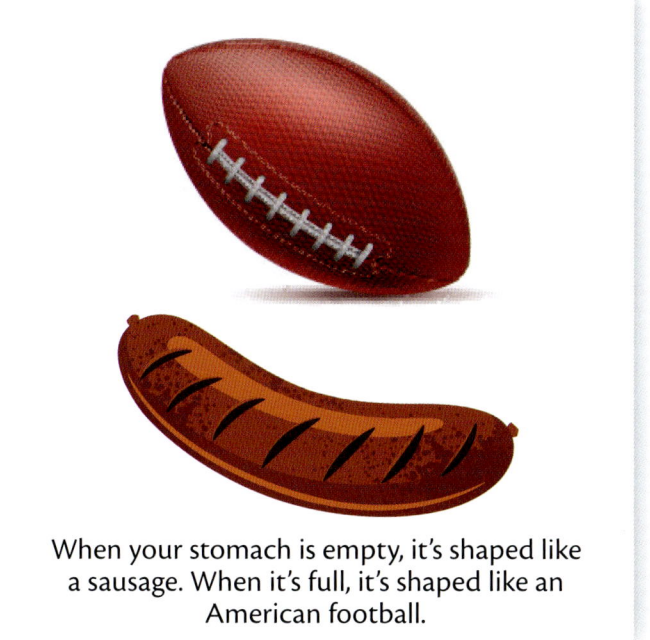

When your stomach is empty, it's shaped like a sausage. When it's full, it's shaped like an American football.

197

GOD MADE ME

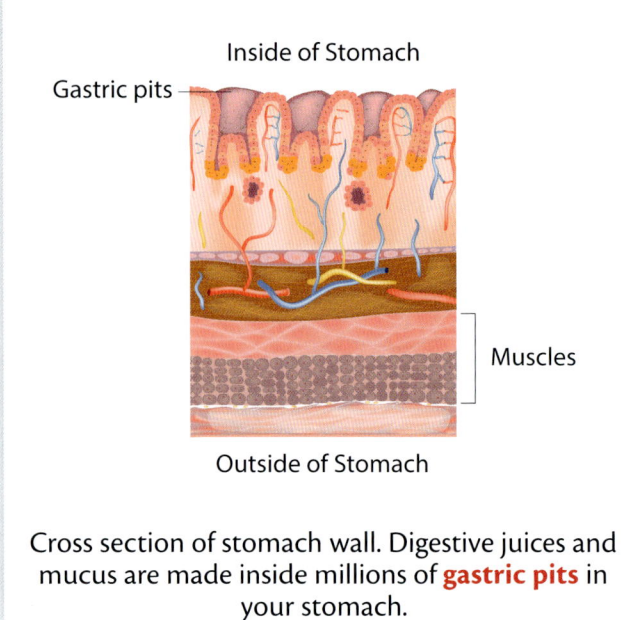

Cross section of stomach wall. Digestive juices and mucus are made inside millions of **gastric pits** in your stomach.

your stomach will make. About one third of your digestive juices are made at this time.

Later, as you are eating, your stomach notices that it's getting stretched. It also notices that it contains partly digested proteins and certain other foods. Now your stomach sends a message to itself, saying, "Get busy and make more digestive juices!"

At the bottom of your stomach is a

> Does thinking about certain foods make your mouth water? If so, it's because your rest-and-digest nervous system is getting you ready to digest by making saliva!

your stomach? It's an enzyme that begins the digestion of protein. This protein-digesting enzyme does its job better when it has acid to help with its job. That's why God put acid and this enzyme together in your stomach.

Digestive juices are what we call your stomach's mixture of acid and enzymes. They are automatically produced by your stomach. Before you even swallow your first bite of food, your stomach gets a message to start making digestive juices. This message comes from your rest-and-digest nervous system. If you taste, see, smell, or even think about food, your stomach gets ready for digestion. The hungrier you are, the more juices

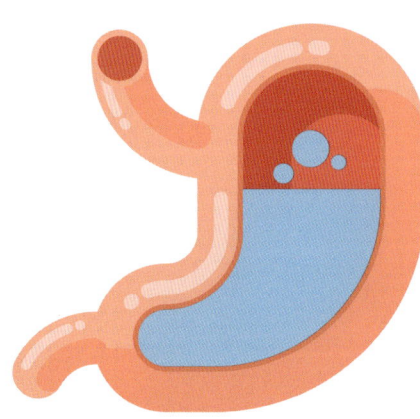

Your stomach makes about one and a half quarts (1.5 L) of digestive juices every day.

sphincter that keeps the chyme in your stomach for the right amount of time. Carbohydrates are ready to leave your stomach in a short amount of time. Proteins stay longer than carbohydrates do. Fats take the longest of all. When the chyme is ready to leave, the sphincter relaxes and lets chyme pass out of your stomach a little at a time.

Every now and then, food goes backwards from the

CHAPTER 18: TO YOUR STOMACH

stomach to the mouth. This is called **vomiting** (vah-mitt-ing). Vomiting is something our bodies do automatically. Vomiting can happen when the stomach is irritated by a sickness or by something bad in the food that's been eaten. A message about the irritation goes to a part of the brain that's in charge of vomiting. During vomiting, the brain does not tell the muscles inside the stomach to squeeze. Instead, it tells the diaphragm muscle and the muscles outside the stomach area to squeeze on all sides. At the same time, the sphincters in the esophagus relax. Thankfully, the trachea automatically closes to keep vomit from going into the lungs. Now the vomit is forced up the esophagus and out the mouth.

Some people's brains send a vomiting message because of dizziness or because they sense something yucky.

Prayer

Dear Heavenly Father, it's wonderful how You give us our teeth and saliva to give our food a start on digestion. It's fun to chew and it's great to swallow, knowing You have made a safe way to keep the food out of our lungs. Thank You for the comfortable feeling in our stomachs when they are full! Help us to eat the proper amounts of the right food. Amen.

Time to do Activity 54 in the Activity Book!

CHAPTER 19
Absorbing in Your Abdomen

Test all things; hold fast what is good. (1 Thessalonians 5:21)

Your Long Small Intestine

Your **small intestine** (in-test-in) is the part of your digestive system that absorbs nutrients from your food and puts them into your blood. Your small intestine is a very long tube that moves your food through itself. It takes up a lot of space in your abdomen. So why is it called your *small* intestine? Your body also has a large intestine which we will learn about later. The small intestine is called *small* because its tube is narrower than your large intestine's tube.

Now let's learn how your digestive system absorbs nutrients so your body can hold onto the good things in your food. This important job reminds me of this verse!

After you eat, your stomach opens and closes its lower sphincter so that your small intestine receives chyme a little at a time.

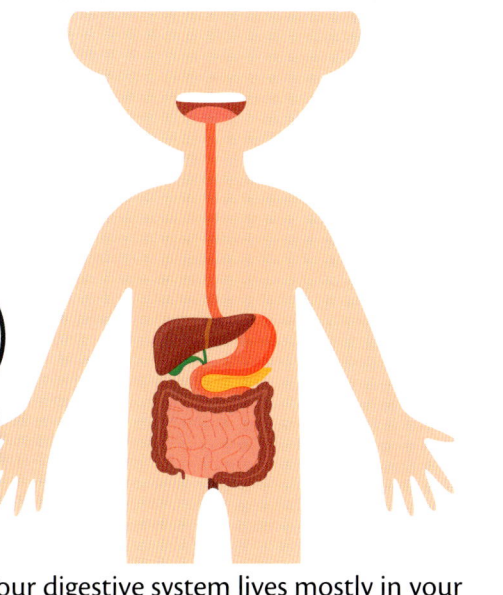

Your digestive system lives mostly in your abdomen.

GOD MADE ME

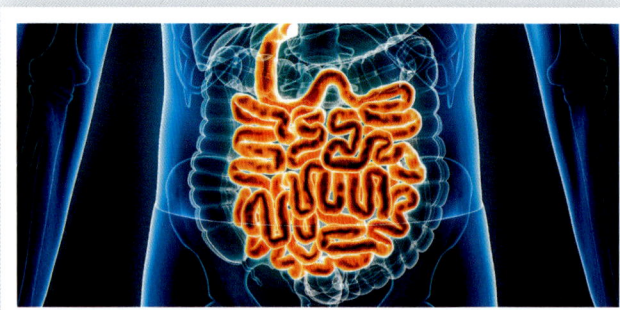
Your small intestine (orange) is about 10 feet (3 m) long as it lives and does its job inside you.

Definitions

Your **abdomen** (abb-duh-muhn) is the part of your body between your diaphragm and the place where your legs begin. It mostly contains digestive organs.

Villi (Vill-eye) are tiny, finger-like projections that cover the inside surface of your small intestine. Villi absorb nutrition from your food. One of these projections is called a **villus**.

Microvilli are even tinier "fingers" than the villi. They live on the outside cells of each villus.

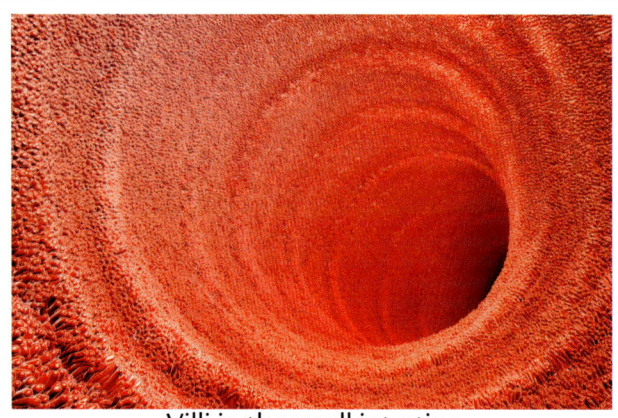
Villi in the small intestine

When we measure the size of a surface, we call this *surface area*. To understand this, let's measure the surface area of a die: Each edge of the die is about one centimeter long. We would then say that each flat side of the die is about one square centimeter. Since the die has six sides, the surface area of the die is six square centimeters.

You can also measure the surface area of the inside of hollow things. Scientists have figured out how big the surface area is inside the small intestine. If your small intestine was smooth inside like a garden hose, its total surface area would be about the same size as the top of a baby blanket. But it is not smooth! Scientists have figured out that the villi and microvilli covering the inside walls of the small intestine give it much more surface area. If we include these in our measurement, the surface area of your small intestine would be about the same as a tennis court!

Think about the amount of food and drink you eat in one meal. Can you imagine spreading that food over a tennis court? That would be very difficult, but God has given your

CHAPTER 19: ABSORBING IN YOUR ABDOMEN

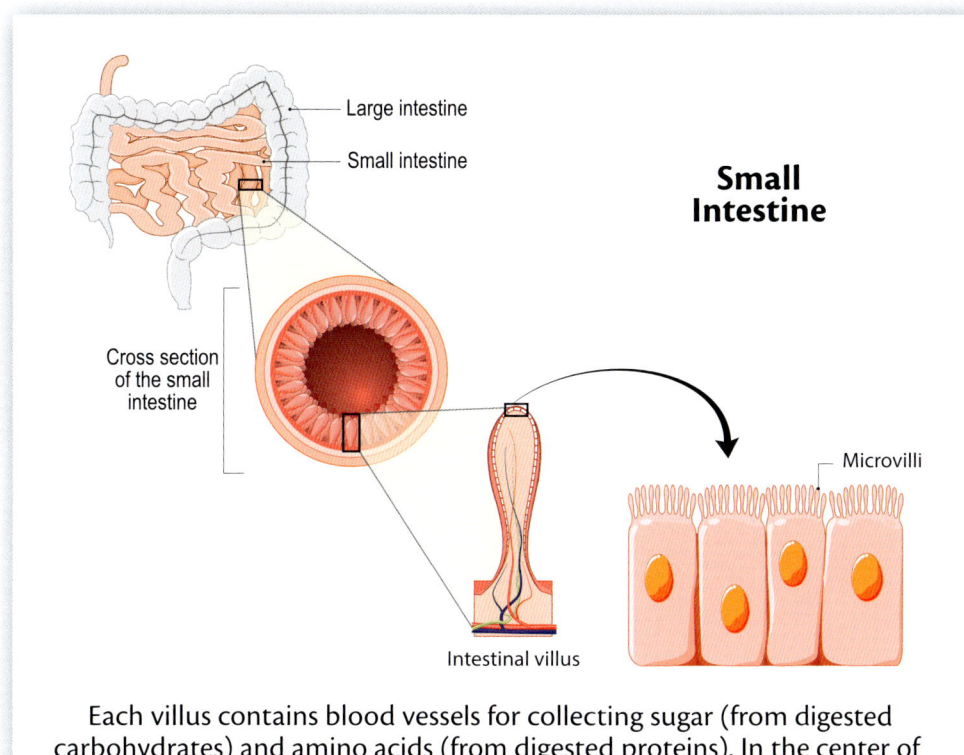

Each villus contains blood vessels for collecting sugar (from digested carbohydrates) and amino acids (from digested proteins). In the center of each villus there is also a lymph vessel that collects broken-down fats.

small intestine amazing ways of spreading your food around inside itself so that the nutrition can be absorbed and used.

Your small intestine has certain motions that mix and spread the chyme. It can squeeze the chyme forwards and backwards. It can also pinch with thin rings of muscles to cut the chyme in pieces over and over. Both these motions help mix and spread the chyme over the villi that absorb nutrition.

While this is happening, your villi move wildly to stir the chyme and mix it with intestine juices.

Intestine juices are a yellow liquid your small intestine makes out of enzymes and water. It makes the juices in tiny glands between your villi. Some of these glands make carbohydrate-digesting enzymes. Others make protein-digesting or fat-digesting enzymes. One enzyme even digests DNA from the food you eat so your body can use those molecules! To make intestine juices, these enzymes are mixed with a lot of water that your small intestine has brought in from your blood.

As your food is breaking down, your small intestine is using peristalsis to move the chyme along inside itself. This peristalsis is weaker than the squeezing of your esophagus. God made it weaker so nutrition will move slowly through your small intestine, giving it time to be absorbed. When the chyme reaches the end of your small intestine, it waits at a sphincter. When the sphincter opens, the chyme moves into your large intestine.

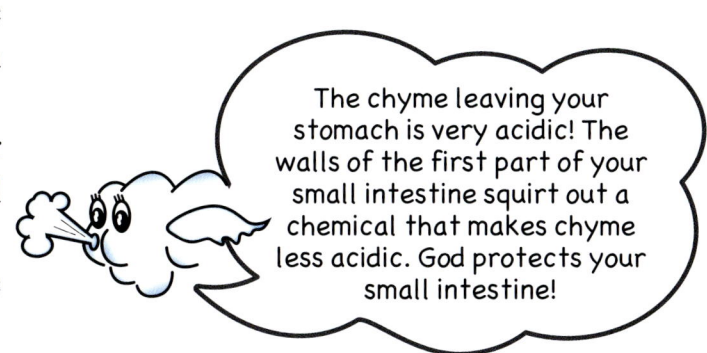

The chyme leaving your stomach is very acidic! The walls of the first part of your small intestine squirt out a chemical that makes chyme less acidic. God protects your small intestine!

203

GOD MADE ME

 **Time to do Activity 55 in the Activity Book!**

Your Pancreas Helps Digestion

God has created a very complicated recipe for energy. He can take any combination of food and "cook" it into the perfect meal for you. He designed your digestive system to add its own ingredients to your food. This turns your food into energy and other things your body needs.

Your digestive system has nearby organs that provide these extra ingredients for your digestion. These organs are attached to your digestive system, but food does not pass through them. Let's learn about your **pancreas** (pang-kree-us) first.

Your pancreas has two jobs:

- **Your pancreas makes about two quarts (2 L) of digestive juices every day.** These juices contain a chemical that helps make the chyme less acidic. They also contain enzymes that help digest carbohydrates, fats, and proteins. The juices travel from your pancreas to the top of your small intestine through a tube called a **duct**.

When you have food in your stomach, your automatic nervous system tells your pancreas to start putting out digestive enzymes. But your pancreas also gets messages through hormones sent out by your small intestine. When your small intestine feels the chyme coming in from your stomach, it makes hormones that tell your pancreas to get busy. These hormones travel through your blood to get their messages to your pancreas. One hormone tells your pancreas to give the juice that will make chyme less acidic. Another hormone tells your pancreas when you have eaten protein and need more of the protein-digesting enzymes.

But God didn't want the enzymes made in your pancreas to digest your pancreas. He wisely made them **inactive** in your pancreas! This means that the enzymes won't do what they were created to do until the right time. The enzymes become **active** when they reach the small intestine and are "switched on" by different enzymes made there.

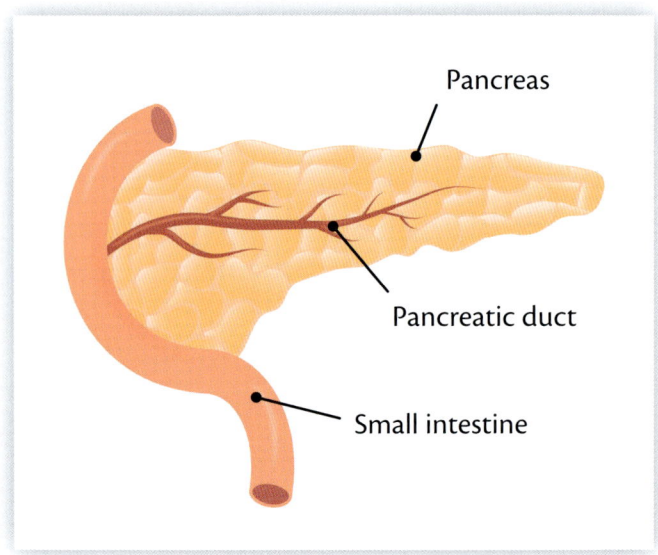

204

CHAPTER 19: ABSORBING IN YOUR ABDOMEN

Your pancreas makes hormones to keep the right amount of sugar in your blood. These hormones are made by small groups of cells that are scattered throughout your pancreas like little islands. One hormone (insulin) keeps your blood sugar from becoming too high. Another hormone keeps your blood sugar from becoming too low. These hormones are released into your blood, not into your digestive system. They take messages to the parts of your body where sugar is stored. They "tell" the sugar to either stay out of the blood or to jump into the blood. This job is not part of your digestion, but God gave your pancreas this job too.

Definitions

The **pancreas** is an organ that makes digestive juices and delivers them to the small intestine. It also makes hormones that make sure blood has the right amount of sugar.

A **duct** is a tube that carries liquids away from a gland or organ.

Remember, a hormone is a chemical that's made in one part of your body and causes another part of your body to do something.

When you're feeling excited, your automatic nervous system tells your pancreas to raise your blood sugar. This is why excitement makes you energetic!

Time to do Activity 56 in the Activity Book!

Your Liver Helps Digestion

Your liver is a large organ. When you see a piece of animal liver at the store, it doesn't look very important. It looks like a reddish-brown blob with no interesting parts. But your liver is full of well-organized cells that do about 500 amazing jobs! Since some of your liver's jobs help your digestive system, let's look at those jobs in this chapter.

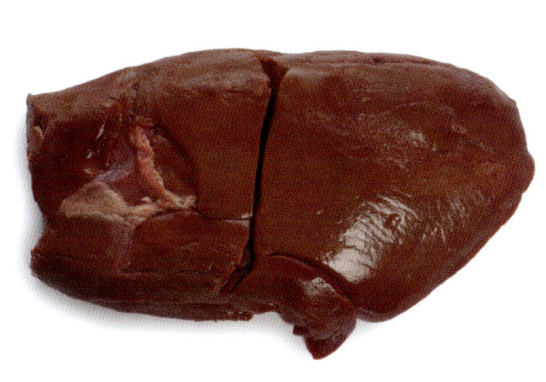

Raw lamb's liver. The liver is a complicated organ—almost as complicated as the brain.

GOD MADE ME

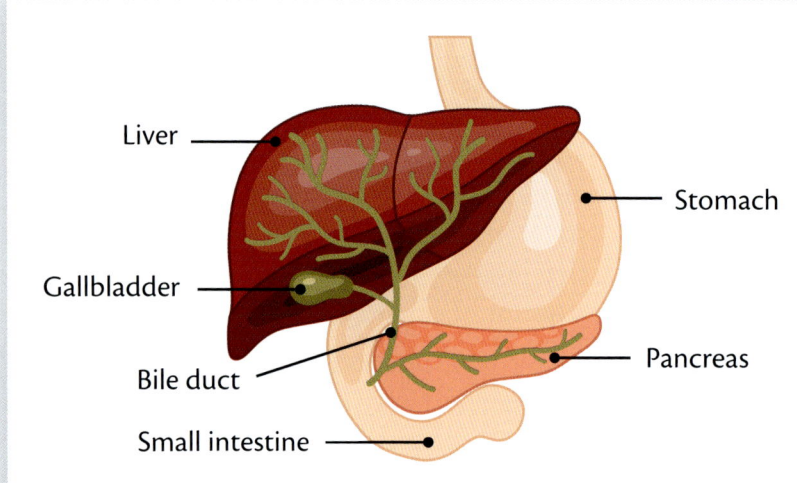

When you eat fats, your gallbladder squeezes bile into your small intestine through your **bile duct**. Digestive juices from your pancreas enter your small intestine at the same place (through a sphincter).

God gives you bile so you can eat things containing fat like this yummy bacon, egg, and avocado sandwich!

Your liver makes an important digestive juice called bile. This juice is made of chemicals, not enzymes. Bile is a sticky, yellow-green liquid that breaks large fat globs into smaller pieces so they can be digested. When bile is made by your liver, it does not go directly into your small intestine. The bile first goes into your **gallbladder** to be stored until you need it.

Your gallbladder is not green. But if you could see your gallbladder, it would look green because of the bile showing through. Bile is green because of an important job your liver does. Your liver helps get rid of dead red blood cells. It changes their red chemical into a yellow-green chemical. Then, as bile travels down your intestines, that green chemical changes to a yellow chemical. Then it changes into the brown color you see in **stools**.

Have you ever tried to rinse butter or oil off your hands with plain water? Fat doesn't rinse off easily. Your parents will probably tell you to use soap. Soap breaks fat into smaller pieces that can be rinsed away. The digestive chemicals in bile act like soap. They break down fat so it can be absorbed!

When you eat fat, your gallbladder squeezes to pour out bile into your small

Definitions

Your **liver** is a large, complicated organ that helps your body digest, absorb nutrients, store important things, remove harmful chemicals, stop bleeding, fight germs, make hormones, and send nutrients and hormones where they are needed.

Stool is a scientist's word for the solid waste that leaves the digestive system when someone uses the toilet.

CHAPTER 19: ABSORBING IN YOUR ABDOMEN

intestine. God did not make your body wasteful. When the digestive chemicals in bile are finished with their jobs, they leave through the walls of your small intestine. They are absorbed into your blood and carried back to your liver so they can be used again.

Puff's List of Bile's Jobs

- Bile makes big blobs of fat smaller so they can be digested better.
- Bile makes chyme less acidic.
- Bile's soapiness kills many of the germs that might be in your food.
- Bile helps your body absorb vitamins A, D, E, and K that God gives to us as fats.

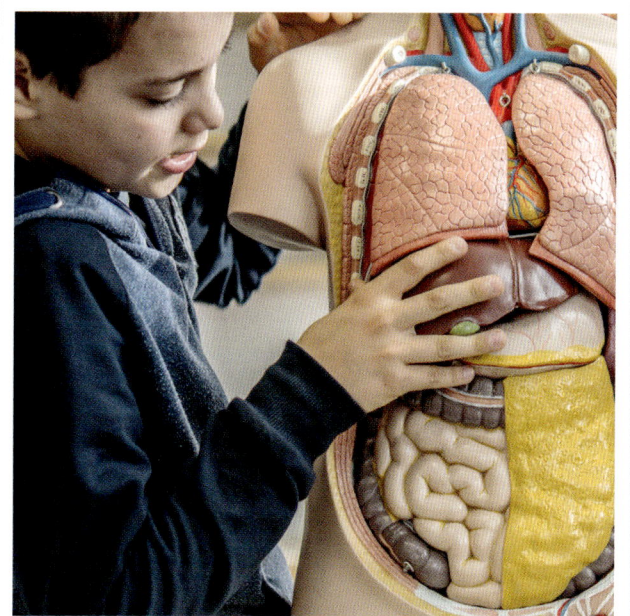

If you put your hand over your lower ribs on the right side of your chest, your hand will be over your liver.

Your liver is the last step in digestion. It does more than just break down food into smaller pieces. It *changes* your food into exactly the right form your body can use. Your liver is like a chemistry lab doing different jobs for different nutrients.

Even though food doesn't pass through your liver, your liver still does its jobs. How does your liver find the food to work on? The food comes to your liver through your blood! Do you remember that your small intestine absorbs nutrition from your food and puts it into your blood? But that blood does not go straight to your heart. Instead, it goes straight to your liver! Then your liver gets busy:

- When your liver gets sugar from your small intestine, it changes the sugar into a kind of starch. Then it stores the starch until your body needs energy. When you need more energy, your liver changes the starch into sugar again.
- Your liver stores fats for a short time. It changes them into simpler molecules your body can use.
- Your liver finishes breaking down proteins into amino acids. If there are any amino acids you don't need, your liver changes them into other things that your body gets rid of in waste.

Once your liver is finished with these jobs, it puts the final nutrients into your blood. They travel to your heart and are delivered all over your body. Your liver also puts wastes into two places—into your blood

GOD MADE ME

and into your bile. If a waste goes into your blood, it comes out in your urine. If a waste goes into your bile, it comes out in your stools.

Inside Your Living Liver

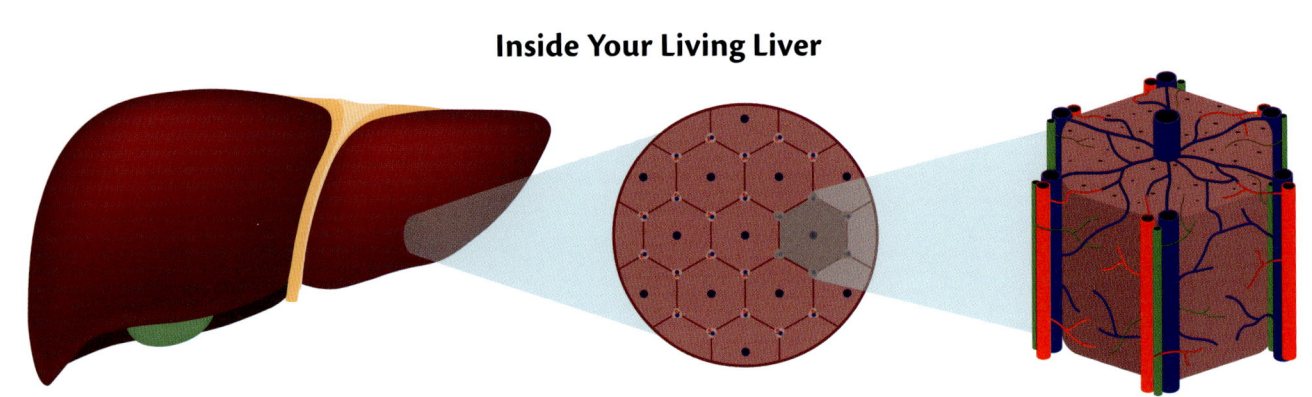

Your liver is made of cells arranged in columns. The cells make bile and send it to your gallbladder through the green vessels. The cells receive blood from your small intestine through the blue vessels. The cells clean and prepare your blood. They deliver the perfect nutrients to your body through the red vessels.

Prayer

Dear God, there are so many complicated jobs our small intestine, liver, gall bladder, and pancreas do to help our bodies use the food You give us. Thank You for giving us every organ, duct, and chemical we need for our digestive tract. We praise You for making our digestive system automatically work so well! Amen.

Time to do Activity 57 in the Activity Book!

CHAPTER 19: ABSORBING IN YOUR ABDOMEN

Puff's Health Hint

You can help your hard-working liver by eating these foods!

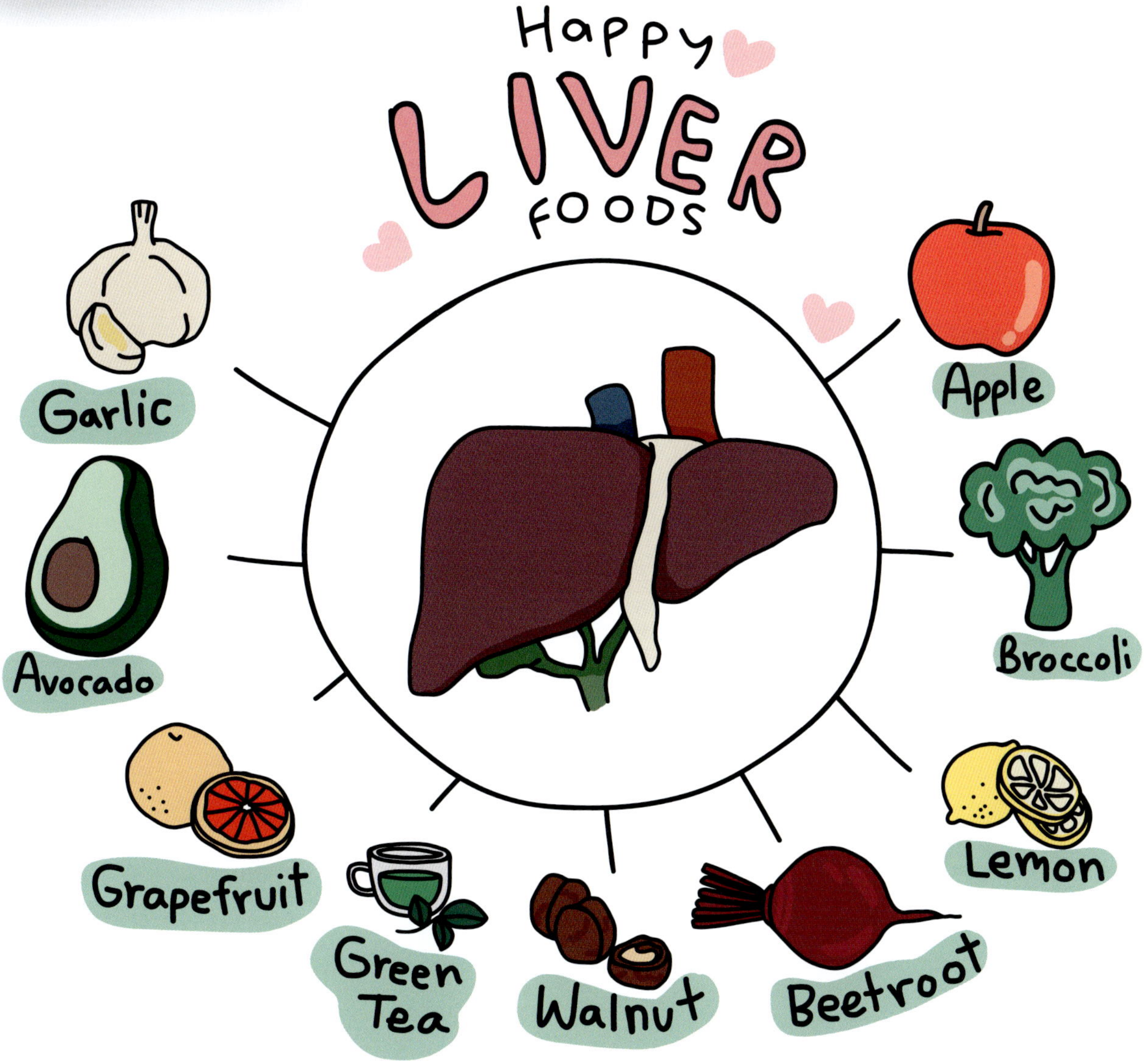

Happy LIVER Foods

- Garlic
- Avocado
- Grapefruit
- Green Tea
- Walnut
- Beetroot
- Lemon
- Broccoli
- Apple

209

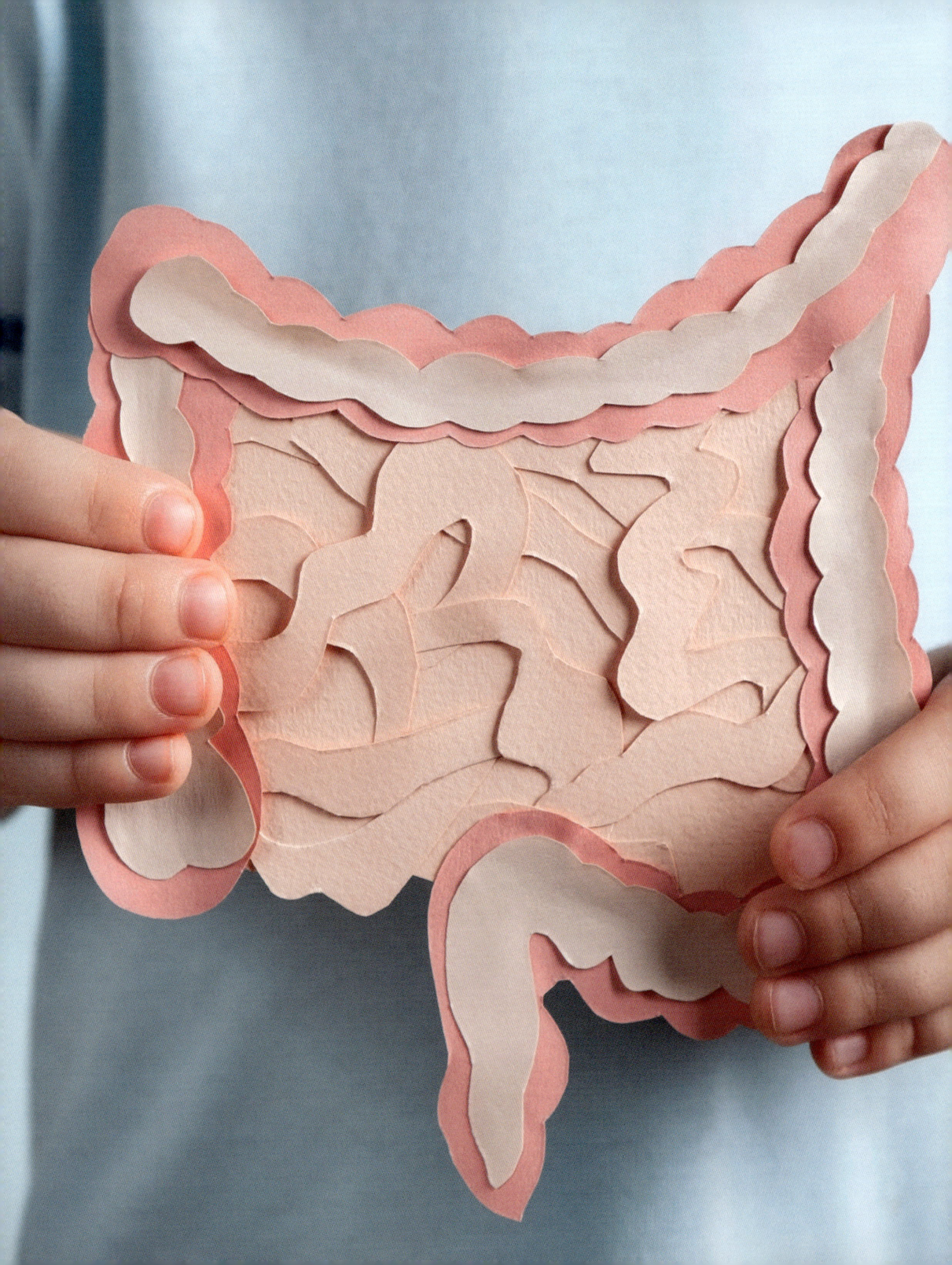

CHAPTER 20
Your Large Intestine Finishes Digestion

In this chapter, we will learn about the last part of your digestive system—your large intestine. Important jobs like digestion and most of your absorption are completed in your small intestine. But that doesn't mean important things don't happen in your large intestine! We'll learn about the jobs your large intestine does.

Your large intestine is filled with trillions of helpful creatures that are too tiny to see without a microscope. We will learn about God's purpose for these almost-invisible good guys. We will also look at the almost-invisible cells your body is made of. We'll learn about the digestion that happens inside them. This verse tells us that all the things we can see and all the things we can't see were created by Jesus and for Him.

> **For by Him all things were created that are in heaven and that are on earth, visible and invisible, whether thrones or dominions or principalities or powers. All things were created through Him and for Him.**
> **(Colossians 1:16)**

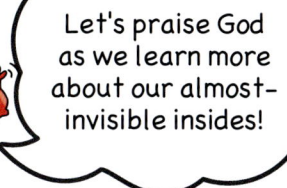

Let's praise God as we learn more about our almost-invisible insides!

Your Short Large Intestine

Your **large intestine** is much shorter than your small intestine. But it's called your *large* intestine because it's wider than your small intestine. Your large intestine does

GOD MADE ME

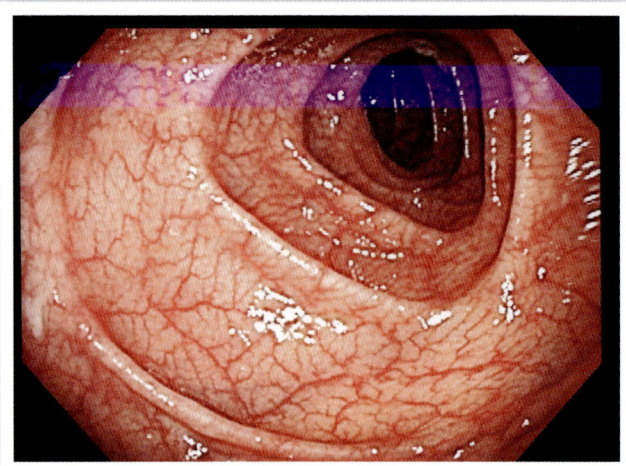
Inside the large intestine

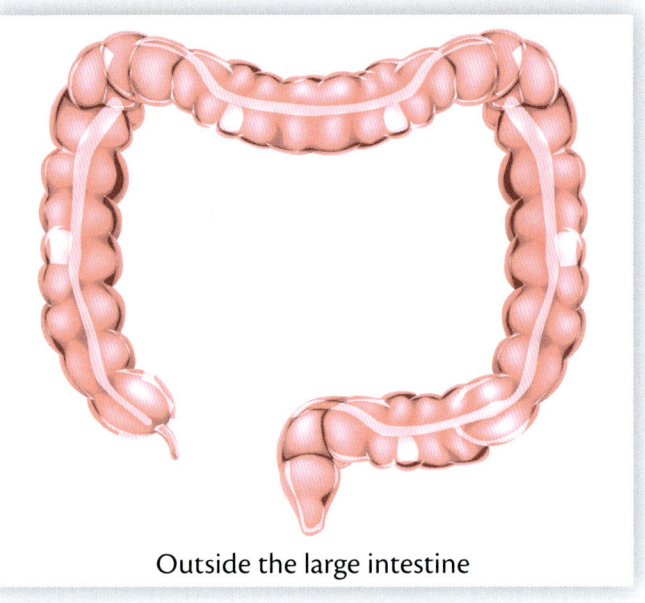
Outside the large intestine

not do any digesting. Its walls are smooth inside because it has no villi. God made your large intestine different from your small intestine because it has different jobs. The main jobs of your large intestine are to absorb water and to remove stool from your body.

After you eat, it takes about six to eight hours for food to pass through your stomach and small intestine. Then the food enters your large intestine. It will take about 36 more hours for it to move through your large intestine. From the time you swallow food to the time the stool leaves your body takes two to five days. This amount of time is different for different people.

Let's learn what happens to the chyme during its time in your large intestine.

- You have a sphincter between your small and large intestine to keep them separate. This sphincter is usually closed. But after you eat a meal, when the meal is still far up in your stomach, this sphincter gets a signal to open and let some chyme into your large intestine. Your small intestine needs to make room for the food that will be coming later.

- Your large intestine looks bumpy because it has pouches inside it. When chyme enters a pouch, some water is

Definitions

Your **large intestine** is the last part of your digestive tract. It removes the useful liquid from chyme and puts it into your blood. Then it gets rid of the solid waste your body doesn't need.

The **rectum** is the last space in your large intestine. It's where stools are stored until it's time for them to leave your body.

The **anus** is the opening at the end of the digestive tract, where stool leaves the body. It's kept closed by a sphincter.

CHAPTER 20: YOUR LARGE INTESTINE FINISHES DIGESTION

squeezed out of the chyme. The water is absorbed into your blood. After about 25 minutes, your intestine moves this chyme along to the next pouch where more water is squeezed out. This movement happens every 25 minutes to all the chyme in all the pouches. You don't usually feel these movements. Your large intestine absorbs about one quart of water (1 L) a day.

- Your large intestine also absorbs certain essential minerals that are in the water it takes from the chyme. God is so wise to make our bodies keep these important nutrients instead of treating them as waste!

- The walls of your large intestine don't make digestive enzymes, but they do make mucus. Mucus protects the walls from undigested fiber that's scrubbing its way through. Mucus also holds pieces of undigested food together to form stool.

- Two or three times a day, the large intestine does a big squeeze to move the chyme toward the **rectum**. This usually happens after a meal. God designed this to make room for the digested food that will come later. You may feel these movements.

- As the rectum fills, its walls get stretched. When it is stretched to a certain amount, a reflex happens. The reflex message tells certain muscles to

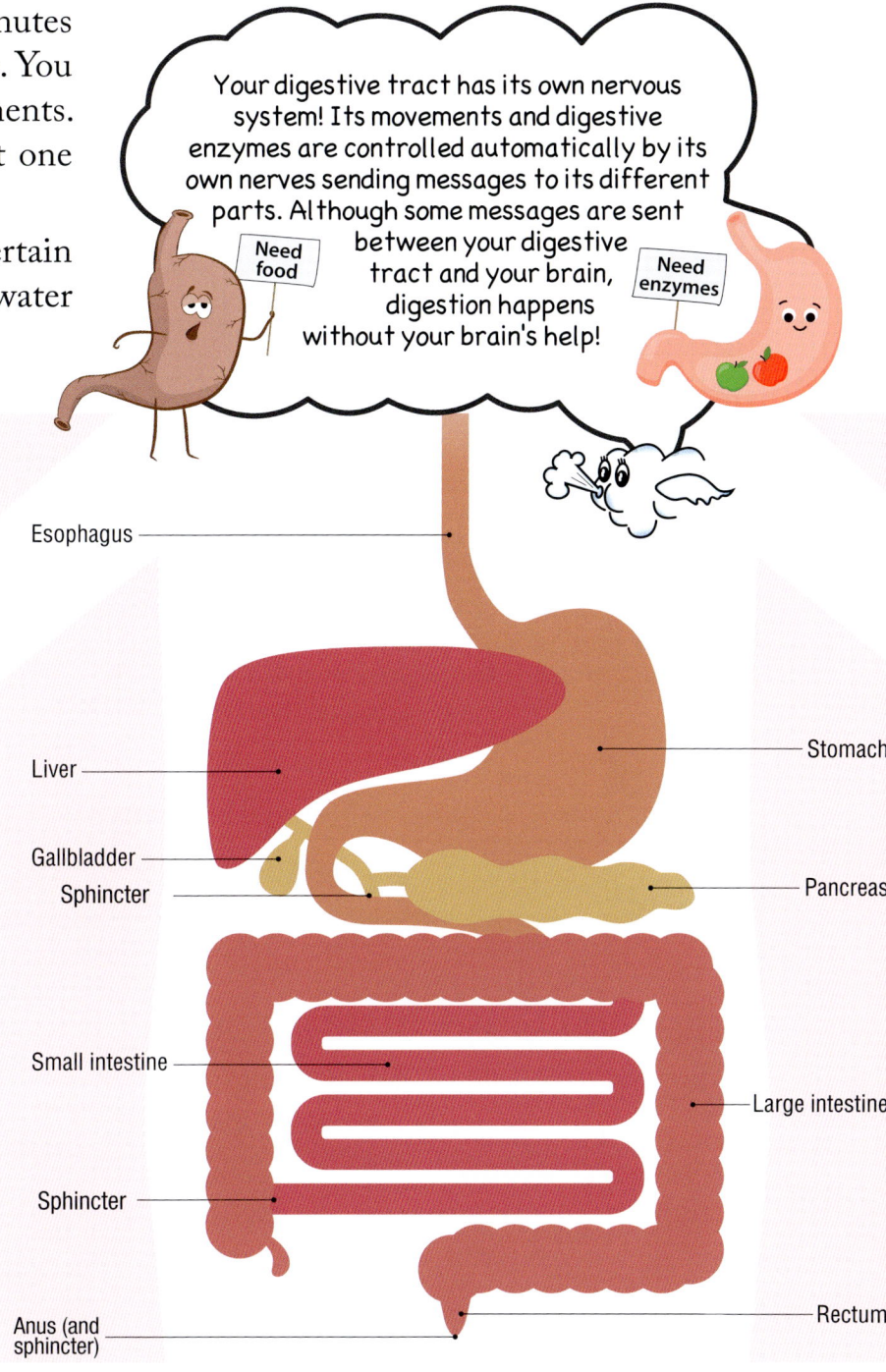

Your digestive tract has its own nervous system! Its movements and digestive enzymes are controlled automatically by its own nerves sending messages to its different parts. Although some messages are sent between your digestive tract and your brain, digestion happens without your brain's help!

Need food

Need enzymes

Esophagus
Liver
Gallbladder
Sphincter
Stomach
Pancreas
Small intestine
Sphincter
Large intestine
Anus (and sphincter)
Rectum

213

Thump's Health Hint

Sometimes intestines can become irritated by germs. But God designed a way to wash away the irritation. The intestines do this by moving the chyme through quickly, before the water can be absorbed out of it. When watery stools happen three or more times a day, we say a person has diarrhea (die-uh-ree-uh).

Sometimes people with diarrhea become dehydrated (dee-hide-rate-ud). This means their bodies have lost a lot of water, which could be dangerous. Scientists who studied digestion discovered a way to help dehydrated people.

They found that when the small intestine absorbs glucose (a kind of sugar), water is automatically absorbed with it. Someone invented a way to quickly fix dehydration by giving patients a special drink containing the right amount of sugar and water. This is called *oral rehydration therapy*. It has saved thousands of lives because it allows water to be absorbed sooner in the intestines than it normally would be.

You can buy special rehydration drinks that are designed for sick people. But don't use sports drinks for diarrhea. Sports drinks have higher amounts of sugar, which can make diarrhea worse.

squeeze and tells the sphincter at the **anus** to relax. This is when the stool is removed from your body.

- When stool leaves the body, it's about three-fourths water. The rest is undigested fiber, mucus, and millions of bacteria (which we will learn about in the next section).

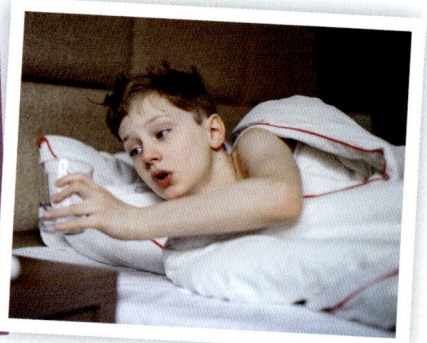

 Time to do Activity 58 in the Activity Book!

The Good Guys inside You

God's world is full of living things that are too tiny to see with your eyes. We say they are **microscopic** because we can only see them with microscopes. We call them **microbes**. Bacteria (back-teer-ee-uh), viruses (vy-russ-uhz), and fungi (fung-guy) are kinds of microbes. But of these, scientists have mostly studied the bacteria.

Although some kinds of microbes can cause sickness, most do not harm you, and many of them help you. In fact, the helpful microbes are very important for your health!

Helpful bacteria are found mostly in your intestines and on your skin. A grown man's body is made of about 35 trillion

CHAPTER 20: YOUR LARGE INTESTINE FINISHES DIGESTION

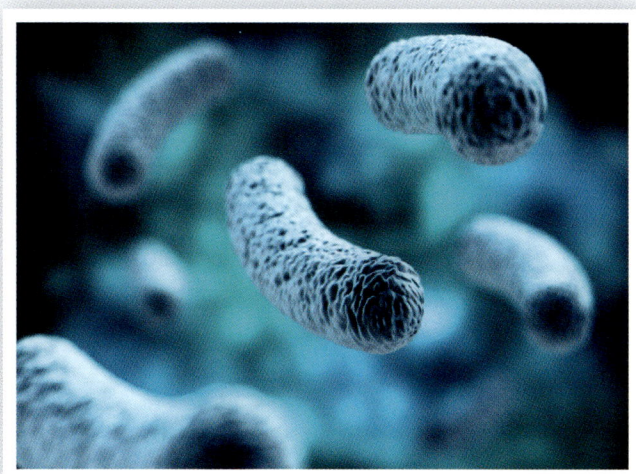

Lactobacillus—one kind of friendly bacteria living in your intestine

Definitions

Bacteria are microbes made of only one cell.

Viruses are tiny particles that can cause disease. Scientists can't decide if viruses are alive or not. They have DNA, so scientists say they are alive in this way. In other ways, viruses do not seem alive. They can only do the jobs of life by invading a cell and having the cell do the jobs for them.

Fungi are microbes that act kind of like plants, but they don't make their own food like plants do. Instead, they get food by absorbing it from other living things.

weigh as much as his brain.

Your body began to gather valuable bacteria as you were being born into the outside world. Then, as you drank milk from your mother and touched the things around you, your body gathered more kinds. Each time you got a new kind of bacteria, its cells would divide and make more of themselves. You became their home.

Your **cecum** (see-come) is a pouch at the beginning of your large intestine. Most of your intestines' 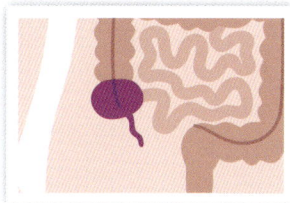 microbes live in your cecum. The cecum has a finger-like pouch attached to it which we will learn about later.

Imagine if we had a bacteria zoo with a cage for each kind of bacteria belonging to you. We would need about 1,000 cages for 1,000 different kinds!

Lactobacillus on villi. Some of your microbes live in your stomach and small intestine.

cells. But that man also has about another 40 trillion bacteria cells living in and on him! How can this work? This works because God made bacteria cells 1,000 times smaller than human body cells. All the bacteria living in and on the man only

215

GOD MADE ME

Thump's List of Your Microbes' Important Jobs

- **To digest for you.** The first kinds of bacteria to grow in your intestines after you were born were ones that could digest the kind of sugar in milk. Certain other bacteria can turn fiber into fatty acids that keep you healthy.

- **To fight germs.** Some microbes make chemicals that kill germs. Others tell your own body when to fight germs. When you have a lot of good microbes, they can crowd out germs. Microbes can also help your large intestine make a lot of mucus to keep germs from getting through to your large intestine's wall.

- **To make molecules that keep your brain happy and working well.**

- **To turn fiber into energy for your large intestine's cells.** When your microbes digest fiber, they get energy for themselves. But their digestion also makes energy to feed your cells!

- **To help you have a healthy heart.**

- **To help you have a healthy weight.**

- **To make vitamins B and K.**

- **To help your liver get back the important ingredients it loses in bile.**

- **To do many more wonderful and complicated things!**

Scientists are still discovering ways that God made the good guys inside you so important. Scientists think that almost half of the different chemicals circulating around your body come from your microbes!

Feed Your Microbes!

CHAPTER 20: YOUR LARGE INTESTINE FINISHES DIGESTION

Puff's Health Hint

Your microbes need food too! But by the time the food you eat gets to your large intestine, your body has removed all the nutrition it can. The only thing left is something you can't digest—fiber. But God made another way to feed your good guys. Your microbes can digest fiber to get energy! Fiber is also called *cellulose*. It's the part of plants that gives them structure. Here's how you can help your microbes:

- Eat a lot of fruits and vegetables, which are high in fiber.
- Eat foods that contain **probiotics** (helpful microbes).
- If you have had diarrhea for a while, or if you have been given antibiotic medicine for an infection, many of your good guys could have died. You can help your intestines build them up again! You can take probiotics as supplements. They should be available where vitamins are sold.

 **Time to do Activity 59 in the Activity Book!**

Cells: the Places We Use Nutrition

We've learned that your small intestine is lined with millions of tiny villi. Each villus is covered with a layer of cells. These cells are responsible for absorbing nutrients from the chyme and sending those nutrients to your body's cells.

Let's learn about the wise way God made these cells absorb things. Have you ever travelled on an airplane? When it's time to get on your plane, the flight attendant checks your boarding pass. If you don't have a boarding pass, or if you are at

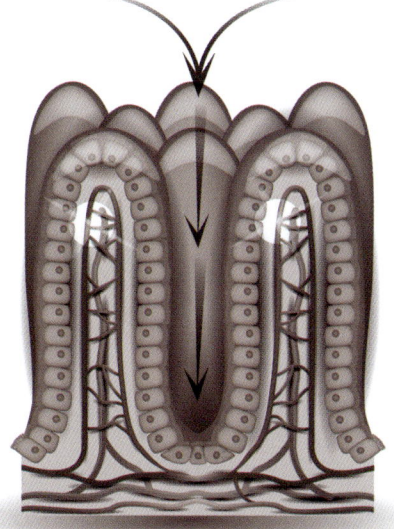

NUTRIENTS

Cells on the surface of a villus absorb nutrients and pass them on to the blood and lymph vessels inside.

GOD MADE ME

the wrong plane, you won't be allowed on board.

Cells have "flight attendants" in their cell membranes! These attendants are molecules that guard openings in their cell. Only the right nutrients will be allowed through. These attendants do their job well. Each one keeps its door closed until the right nutrient comes along. Then the nutrient is welcomed in!

Digested carbohydrates, proteins, and fats are brought into your villi cells through

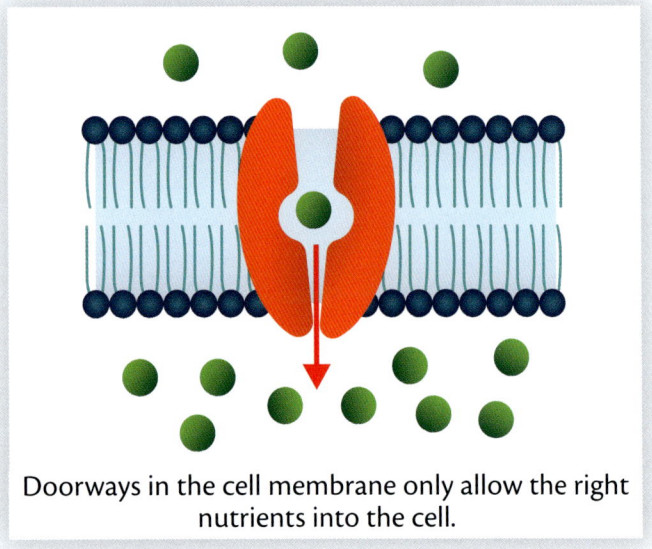

Doorways in the cell membrane only allow the right nutrients into the cell.

these special doorways. Vitamins, minerals, and other essential nutrients are brought in the same way. Water can enter these cells quickly along with certain nutrients when the doors open. But water can also come in slowly through a different kind of hole that doesn't need a "flight attendant."

Once these nutrients are absorbed by your villi cells, they will be put into your blood stream. Most nutrients get into your blood by first going into the capillaries inside each villi. But the digested fats don't go into these capillaries. First, your villi make digested fats into new fat molecules. These new fat molecules are too big to go through a capillary wall into your blood. Instead, they will go into the lymph vessels in each villi. Then they will travel with your lymph to where it empties into your blood.

Zoom in to the Cell

Do you remember that cells are kind of like little creatures? They are alive, they take in nutrition, they get rid of waste, and they have different parts with different jobs. In your body, you have different parts with different jobs. We call these parts *organs*. Since a cell is like a little body, we call its tiny parts **organelles** (ore-gan-ellz).

Since cells are like little bodies, we might expect them to "eat." If they eat, they might need to digest and get rid of waste too. God has given cells tiny organelles that help them do these things! Let's learn about them.

Endoplasmic reticulum (end-oh-plaz-mick ree-tick-you-lumm) is a fun name

CHAPTER 20: YOUR LARGE INTESTINE FINISHES DIGESTION

to say. But if that's too hard, you can just call it the **ER**. The ER looks like a maze. The maze's passageways connect with openings in your cell membrane so products can be sent out into your body. The ER also provides structure in your cells. Let's see what else it does!

There are two kinds of ER:

- **Rough ER** looks rough under a microscope because it has lots of ribosomes living on it. These are different ribosomes than the ones we learned about earlier. Those ribosomes live in your cells' liquid, but both kinds of ribosomes help make protein. Some of the proteins made by the ribosomes on your rough ER become the "flight attendants" in your cell membranes. They let nutrition into your cells. Other proteins made on your rough ER are your digestive enzymes. These enzymes will be sent out of the cell and given to your digestive system.

- **Smooth ER** has no ribosomes and does not make proteins. It looks smooth. Smooth ER makes special types of fats you need. It also acts a little like your liver because it breaks down molecules that would be poisonous to your cell. And, like your liver, it can store nutrients.

Lysosomes (lie-so-sohmz) come in a lot of different shapes and sizes. Usually they look like tiny water balloons. Lysosomes do the digesting jobs for their cell.

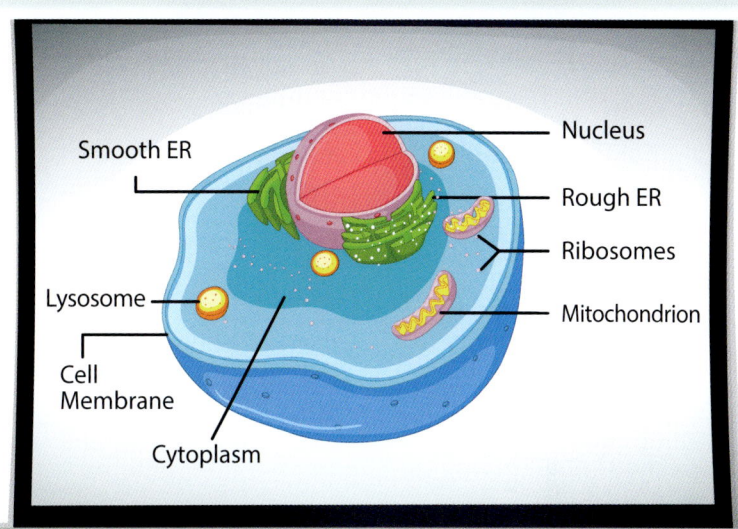

Zoom in to the Cell

- Lysosomes are like stomachs. They're filled with an acidic liquid.
- Lysosomes contain about 60 digestive enzymes that were made in the ER. These enzymes break down large molecules of fat, carbohydrates, and proteins that are not needed anymore.
- Lysosomes break down organelles that don't work anymore.
- Lysosomes help you fight germs. Some cells in your body capture germs and "eat" them by engulfing them. The engulfed germ is then delivered to one of the cell's lysosomes. The lysosome kills the germ by digesting it.
- The digested products are used again by the cell.

GOD MADE ME

- When cells get old, they stop doing their jobs. Your body needs to make new cells, and the old ones will die. God designed cells to die at a certain time. When it's that time, the lysosomes begin digesting the cell from the inside to kill it. A child your age will lose 20 to 30 billion cells a day this way.

Prayer

Lord, we don't understand how You made so many things in this world. And each thing is so complicated. Thank You for our complicated large intestines! Thank You for our complicated cells! Thank You for our complicated microbes! You made these complicated things work together in complicated ways. They need the fiber in our large intestines for their food, and we need their many helpful services. Help us take care of our wonderful bodies, cells, and microbes. Amen.

Time to do Activity 60 in the Activity Book!

CHAPTER 20: YOUR LARGE INTESTINE FINISHES DIGESTION

God purifies water as it travels through His amazing water cycle. This mountain waterfall is probably very pure because it's just beginning its downhill journey. God also designed amazing ways to purify your body.

UNIT 6
God Purifies You

Every living part of your body, from your blood to your cells, needs to be cleaned and purified (pure-if-ide). To purify means to take out the things that are dirty or not needed.

Our memory verse and hymn talk about purifying the part of you that's not your body. The verse means that your hope in God can make your heart pure as God is pure. The hymn reminds us that Jesus can cleanse you from your sin and make you pure within your soul. In this unit we'll learn how God keeps your body clean!

 ## Memory Verse

We know that when [God] is revealed, we shall be like Him, for we shall see Him as He is. And everyone who has this hope in Him purifies himself, just as He is pure. (1 John 3:2-3)

 ## Hymn to Sing: Lord Jesus, Think on Me

You can listen to it by searching for "Lord Jesus, Think on Me, adapted by Gregory D. Wilbur" on the internet.

Lord Jesus, Think on Me

Lord Jesus, think on me,
And purge away my sin;
From earth-born passions set me free,
And make me pure within.

Lord Jesus, think on me,
That I may sing above
To Father, Spirit, and to Thee,
The strains of praise and love.

The words to this old hymn were written in the 300's.

Kidneys live inside your back, under the place where this girl's thumbs are resting.

CHAPTER 21
Your Kidneys Clean Your Blood

Your word is very pure; therefore Your servant loves it. (Psalm 119:140)

When you talk about the things around you, do you ever use the word *stuff*? You might ask your brother to bring in all his stuff from the car. You might say you're going to eat a banana because of all the healthy stuff in it. When scientists want to talk about stuff, they use the word **matter**.

Matter is anything that has mass (which is kind of like weight). Matter must also take up space. A chocolate cake is matter. A cake takes up space and weighs something. A sound you hear is not matter. A sound doesn't take up any space, and it doesn't weigh anything.

Matter can exist in three different forms. Matter can be solid, liquid, or gas. Water is matter. Water is solid when it's ice. If the ice melts, the water will be liquid

When we read the Bible, we know God is giving us His pure words that will help our hearts be clean. We can love it!

you can drink. If you leave liquid water in an open container, it will evaporate into its gas form. We call these three forms the *three states of matter*.

God uses each of the three states of matter to remove waste from your body:
1. You have already learned how your **respiratory system** removes *gas waste* (carbon dioxide) from your body.
2. You have also learned how your **digestive system** removes *solid waste* (stool) from your body.

225

GOD MADE ME

3. Now we will learn how your **urinary** (yur-in-air-ee) **system** removes *liquid waste* from your body!

Your Blood Needs to Be Filtered

God gave your blood the job of circulating around inside your body. Because it circulates, blood can be used to pick up and deliver things. We have already learned about a few important pick-up-and-delivery jobs your blood has. One job is to deliver oxygen to your cells. Another is to gather carbon dioxide waste from your cells. You also know that your blood has the job of picking up nutrients from your digestive system and delivering them around your body.

But did you know that your blood also picks up waste from around your body? Your blood has the job of taking this waste to your **kidneys** (kid-neez). Your kidneys are the busiest part of your urinary system. They are perfect for the job of removing wastes from your blood. Let's look at the kinds of waste your kidneys remove from your blood.

> Your kidneys work hard! Your kidneys use almost one fourth of all the oxygen you breathe as they work to purify your blood!

Wastes That Your Kidneys Remove from Blood

- Chemicals left over after you digest amino acids
- Chemicals left over after you digest DNA in foods
- Chemicals left over after your muscle cells make energy
- Extra salts and vitamins your body doesn't need
- Extra water

Your urinary system uses liquid to do its job. These wastes leave your body in a liquid called **urine** (yur-in).

Do you remember that each edge of a die is about one centimeter long? Since dice are shaped like cubes, we can say that the amount of plastic the die is made of is one *cubic* centimeter. Now, imagine that 125 hollow dice are filled with your blood. That would be 125 cubic centimeters of blood. Imagine that your kidneys filtered that

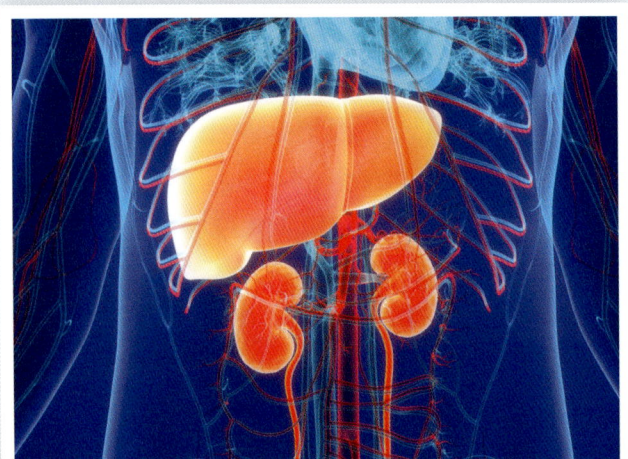

Kidneys, from the front. Your kidneys are two dark-red, bean-shaped organs. Your right kidney is lower than your left kidney because of the shape of your liver.

226

CHAPTER 21: YOUR KIDNEYS CLEAN YOUR BLOOD

> **Definitions**
>
> **Matter** is anything that has mass and takes up space.
>
> **Urine** is the liquid put out by your kidneys. It contains wastes and unneeded water and salt it has filtered from your blood.
>
> The **urinary system** is your kidneys, bladder, and all the tubes used to remove urine from your body.

Purified blood leaves the kidney

Blood that contains waste enters the kidney

Urine leaves the kidney

Your two kidneys filter all the blood in your body every 30 minutes!

blood and removed all the waste from it. Afterwards, 124 of the dice would be filled with clean, purified blood. The one die that's left would be filled with urine.

For every 125 cubic cm of waste-filled blood that enters your kidneys, 124 cubic cm of purified blood is returned to your bloodstream. That leaves only one cubic cm of liquid as urine!

 Time to do Activity 61 in the Activity Book!

Inside Your Kidneys

To help us understand kidneys, let's compare them to a clothes dryer. Your body cells automatically make waste as they do their work like the tumbling clothes make lint. That waste goes into your blood like lint goes into the air. Your kidneys clean blood by trapping the waste just like the dryer filter traps lint. Your kidneys

If your family uses a clothes dryer, you can help by cleaning lint out of its filter. Without a filter, lint would blow out the dryer and collect somewhere it shouldn't.

227

put the clean blood out into your circulatory system like the dryer blows out clean air. Then urine removes the waste from your kidneys and out of your body just like you remove the lint from the dryer's filter.

Let's look inside a kidney and learn how God makes it work!

The Path of Liquid through One of Your Kidneys
See picture on next page

1. Blood that contains waste comes toward your kidneys through the big artery that leaves your heart. Two smaller arteries branch off, one going into each kidney.
2. The artery entering each kidney branches into smaller and smaller vessels that go to the deep part of your kidney.
3. There, blood becomes purified in about a million tiny **nephrons** (neff-ronz). Let's learn what happens in a nephron.
4. When blood first enters the nephron, it goes through a tangle of capillaries living inside a capsule. In this tangle, blood cells are separated out and returned to the bloodstream. The liquid that's left is called **filtrate**. This filtrate still contains wastes. It also still contains the good things in blood plasma. The filtrate is collected by the capsule.
5. The filtrate leaves the capsule and travels along a wiggly tube. The wiggly tube has wiggly blood capillaries wound through its wiggles. These capillaries and the filtrate tube will trade things with each other right through their walls. The tube will give the blood good things that were taken out earlier in the tangle. The capillaries will add other wastes and extra water to what's already in the tube.
6. Now the filtrate is called **urine**. The urine moves out of the nephron into a collecting tube. The collecting tube joins with other nephrons' collecting tubes to make bigger collecting tubes.
7. Your collecting tubes empty their urine into a large space in your kidney. This space becomes the tube that removes urine from your kidney. Later, we'll learn where the urine goes next.
8. What happens to the purified blood? The purified blood from your nephrons travels out through little veins. These veins join together and leave your kidney. From there, the purified blood empties into the big vein that goes right to your heart.

Time to do Activity 62 in the Activity Book!

CHAPTER 21: YOUR KIDNEYS CLEAN YOUR BLOOD

The Path of Liquid through One of Your Kidneys

Each kidney has over one million tiny nephrons. The nephron above was drawn extra large to show us how it works.

Your Kidneys Do Other Amazing Jobs!

Your kidneys are organs that do several jobs in your body. Since blood circulates through your kidneys, God put them in charge of noticing certain things about your blood. And when they notice something that could be a problem, they do something about it! They also don't mind doing job assignments delivered by messages from other parts of your body.

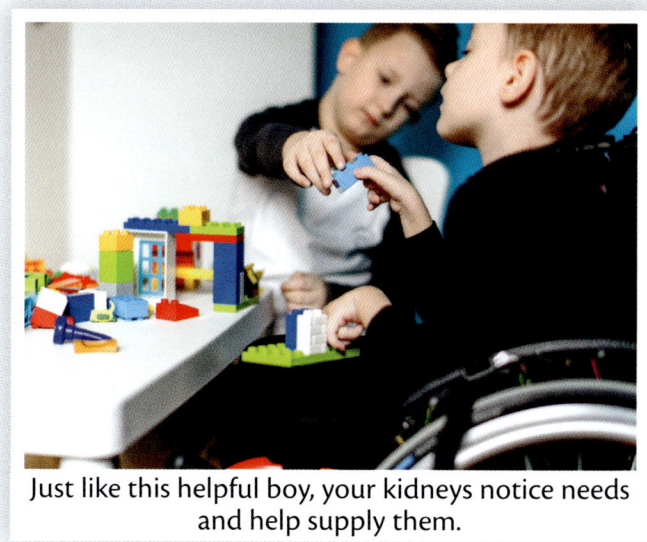

Just like this helpful boy, your kidneys notice needs and help supply them.

Like these children, your kidneys are happy to do the jobs they are asked to do.

Extra Jobs Your Kidneys Do

 Your kidneys tell your bone marrow to make red blood cells. Do you remember that red blood cells are the part of blood that carries oxygen around your body? If your kidneys notice that the blood passing through them needs more oxygen, they'll fix the problem. Your kidneys will make more of a hormone that tells your bone marrow to make more red blood cells. This will help your body get the oxygen it needs.

 Your kidneys keep the right amount of acid in your blood. Your kidneys notice if your blood has too much or too little acid. They can make the right amount of a certain chemical that makes your blood normal again.

 Your kidneys help strengthen your bones. You need vitamin D for strong bones. But most of the vitamin D traveling around in your blood is inactive (it can't be used). Your kidneys help change that vitamin D so you can use it. This active vitamin D helps your body absorb calcium from your food to keep your bones strong.

 Your kidneys help keep the right amount of pressure in your blood vessels. If blood pressure is too low, parts of the body might not get enough oxygen. Kidneys notice if blood pressure becomes too low. They can make an enzyme and a hormone that raise blood pressure two different ways.

 Your kidneys can change the amount of water they put into your urine. If your brain notices that your blood is losing water, it sends a message to your kidneys. It tells them to take less water from your blood when they make your urine. When you have plenty of water in your blood, your brain sends your kidneys a message to add more water to your urine again.

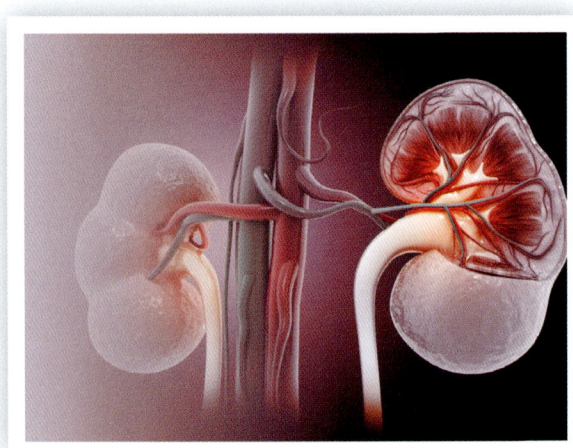

 Your kidneys can make sugar if your blood sugar gets too low. Your kidneys notice if the sugar in their own cells gets low. They also notice when there are messages in your blood calling for more sugar. Then they solve the problem by making sugar.

CHAPTER 21: YOUR KIDNEYS CLEAN YOUR BLOOD

Jobs of a Kidney

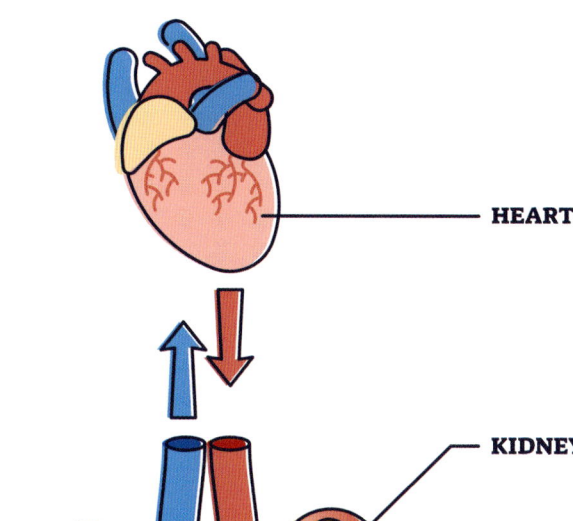

HEART

KIDNEY

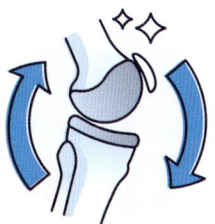
Helps body absorb calcium for strong bones

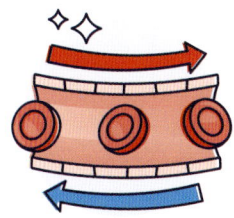
Tells bone marrow when to make more red blood cells

Filters blood to remove wastes and extra water

Makes sugar when blood needs it

Helps blood have the right amount of acid

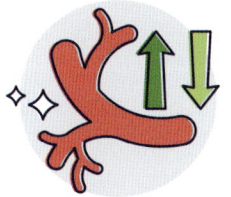
Controls the amount of liquid in blood

Prayer

Dear God, we are thankful for our kidneys. We don't feel anything while they do their jobs, and we hardly ever think about them. But we praise You for giving us kidneys to purify our blood and to do so many other important jobs. We know You want us to have pure hearts too. Thank You for helping us be pure by giving us Your pure words in the Bible! Amen.

Time to do Activity 63 in the Activity Book!

231

CHAPTER 22
Your Bladder and Its Tubes

... Christ also loved the church and gave Himself for her, that He might sanctify and cleanse her with the washing of water by the word. (Ephesians 5:25-26)

Now that you've learned how your kidneys so carefully remove waste from your blood, let's see how your body gets rid of this liquid waste.

Other Parts of Your Urinary System

Your urinary system has other parts besides your kidneys. These parts do not purify your blood, but they are needed to take the urine out of your body.

Urine leaves your kidneys through two long tubes called your **ureters** (you-readers). When you were seven years old, your ureters were about eight inches (20 cm) long. When you become an adult, your ureters will be 10-12 inches (25-30 cm) long.

Ureter walls are thick. If you could look at a short piece of a ureter, it might remind you of a macaroni noodle. God put muscles all along the walls of your ureters. This way, they can use **peristalsis** to squeeze urine along just like your digestive system uses peristalsis to move your food along.

Your **bladder** is the place where your body stores urine until it's time to empty it. Both ureters deliver little spurts of urine into your bladder. The urine enters your bladder through openings in back—one opening on the left and one on the right.

> God made water, and He loves to use it! In this verse, He talks about water so we can understand how His word cleanses His people. God gives water all over the earth and all through our bodies to keep things clean and working well!

GOD MADE ME

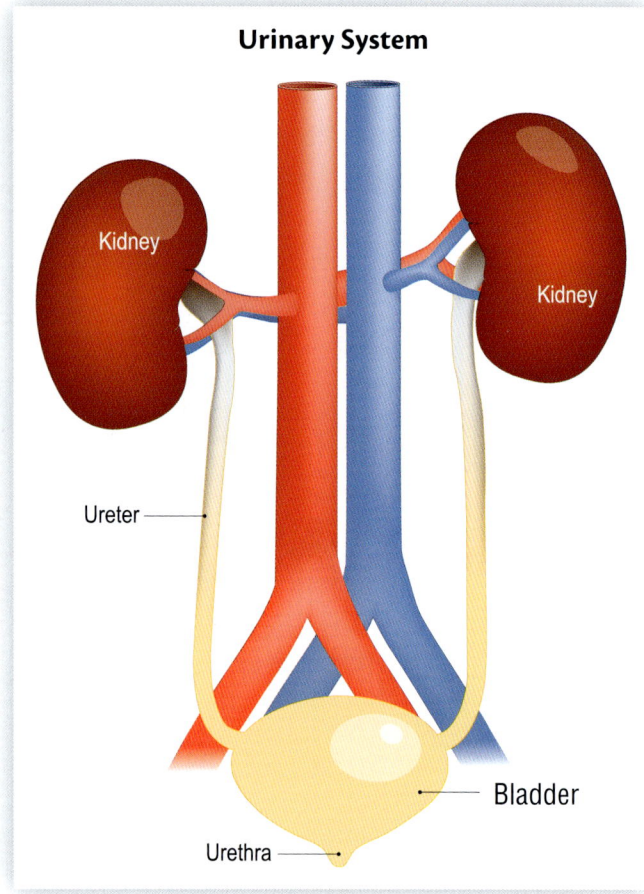

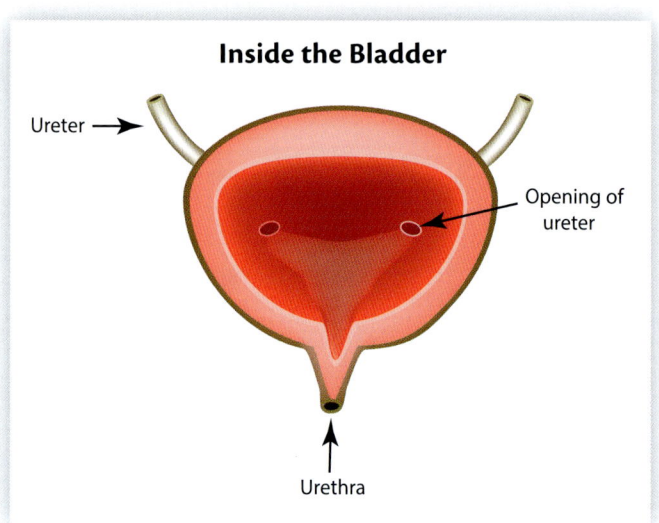

These openings have little flaps that act like the valves in your heart or veins. They allow urine to enter your bladder, but they keep it from traveling backwards into the ureters again.

Your bladder might get heavy as it fills with urine, so God has given it a cord to hang from. One end of this cord is attached to the top of your badder. The other end is attached up behind your belly button.

When it's empty, the inside walls of your bladder scrunch into folds. Your bladder becomes smaller. When it's full, your bladder walls become smoother and its top grows rounder. Your bladder can stretch. It can also squeeze the muscles in its wall to help empty out the urine.

Your **urethra** (you-wreath-ruh) is the tube that takes urine from your bladder to the outside of your body. The urethra has a sphincter about one inch (2 cm) down from your bladder. This sphincter is normally closed. It opens when you decide it's time to empty urine.

Your bladder and its tubes make mucus to fight germs inside your urinary system.

Your bladder stretches like a water balloon, but its walls are much tougher than a balloon.

CHAPTER 22: YOUR BLADDER AND ITS TUBES

 Time to do Activity 64 in the Activity Book!

How You Know When It's Time to Go

If you are eight to 10 years old, your bladder can probably hold 10-11 ounces (300-325 ml) of urine. But you probably feel a desire to **urinate** (empty your bladder) when your bladder contains about 8 ounces (235 ml) of urine.

Do you remember you have certain receptors in your skin? These receptors are parts of neurons that give you the sense of touch. Some of these skin receptors tell you when your skin is stretching. Your bladder walls have stretch receptors too! As your bladder fills with urine, it stretches. When it stretches a certain amount, a message is sent. This message is a *reflex* message.

You may remember what reflexes are. A reflex is when a message travels to your spinal cord. Then, without traveling to your brain, another message travels out from the spinal cord to solve the problem. If you touch a hot stove, your arm is quickly told to jerk your hand away. This is a reflex. In a reflex, your nerves send these signals only to your spinal cord, without asking your brain for help.

The reflex your bladder uses sends a message from your bladder to your spinal cord, saying that your bladder has stretched to that certain amount. Your spinal cord sends a message back to your bladder. This message goes to your bladder's muscles and tells them to tighten. The nerves that send your bladder's reflex messages are automatic. You don't have to tell them when to send their messages.

When your bladder muscles tighten, your bladder feels uncomfortable, but you don't urinate yet. It's time for your brain to help. The messages are no longer reflexes. They are also no longer automatic. Your brain gets the message that your bladder is uncomfortable. Your brain tells you to start looking for a place to urinate.

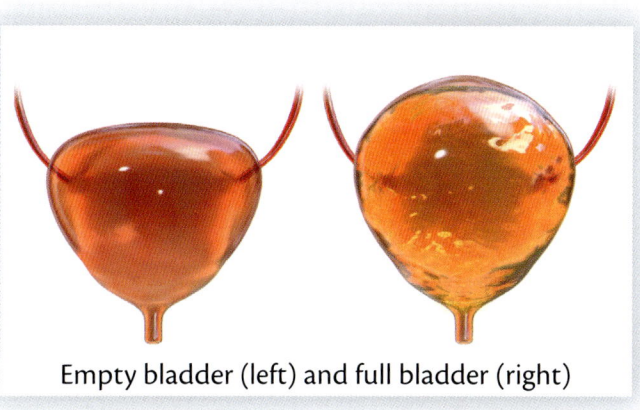
Empty bladder (left) and full bladder (right)

There are other interesting things your nerves do once you have found a place to urinate. As you are moving toward the place, your brain is automatically telling the sphincter in your urethra to stay squeezed shut. Then, when you can urinate, your brain tells you to purposely relax your sphincter. At the same time, other parts of your brain make your bladder muscles squeeze. Your bladder squeezes automatically—you cannot purposely make your bladder squeeze. But you can help a little by purposely squeezing different muscles in your abdomen. Now the urine leaves your body through your urethra.

Once you have urinated, your bladder muscles stop squeezing. But the sphincter in your urethra must now squeeze shut and stay shut. And your bladder starts filling again!

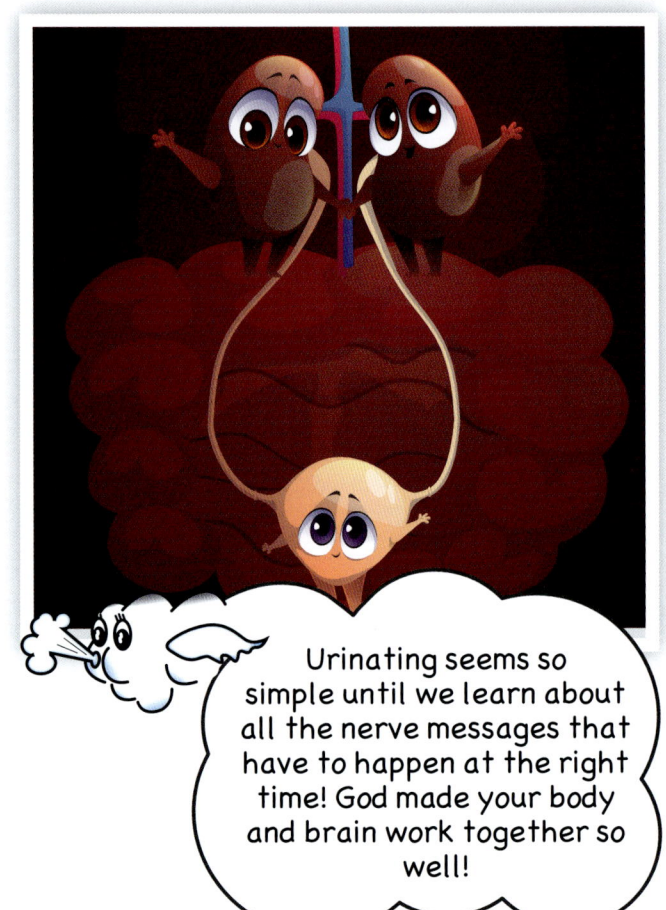

Urinating seems so simple until we learn about all the nerve messages that have to happen at the right time! God made your body and brain work together so well!

 **Time to do Activity 65 in the Activity Book!**

Liquid In, Liquid Out

A certain part of your brain notices when your blood is low in water. That's the same part that tries to solve the problem. It's the part of your brain that makes you feel thirsty! It's amazing how God made things in our bodies communicate with each other to solve problems and keep us working well.

All the jobs that happen in your body happen in liquid. That liquid comes from water. You lose water through your skin as sweat. You lose it through your nose and mouth when you breathe. You also lose water through your stools and urine. Your kidneys need to put wastes in water in order to remove them. You can help your kidneys by drinking enough water!

An eight-year-old child should drink about seven cups (1.5 L) of liquid a day. No more than one cup of that amount should be sweet liquids like juice or flavored milk.

Let's learn why urine is yellow. Do you remember that bile from your liver and

CHAPTER 22: YOUR BLADDER AND ITS TUBES

gallbladder is yellowish green? It gets that color from the waste that's made as your liver breaks down old red blood cells. Your body changes that yellow-green chemical to a yellow chemical in

Use this chart to tell if you are drinking enough. If your urine has no color, you are drinking too much.

your digestive system. A little of the yellow chemical enters your blood from your intestines. Soon your kidneys filter the yellow chemical out of your blood and put it into your urine.

Thump's Health Hint

Babies and older people can become dehydrated sooner than other people. Babies younger than 18 months and adults older than 60 years do not get much help from their kidneys when they start to get dehydrated. Their kidneys can't change the amount of water they use when they make urine.

These people may not know they are getting dehydrated and need to drink more water. Now that you know this, you can help little ones and older ones get enough to drink!

Prayer

Father, thank You for giving us water to drink! People all over the earth are glad You bring fresh water down from Your clouds. We're also glad You have given water so many jobs to do in our bodies. Thank You that water helps purify us in such amazing ways! Amen.

 Time to do Activity 66 in the Activity Book!

CHAPTER 23
Your Body Has Many Parts That Purify

> But now, God has set the members [body parts], each one of them, in the body just as He pleased. (1 Corinthians 12:18)

Your body has many organs and parts that work to purify. We have learned about these organs in earlier chapters. In this chapter, we will be looking at what these parts do to keep you clean inside. Let's be amazed at the wonderful way God put us together and keeps our bodies pure!

This verse says that God put all your body parts inside you! He put them in the right places and gave them their jobs so your whole body would work well. God put this verse in the Bible to show that He has made you a special part of Christ's body-His people!

Large Organs That Purify

Your lungs, liver, intestines, and kidneys are large organs that have several jobs. But the job of purifying takes up a lot of their energy! Let's look at these organs again and review the way they clean you inside.

Hi! I'm Scrub. I LIKE to keep things clean! Don't you? Let's review some of the ways God purifies your body.

GOD MADE ME

Lungs: As you breathe, your lungs bring in oxygen for your body to use. But your lungs also collect waste that your body must get rid of. This waste is the gas we call carbon dioxide. Your blood brings carbon dioxide to your lungs and drops it off. Your lungs blow it out each time you exhale.

If we could collect all the carbon dioxide an adult exhales each day, it would weigh a little over two pounds (1 kg). It's very important that your lungs purify your body by removing carbon dioxide!

Liver: Sometimes you might have to take medicine that helps a problem. But then your body needs to get rid of the medicine because it could make you unhealthy if it stayed inside you. Thankfully, your liver is able to take these chemicals out of your blood.

Sometimes we may accidentally eat food that contains harmful chemicals. We might eat the green part of a potato, which has a little poison in it. Or we may accidentally eat spoiled food. Spoiled food can contain chemicals made by the germs that cause mold or rottenness. Thankfully, God made our livers able to remove these chemicals too.

Your body makes its own poisons! Your body breaks down the food you eat, and

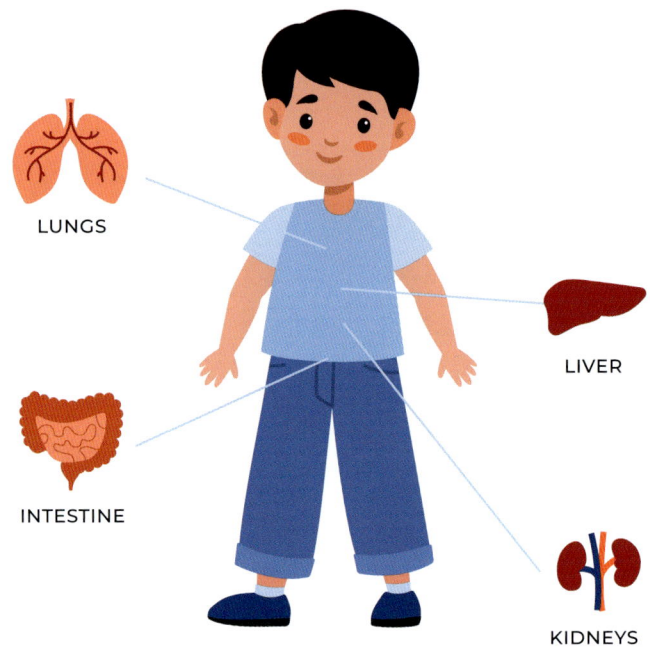

Your lungs, liver, kidneys, and intestines do much of the purifying in your body. But if more than half the cells in these organs stopped working, a person could still live. God is very generous to give you extra!

Your liver puts some wastes, like urea, into your blood so your kidneys can remove them from your body. Your liver puts other wastes into bile for your intestines to remove.

All toxins are poisons. All poisons are not toxins.

CHAPTER 23: YOUR BODY HAS MANY PARTS THAT PURIFY

Definitions

Poison — a chemical that could harm a living thing.

Toxin — a poison that's formed by a living thing. We can say that a poison dart frog makes a poison, or we can say it makes a toxin. We can say gasoline is a poison. We cannot say gasoline is a toxin because it's not made by a living thing.

Detoxify (dee-tox-if-eye) — to remove a poison or toxin from somewhere it shouldn't be. You have several organs that detoxify your body.

Even though ammonia is poisonous inside our bodies, it's a helpful ingredient in many glass cleaners.

your cells use the food. As your body does this, wastes are created. One of these wastes is **ammonia** (uh-moan-yuh). Ammonia is produced when your body breaks down protein. But ammonia is poisonous. Your liver changes ammonia into **urea** (yur-ee-uh) and puts it into your blood. Urea is not poisonous.

Kidneys: After certain wastes leave your liver, they enter your blood. The wastes travel with your blood until they arrive at your kidneys. Your kidneys remove these wastes from your blood and make them a part of your urine. Soon the urine travels out of your body.

Intestines: During digestion, your digestive tract saves all the things your body needs from food. It puts those good things into your blood. All that's left is waste you can't use. Your intestines help purify your body by removing that waste as stool.

Chocolate is poisonous to dogs. A dog's liver is different than a human liver. Dog livers can't break down certain chemicals in chocolate quickly enough, so the chocolate will make the dog very sick.

The main waste in urine is urea. Do these two words sound similar to you?

241

GOD MADE ME

 Time to do Activity 67 in the Activity Book!

Other Parts That Purify

Now let's look at other parts of your body that help keep you pure. These parts are not big organs, but they still do important jobs. We're going to review your lymph, blood, skin, and other body parts to see how they help purify.

Blood: Your blood is an important part of keeping you pure. Your blood carries harmful chemicals and wastes away from your cells. It carries this waste to the places that know how to get rid of it.

Your blood collects carbon dioxide waste from cells. Then it takes the carbon dioxide to your lungs so your lungs can get rid of it.

Your blood also travels directly from your small intestine to your liver before it goes to your heart and the rest of your body. God designed it to work this way so unhealthy chemicals you may have eaten can be **detoxified** in your liver before they travel anywhere else.

As your body changes food into energy in your cells, poisonous wastes are left over. Your blood delivers these wastes to your kidneys so they can

A trash service takes neighborhood trash and gets rid of it at the right place. Your blood also takes waste from your cells and gets rid of it at the right place.

be removed from your blood and sent out of your body in urine.

Lymph: Remember, the liquid in your blood is called *plasma*. The liquid surrounding your cells is called *lymph*. These liquids have different names in the two places, but they are the same liquid. This liquid is constantly being traded back and forth between your body cells and your blood. It's traded though the capillary walls.

A small amount of your lymph is not traded back into your capillaries. It has the job of collecting certain wastes that your capillaries can't take. This

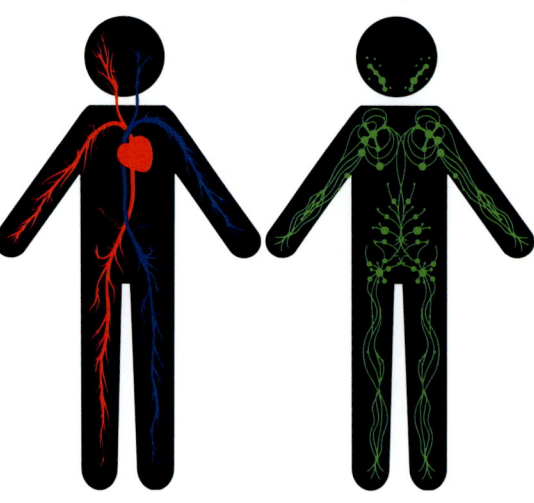
Circulatory and Lymphatic Systems

242

CHAPTER 23: YOUR BODY HAS MANY PARTS THAT PURIFY

lymph collects germs and damaged cells and takes them to your lymph nodes. There, they are destroyed and turned into waste. Then your lymph system empties this waste into your blood. Your blood takes the waste to your liver or kidneys to be removed and sent on its way.

Sweat glands: God formed sweat glands on the palms of your hands and soles of your feet when you were four months old inside your mother. They did not start working until soon after you were born.

One of the jobs God gave sweat glands is to remove wastes. Sweat is mostly water, but it also has salt and a small amount of wastes. Your sweat glands help your kidneys by removing a little urea. Your sweat glands also help your lungs by removing a little carbon dioxide from your body. There are even some poisonous chemicals that sweat glands can remove more quickly than the kidneys and lungs can.

Tears: Tears contain enzymes that can break down harmful things and wash them out of your eyes. Tears trap small things like dust and dirt that accidentally get in your eyes. Your eyelids can then sweep them away through the tiny tube that leads from the inside corner of your eye into your nose.

While you sleep, tears keep moving toward the corners of your eyes. Your eyes are drier at night, and you may wake up with dried tears in those corners. Scientists don't have a name for these eye globs or crumbs. Does your family have a name for them?

Your eyes make a small amount of tears normally. But if something irritates your eyes, or if you cry, they make a large amount. In one year, your eyes make about 15-30 gallons (55-110 L) of tears.

**Put my tears into Your bottle;
Are they not in Your book?
(Psalm 56:8)**

This verse shows us that God keeps track of our troubles. It's as though He has a special bottle for measuring tears!

GOD MADE ME

Mucus: Mucus protects many of your body's openings. It's thick and sticky. Mucus helps trap dirt, germs, and other things that shouldn't be inside you. And if something is already in (like germs that have reproduced), mucus helps move it out. You probably notice nose mucus more than mucus from other places. Mucus is also made in your eyes, mouth, urethra, and digestive tract.

Mucus helps move pollen out of your nose when you sneeze.

Earwax: Earwax starts with a mixture of products made by tiny glands in your ear canal. This mixture cleans your outer ear by trapping dust, dirt, and dead skin cells. The movement of your jaw helps earwax move toward your ear canal's opening, where it can fall out of your ear or be removed.

Time to do Activity 68 in the Activity Book!

A New Cleaning Service Has Been Discovered!

Scientists have recently discovered a body system that cleans waste out of your brain! They named it the **glymphatic** (glimf-attic) **system**. The glymphatic system uses a liquid called **cerebral spinal fluid**. We'll just call that fluid CSF. Your brain and spinal cord are surrounded by CSF.

People used to think that the only job of CSF was to cushion the brain and spinal cord and protect them from bumps. But scientists have recently found that CSF also moves among your brain's neurons, collecting wastes.

Your glymphatic system has no pump like your circulatory system has. And it doesn't have vessels with muscles inside like your lymph system has. God designed a different way for CSF to move and purify your brain. Scientists still don't know everything about it.

Your CSF is made deep within your brain. Then it uses an interesting way to travel among your neurons. Picture one of your arms as a blood vessel

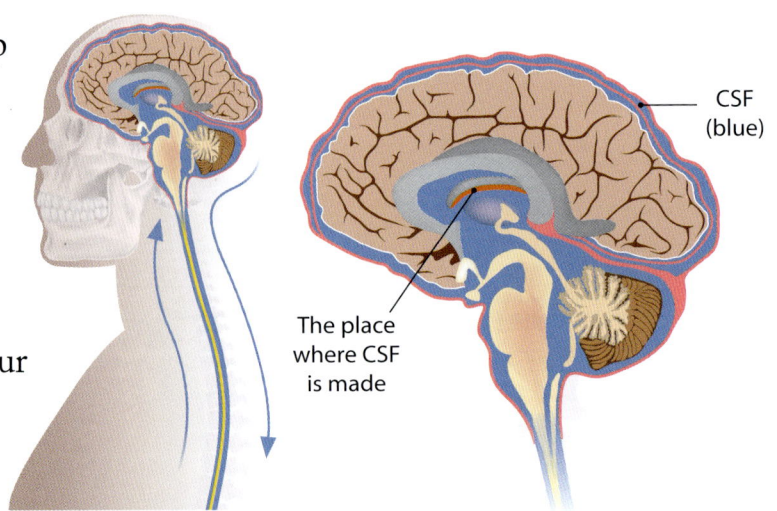
CSF (blue)
The place where CSF is made

244

CHAPTER 23: YOUR BODY HAS MANY PARTS THAT PURIFY

in your brain. Now picture that arm wearing a long sleeve that's big and loose. Your brain's blood vessels are surrounded by tubes that are loose like big sleeves. Just like there would be air inside the big sleeve, there is CSF inside the tubes. And, just as the air can move through the sleeve's cloth, CSF can move through the walls of its tubes very easily.

As the CSF moves out of its "sleeves," it flows among your brain's neurons. It's busy picking up wastes. At the same time, it's delivering nutrients to your neurons. Then the CSF moves into a different sleeve, taking the wastes with it. The wastes will end up being dumped into your lymph system, and your lymph system will get rid of them!

Scientists think CSF flows by the help of the blood that moves in spurts through the vessels sharing its "sleeves." This may be why God put CSF's loose tubes around your brain's blood vessels.

Scientists are pretty sure that your glymphatic system is mostly active when you are sleeping. It seems to shut off when you are awake. They believe it cleans out more waste during *deep* sleep and when you sleep on your side instead of your back or stomach. Scientists also believe that wastes can be moved out of your brain during exercise.

Not long ago, no one knew about the glymphatic system. But even though we didn't know about it, God has had this wonderful system working secretly and wonderfully, deep in brains, ever since creation!

Oh, the depth of the riches both of the wisdom and knowledge of God! How unsearchable are His judgments and His ways past finding out! (Romans 11:33)

Now you see another reason why sleep is important! Be sure to get a good night's sleep every night.

Prayer

Dear God, it's nice to have a clean, pure body. We see that our bodies need to be purified in so many ways so we can live. Thank You for thinking of all these wise systems and processes in Your deep wisdom and knowledge! Help us to always be learning about You. Amen.

Time to do Activity 69 in the Activity Book!

CHAPTER 24
God Purifies Your Cells

We have been learning about the many ways our bodies purify themselves. Your body has many places that work hard to remove wastes and harmful chemicals. Each of your cells is like a tiny working body. That means your cells make waste too. They have parts that get old and need to be broken down into waste. They also have parts that need repairing. Let's see how God keeps your cells pure and repaired!

Your Cells Can Reuse Waste

Do you remember that your cells have organelles called lysosomes? Lysosomes are the stomachs of your cells. Lysosomes help keep your cells clean inside by digesting old organelles and other things your cells don't need.

We've learned that God has given the ribosomes in your cells the big job of making proteins. Your lysosomes help out! As lysosomes digest waste that's made of protein, they break it into smaller pieces. Your ribosomes reuse these small pieces to make new proteins!

If any wastes cannot be digested by

Making something new out of old things saves time and energy. God made your cells reuse things to be efficient!

GOD MADE ME

lysosomes, the cell can sometimes spit them out. Once outside the cell, the waste might be broken down by enzymes. If not, the waste may just clump together like trash at a dump. This is one reason it's important to put only good things in your body.

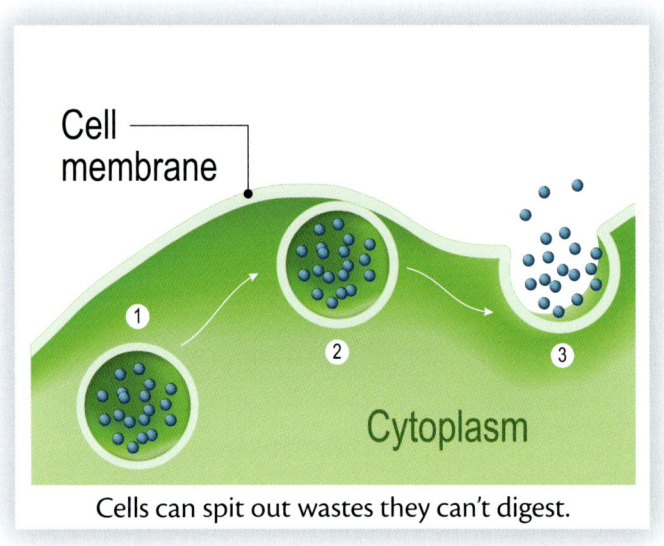

Cells can spit out wastes they can't digest.

Some of the Proteins Made by Cells

- Proteins that carry oxygen from blood to cells
- Proteins for muscle movement
- Digestive and other enzymes
- Proteins for fighting germs
- Hormones
- Proteins that make hair and nails

 Time to do Activity 70 in the Activity Book!

Your Cells Have Tiny Shredders

You know that, inside your tiny cells, you have tinier organelles. But your cells have structures that are even tinier than organelles. These structures are so tiny that they are made of only a small number of molecules! Let's learn about **proteasomes** (pro-tee-uh-sohmz).

Proteasomes look like little trashcans. In fact, their job is to get rid of trash in your cells, like damaged and unwanted protein. Each proteasome has a lid on top and another lid on the bottom. Shredding happens in the "can." Let's have Scrub tell us how this kind of cleanup works.

Most of the protein waste in your cells is taken care of by your proteasomes! And, like your lysosomes, they help your cells reuse proteins!

CHAPTER 24: GOD PURIFIES YOUR CELLS

Proteasome Meets Protein Trash

1. Molecules called "flags" look for certain kinds of protein trash in the cell.
2. When they find what they are looking for, one or more "flags" attach themselves to the protein trash. This helps the proteasome know which proteins to destroy and which ones to leave alone.
3. God doesn't want the wrong proteins destroyed, so He made another safety step. Not only does the proteasome recognize the flags, it also can recognize damage on proteins. It checks for both these things before destroying the protein.
4. The proteasome unfolds the flagged protein so it will fit through a small opening into the shredding area (the "can").
5. When it's in there, the protein is shredded into smaller pieces.
6. The pieces are let out of the bottom of the "can" into the cell liquid.
7. Enzymes in the cell liquid break the pieces into amino acids.
8. Ribosomes use the amino acids to make more proteins.

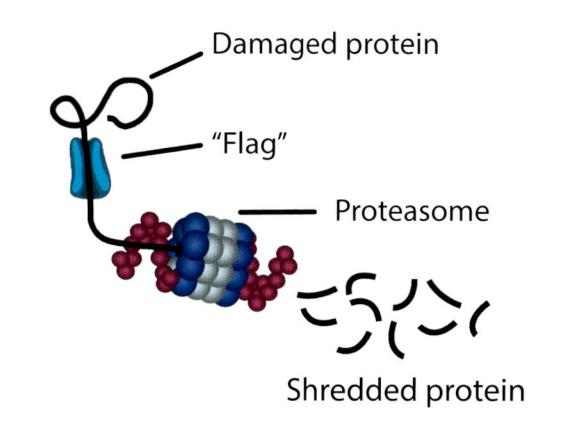

 **Time to do Activity 71 in the Activity Book!**

God Keeps Your DNA Pure

Did you know that damage is happening to your DNA all the time? This damage can cause mistakes to happen when your cells make more cells (mitosis). Remember, mitosis is when copies of your DNA are made. Damage to your DNA can be caused by toxins that enter your cells. It can also be caused by toxins your cells naturally make as waste. Damage happens to the DNA in each one of your cells about 10,000 times every day!

But God has provided a way for your damaged DNA to be fixed! Your cells have 17 different enzymes to repair 17 different

GOD MADE ME

kinds of DNA damage. Each one of these enzymes has two jobs. One job is to fix the damage. The other job is to kill the cell if the damage can't be fixed.

Now let's learn about how God keeps your DNA pure!

- At certain times during mitosis, enzymes check your DNA to make sure it's being copied correctly. If something is wrong, mitosis pauses until your DNA copy is fixed.
- God is very wise. He made DNA in twin strands. When part of one strand is damaged, certain enzymes notice this

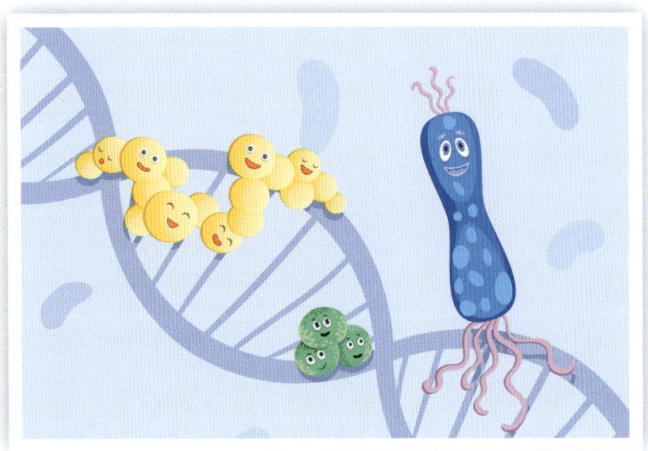

and cut out the damaged area. Then different enzymes put in new DNA. They know just what to put in by looking at the same place on the twin strand.

- Sometimes both strands of DNA are broken off at the same time. This means that there is no copy for the repair enzymes to look at. When this happens, enzymes work to put each strand back together. Usually, a little bit of information is lost from each strand. But that's better than losing all the information from the broken off pieces.
- Sometimes damaged DNA cannot be fixed. That's when enzymes kill the cell. Why is it important to get rid of damaged cells?

1. **DNA tells your cells how to make proteins your body needs.** If a cell's DNA is damaged too much, the cell can't make important proteins. If the cell can't do its job but it stays alive,

250

CHAPTER 24: GOD PURIFIES YOUR CELLS

your body would still have to feed it and take care of its waste. This would use up extra energy.

2. **Damaged DNA could harm your body.** The damage might cause part of the DNA instructions to be missing. If the cell divides into new cells, these cells could be missing important DNA instructions too. The missing information could cause problems in the rest of your body.

God wants your DNA information to stay pure and without mistakes. But more importantly, God will always make sure that all of His Word stays accurate. This verse shows that!

The sum of your word is truth, And every one of your righteous rules endures forever. (Psalm 119:160 ESV)

Prayer

Lord, You are so wise to have made such tiny, complicated structures in our cells. They have important jobs, like breaking down the right proteins or checking and fixing DNA. We're glad to learn about these things so we can thank You for more ways You take care of us. Thank You for keeping us pure. Amen.

Time to do Activity 72 in the Activity Book!

UNIT 7
God Gives You Strong, Beautiful Motion

Are you ready to move? Let's learn about your body's muscle and bone team! Together, your muscles and bones work so you can move. This team is called your musculoskeletal system. Our memory verse reminds us that God is the One who gives us motion and life!

When we sing this hymn, we offer our lives to God for Him to use. Let's learn how our muscles and bones help us be swift and beautiful for the Lord!

 Memory Verse

"For in Him we live and move and have our being." (Acts 17:28)

You can listen to a few different melodies of this hymn by searching for "Take My Life and Let it Be hymn" on the internet.

 Hymn to Sing: Take My Life and Let It Be

Take my life and let it be
Consecrated, Lord, to Thee.
Take my moments and my days;
Let them flow in ceaseless praise.

Take my hands and let them move
At the impulse of Thy love.
Take my feet and let them be
Swift and beautiful for Thee.

Take my love; my Lord, I pour
At Thy feet its treasure store.
Take myself, and I will be
Ever, only, all for Thee.

CHAPTER 25
God Gives You Strong Bones

[God said,] "So I will comfort you; . . .
Your heart shall rejoice,
And you bones shall flourish like grass."
(Isaiah 66:13-14)

In this verse, God speaks comfort to His people. The Lord is the One who blesses us with joy and health. This includes good health for our bones!

When God made octopuses, He didn't give them bones. Since they don't have bones, octopuses can squeeze through very small spaces to hide or find food. God gave them exactly what they need to live in the ocean. God gave you bones so you can live on land!

Do you think bones are hard and dry? That is the way bones look when you see them in cooked meat. But the bones inside you are different than that. The bones inside your body are living. Your bones are strong, but they are not hard. They can actually bend a little. Also, your bones are not dry because they are ⅓ water! Bones have blood vessels to deliver nutrition and nerves so you can feel things. Your bones are living organs!

Why Are Bones Strong?
God made bones strong and stiff, but still flexible (able to bend). Bones are made up of these different things:

GOD MADE ME

 Hard minerals such as calcium

 Proteins that are flexible

What would your bones be like if they were only made up of hard minerals? The minerals would make them very stiff. Of course, you would want sturdy bones. But, if your bones weren't flexible, and you fell down, your stiff bones could shatter like glass.

What if your bones were only made of flexible proteins? If you fell, your bendy bones wouldn't break. But, if your bones were that flexible, you couldn't stand up in the first place! You would probably be lying on the ground since bendy bones couldn't hold you up. You would be like an octopus stuck on the beach. God wisely made your bones strong and flexible with the perfect mixture of minerals and proteins!

Your bones are also strong because of their structure. Most bones are made of two kinds of bone tissue:

- **Compact bone** is bone tissue that's tightly packed and very strong. The outer shell of most bones is made of compact bone tissue.

- **Spongy bone** has big spaces inside. Even though it looks like a sponge, this bone tissue is still very stiff and strong. The spaces in spongy bone are where your body makes blood cells. Spongy bone is only found inside certain parts of some of your bones.

There is another reason your bones are both strong and flexible. Your bones are strong and flexible because they are not solid all the way through. Besides the small spaces in spongy bone, your bones also have large, hollow spaces inside. These spaces allow your bones to twist and bend a little without breaking.

Your bones can grow stronger when you need more strength. When you lift, push, or pull heavy things, your muscles must pull

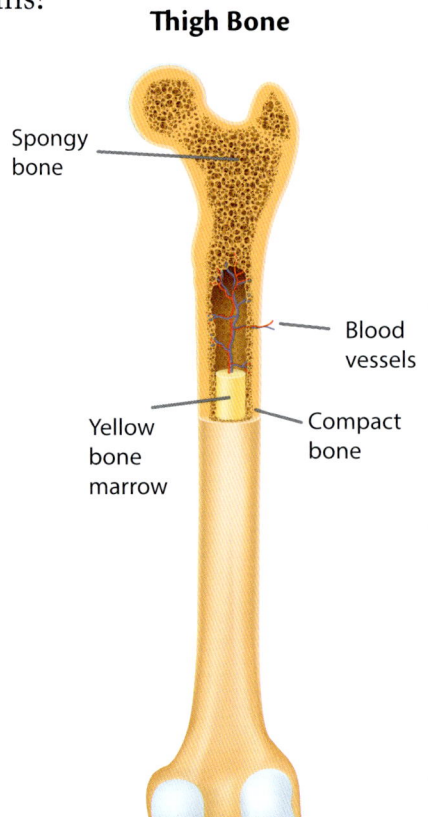

Thigh Bone — Spongy bone, Blood vessels, Yellow bone marrow, Compact bone

Eiffel Tower, Paris, France. The spongy bone inside these brothers is built with a strong criss-cross design. The Eiffel tower behind them is also built for strength with a criss-cross design.

CHAPTER 25: GOD GIVES YOU STRONG BONES

A Broken Bone Made New

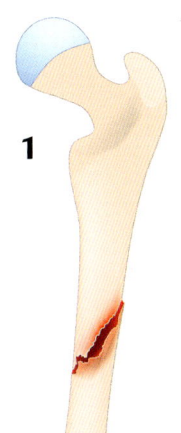 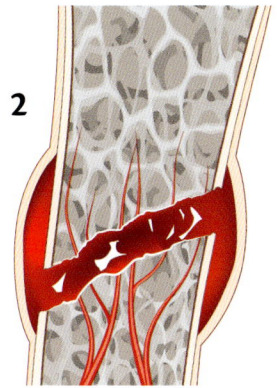 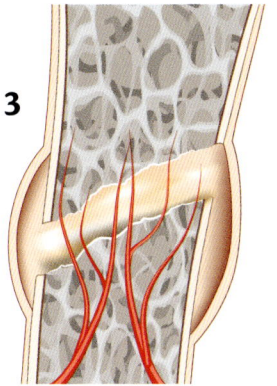 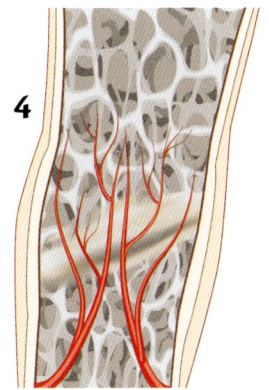 

hard on your bones. Your bones notice this. Over time, they grow thicker and stronger where the muscles are attached.

Bones stay strong by replacing old tissue with new tissue. If a bone breaks, the bone can fix itself. Your body makes new bone to join the broken pieces together again and fill in the break. The new bone material will be stronger than the old material! What an amazing system God made!

Man-made things like bridges and cars get worn out and weaker the more they are used. But bones don't wear out easily. They get stronger the more they are used. Bones are one of the strongest things we have discovered in God's creation!

 **Time to do Activity 73 in the Activity Book!**

How Your Body Builds Bone

When you were about one week old in your mother's womb, cartilage began to grow where most of your bones would someday be. About eight weeks later, that cartilage began to be replaced by bone. Ever since then, your body has been growing more and more bone material, and you've been losing cartilage. When your bones stop growing around age 25, you won't have any more cartilage that will become bone.

Remember, cartilage is strong, bendy tissue that gives structure. Your nose and outer ears are given their shape by cartilage under your skin.

GOD MADE ME

Right now, your bones are growing at areas called **growth plates**. Growth plates are made of cartilage. As new cartilage grows, the bone tissue next to it is pushed away. This makes the whole bone larger or longer. At the same time, the older cartilage gets replaced with new bone. Your bones will stop growing by age 25 because your growth plates will all become bone material and disappear.

Bone tissue is built by special cells. We will call these **bone-building cells**. These cells make new bone and put it out into the empty space around them. As more bone forms around the cells, the space around them isn't empty anymore. It fills up, making it harder for nutrition to get to the bone-building cells. To solve this problem, God changes them.

The bone-building cells change by growing arms that connect to blood vessels within the bone. The cells can now get nutrition (and remove waste) through the blood vessels. The cells' arms also grow to connect with the other bone-building cells.

These cells with arms now have the job of making sure you keep your bone material. We can call them **bone-keeping cells**.

God designed bone-keeping cells to notice things about the minerals in your body and in your bones. They can send messages to new bone-

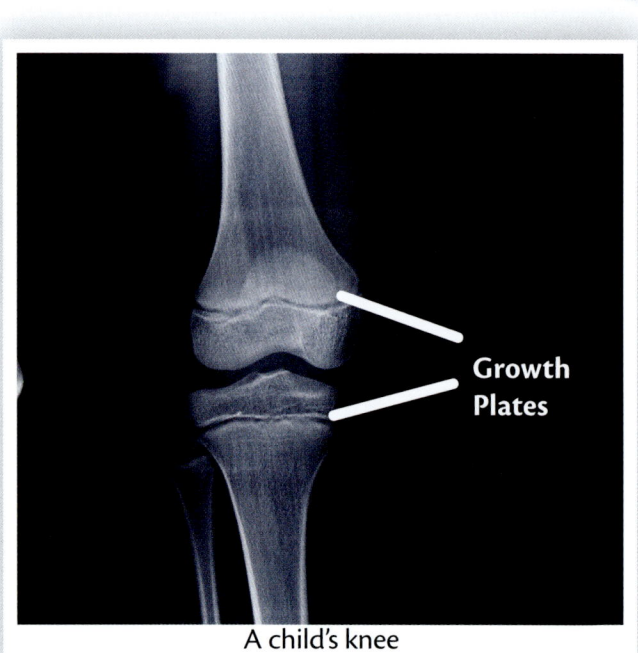
A child's knee

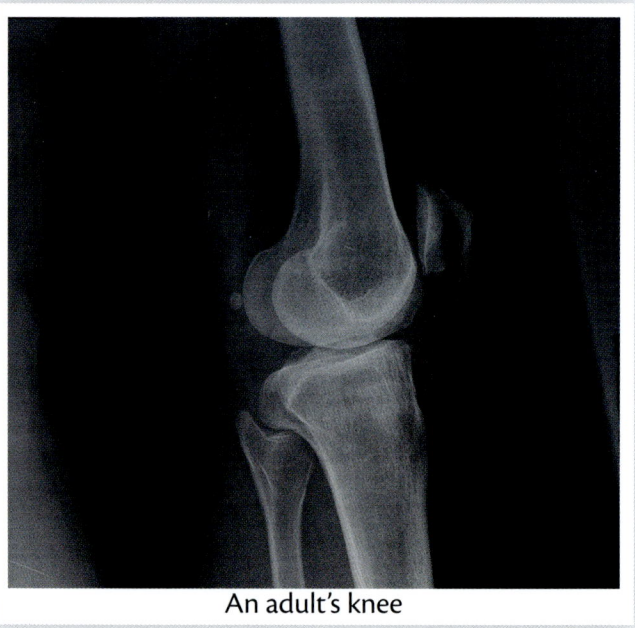
An adult's knee

Knee **X-rays**. X-rays are pictures taken by making energy beams pass through the body. Hard things appear white on an X-ray.

CHAPTER 25: GOD GIVES YOU STRONG BONES

Your bones are made of many small columns. The columns stand next to each other like the sticks of chalk in this boy's hands.

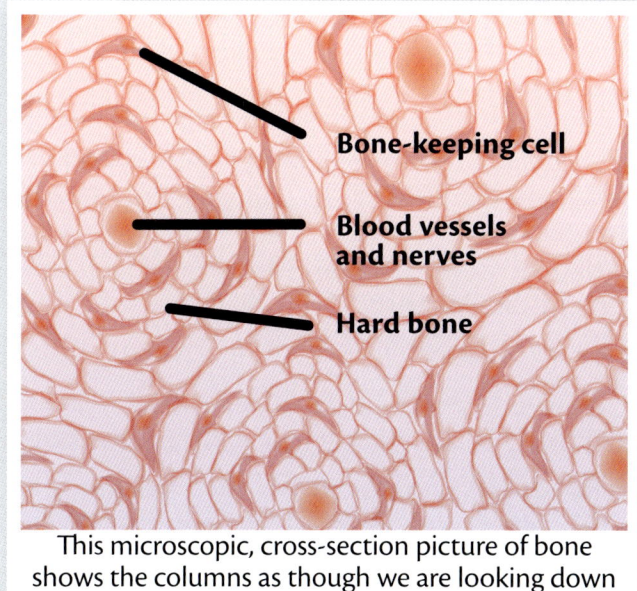

This microscopic, cross-section picture of bone shows the columns as though we are looking down on them.

building cells, telling them when and where to make bone. One bone-keeping cell can stay alive your whole life!

 **Time to do Activity 74 in the Activity Book!**

Bone's Extra Space for Bone's Extra Jobs

God made your bones with space inside. Your bones weigh less than they would if they were solid all the way through. Having lighter bones means that you can move your body more easily, using less energy. God's design for bones gives them the most strength with the least amount of bone matter.

But God did not design your bones with any wasted space! Your bones have two kinds of spaces and two kinds of bone marrow living in the spaces. Marrow looks like soft jelly. It has very important jobs.

 Red marrow is found in spongy bone. Red marrow makes almost all your blood cells. Right now, you have red marrow in many of your bones. By the time you are 25 years old, your blood cells will only be made in your spine bones, breastbone, collarbones, hipbones, and skull.

 Yellow marrow is found in long bones. The long, thin parts of your leg bones and arm bones are hollow.

259

They are hollow but not empty. God has filled that space with yellow marrow. Yellow marrow is mostly fat. This means that these bones are places that store energy!

Your yellow marrow is also the place where certain stem cells live. Remember, stem cells can become several different kinds of cells. Many of your body's cells do not use mitosis to make more of themselves. Instead, they come from stem cells. The stem cells in your yellow marrow can make bone, cartilage, fat, and muscle cells if you need them.

Your bones also have the extra job of storing minerals. Your bones are made of calcium and other minerals that give them strength. These same minerals are also needed for many of your body's other jobs.

When your blood gets low on minerals, your bone-keeping cells send two messages. One message tells your bone-building cells *not* to make bone matter. The other message tells a different kind of bone cell to dissolve some of the bone matter you already have. We'll call these cells **bone-dissolving cells**. The dissolved bone goes into your blood and is delivered where it's needed.

When your blood has enough minerals, bone-keeping cells again send two messages. They tell the bone-dissolving cells to stop dissolving bone. They also tell the bone-building cells to make more bone.

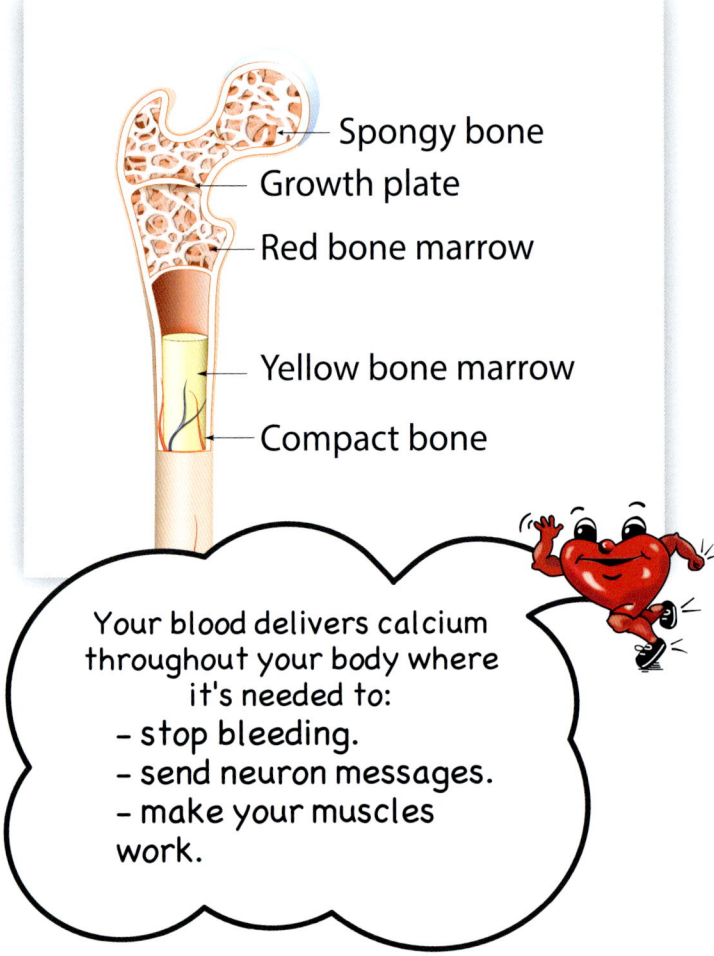

- Spongy bone
- Growth plate
- Red bone marrow
- Yellow bone marrow
- Compact bone

Your blood delivers calcium throughout your body where it's needed to:
- stop bleeding.
- send neuron messages.
- make your muscles work.

If bone matter gets old, it can become fragile. Bone-dissolving cells remove old bone, and bone-building cells make new bone to replace it!

CHAPTER 25: GOD GIVES YOU STRONG BONES

These bone-dissolving cells (red) put out chemicals to dissolve bone as they move slowly along. Notice the shallow grooves behind them where they've removed bone material.

Prayer

Thank You, God, for the strength You give our bones inside us! We love to depend on their sturdiness as we play and work. We see Your wisdom and power in the strong, flexible bones You created. Help us use our strength to serve others. In Jesus' name, Amen.

Time to do Activity 75 in the Activity Book!

261

CHAPTER 26
God Gives You Structure

**For He knows our frame;
He remembers that we are dust.
(Psalm 103:14)**

When God first created everything, He used dust to make people. This verse from Psalm 103 says God remembers that. And it says He knows our frame (shape). It's important to remember that we came from dust. Our frames are fragile. Yet, God has still given us a wonderful, tightly-connected structure to give us shape. Our bones, connected together, are called our **skeleton**.

Your Skeleton

When you were born, you had more than 300 parts that we call bones. When your bones are finished growing, you will only have 206 bones. Your baby skeleton looked almost the same as your adult skeleton will. So why is the number of bones different? Here's the reason. A baby's skeleton starts as cartilage in the womb and slowly changes to bone until the baby becomes an adult.

> God made bones out of strong stuff. He also put each bone in the right place and gave it the right shape. This makes your whole body strong!

GOD MADE ME

When you were a baby, not every bone had changed its cartilage into bone yet. Something that looks like one bone in a baby might really be two pieces of bony matter separated by a piece of cartilage. This is why it's counted as two bones in a baby.

God made your skeleton with different bones for two main purposes:

1. **God made some bones to protect your insides.** These bones make protective shells around your organs.

   - Your **skull** protects your brain.
   - Your **ribs** protect your heart and lungs in your chest.
   - Your **backbone** protects your spinal cord.

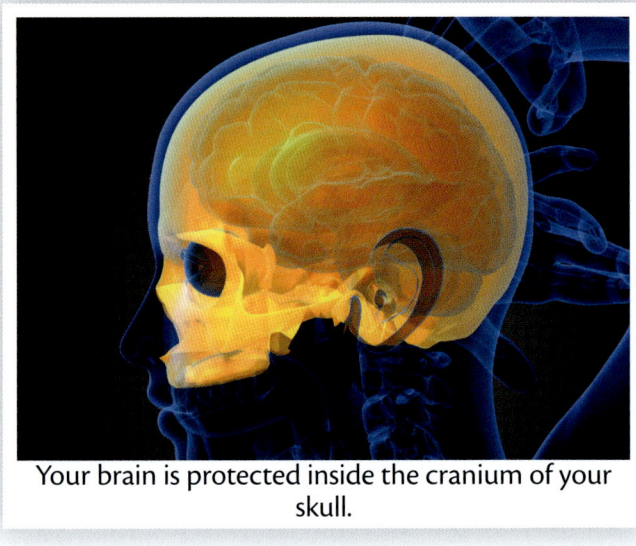
Your brain is protected inside the cranium of your skull.

2. **God made other bones to help you go places and do things.** These bones include your arm and leg bones. They also include the bones that your arms and legs are attached to.

   - Your **shoulder** bones connect your arm bones to your upper body.
   - Your **hip** bones connect your legs bones to your lower body.
   - Your hip bones also protect some of the organs in your abdomen.

   Let's look at some of your protective bones more closely! We will learn more about the bones that help you go places and do things later.

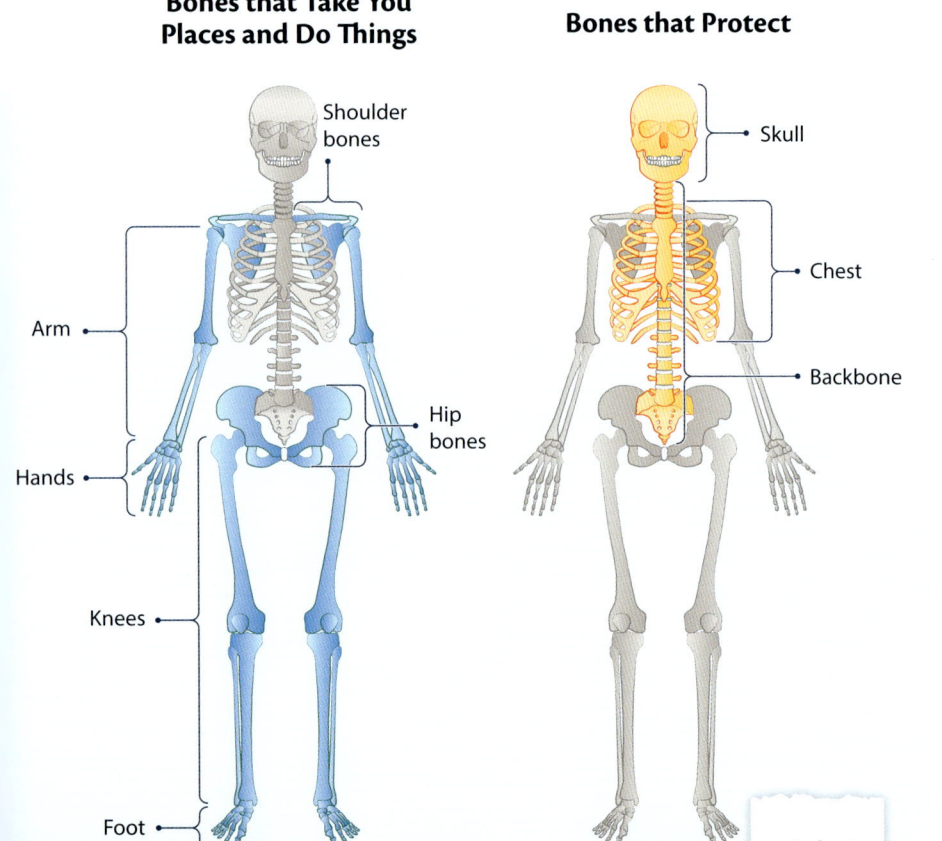

CHAPTER 26: GOD GIVES YOU STRUCTURE

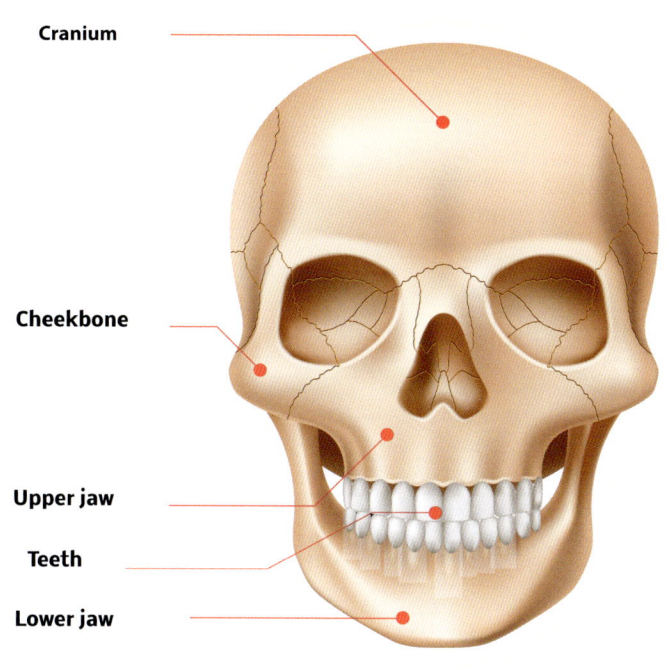

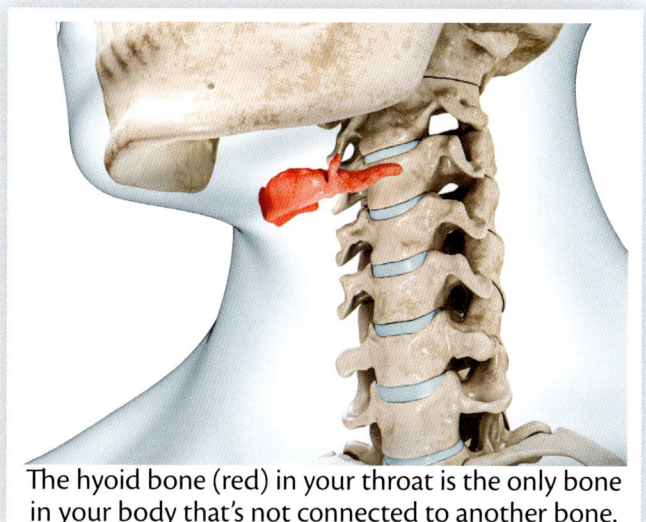

The hyoid bone (red) in your throat is the only bone in your body that's not connected to another bone.

Skull

Your skull is made of 29 bones. Many of these bones are joined together to make the hollow shell where your brain lives.

- The **cranium** (kray-nee-um) is the round part of your skull at the top, sides, and back of your head.
- The bones of your face have large holes where some of your sense organs live and where breathing and eating tubes pass through.
- Your **lower jawbone** moves more than your other skull bones because it helps you chew and talk.
- Your skull also contains three tiny ear bones in each ear.
- Your **hyoid** (hi-oid) bone helps you swallow, cough, make sound, move your tongue, and keep from choking.

Chest

Your chest is protected by your **ribs** and your **breastbone**. Instead of a solid shell like your skull, your chest bones are separate bones that can move a little bit. Your ribs move to make your chest larger and smaller as you breathe.

- You have 12 ribs on each side. Each rib is narrow, curved, and flat. Each rib has one end that's attached to your backbone. In front, the top seven ribs on each side are attached to your breastbone. The lower ribs do not attach directly to your breastbone, but some of them have cartilage that's attached to other ribs. The bottom two ribs on each side are called floating ribs because they don't attach to anything in front.
- Your breastbone gives a place for your muscles, ribs, and collarbones to attach.

265

GOD MADE ME

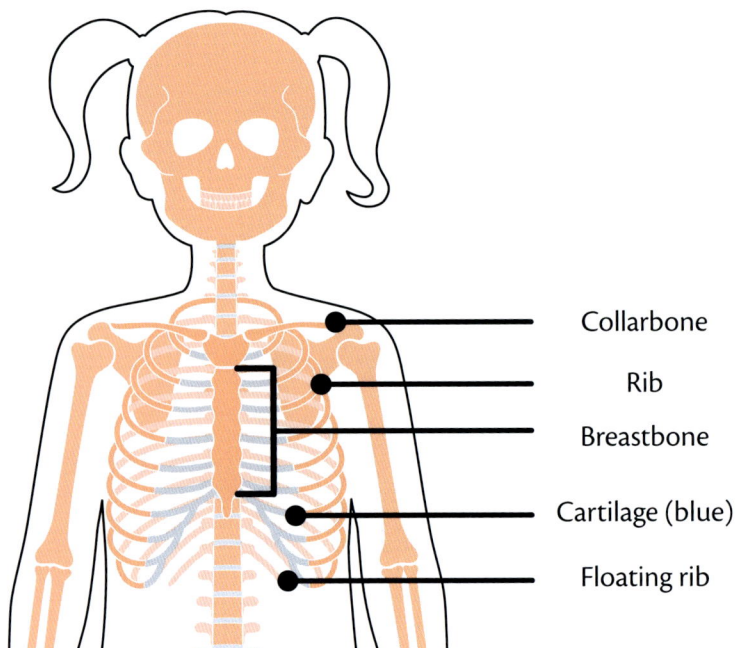

Collarbone
Rib
Breastbone
Cartilage (blue)
Floating rib

The curved springs inside a mattress can bend to keep you from landing too hard when you fall into bed. God gave your backbone several curves that work the same way. The curves can bend a little, so your motions feel gentle each time you take a step.

Backbone

Your backbone (spine) is made of vertebrae. When you were born, you had 33 separate vertebrae. By the time you are an adult, your lowest vertebrae will have grown together into one piece. You will then have 24 separate vertebrae. God made your vertebrae separate from one another so you can twist and bend your body. Your backbone protects your spinal cord and holds you up!

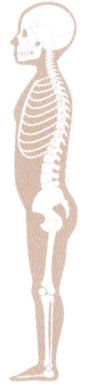

 Time to do Activity 76 in the Activity Book!

Joints Help You Bend Between Bones

Bones give you structure. But God didn't make your skeleton one single, big bone. If your skeleton was one solid piece, you wouldn't be able to move. This is why God gave you smaller, separate bones with **joints** between them. A joint is a special place that separates parts of your skeleton.

Joints join!

266

CHAPTER 26: GOD GIVES YOU STRUCTURE

These brothers' elbows are bending with hinge joints.

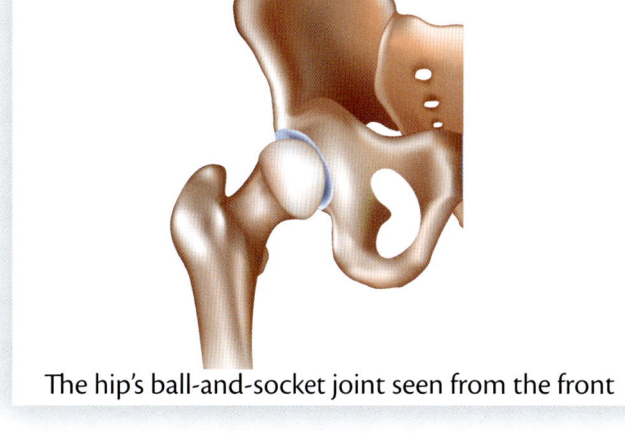

The hip's ball-and-socket joint seen from the front

But, even though joints separate your bones, they also join your bones together in a way that lets you move. Let's learn how your joints work. Be ready to stand up and try out some of your joints!

Hinge joint

Your knees and elbows are **hinge joints**. Hinge joints work like a door as it opens and closes. Hinge joints allow your arms and legs to fold almost in half and then "open" all the way. Hold up one leg and swing your foot forward and back again. This is your knee's hinge joint at work.

Ball and socket joint

Your hips and shoulders are **ball-and-socket joints**. Your leg bone has a ball-shaped top that fits into a cup-shaped socket in your hip bone. Your arm bone also has a ball-shaped top. It fits into a socket in one of your shoulder bones.

Ball-and-socket joints allow your arms and legs to lift in any direction and to swing in a circle! They can even twist inside the socket. Try some lifting, swinging, and twisting motions with your arms and legs. Now see if you can swing one arm and one leg in circles at the same time? How about

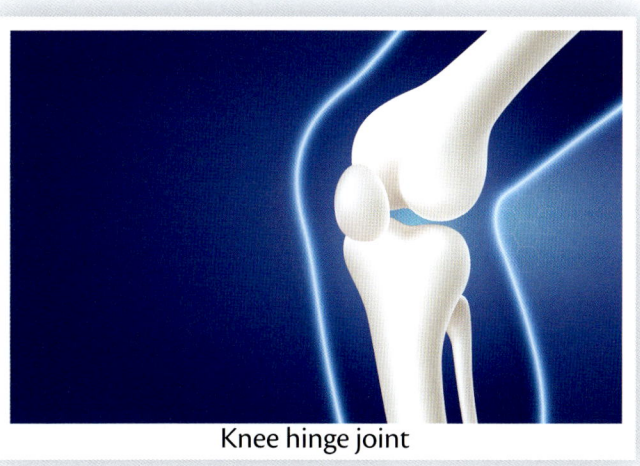

Knee hinge joint

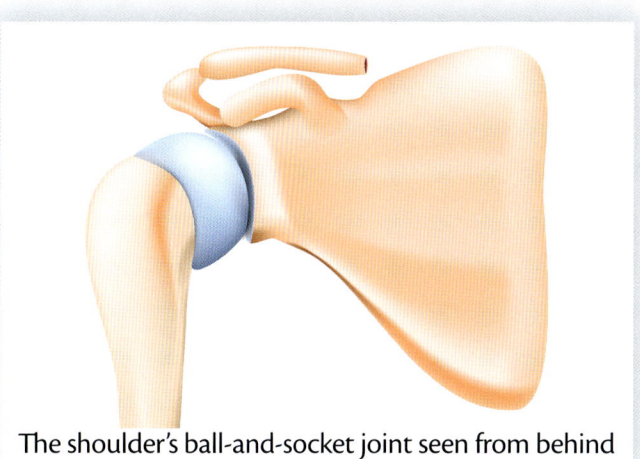

The shoulder's ball-and-socket joint seen from behind

267

GOD MADE ME

two arms and one leg at the same time?

Pivot joint

A **pivot joint** allows a bone to turn (or rotate). The reason you can shake your head to say "no" is because of the pivot joint between two vertebrae below your head. Most of the vertebrae in your backbone only move a little bit against the ones next to them. But your neck's pivot joint allows you to turn your head quite far to each side.

The lower vertebra of this pivot joint has a peg-like bump that sticks up. The bump fits inside a special hole that God made in the vertebra above it. This perfect design lets you turn your head steadily and even quickly without your heavy head getting out of its place. Try turning your head while you picture one bone twisting on top of the other, held together with a peg.

A ball-and-socket joint can also allow one bone to rotate against another bone. But imagine how wobbly your head would be if your neck's pivot joint was a ball-and-socket joint! God made your joints to work just right in each part of your body.

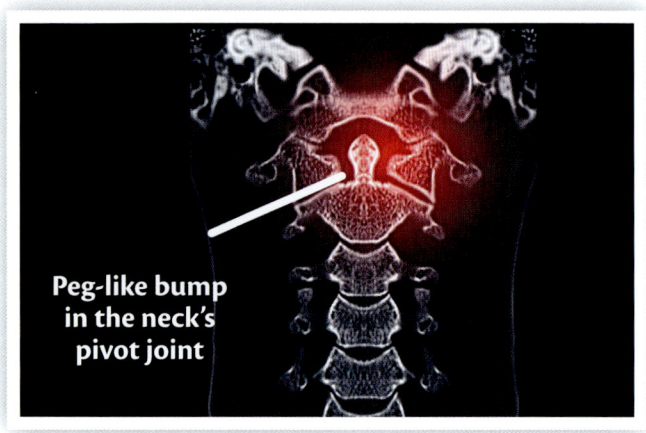

Peg-like bump in the neck's pivot joint

Gliding joint

Your wrists and ankles have **gliding joints** that allow sideways motion as the bones glide across each other.

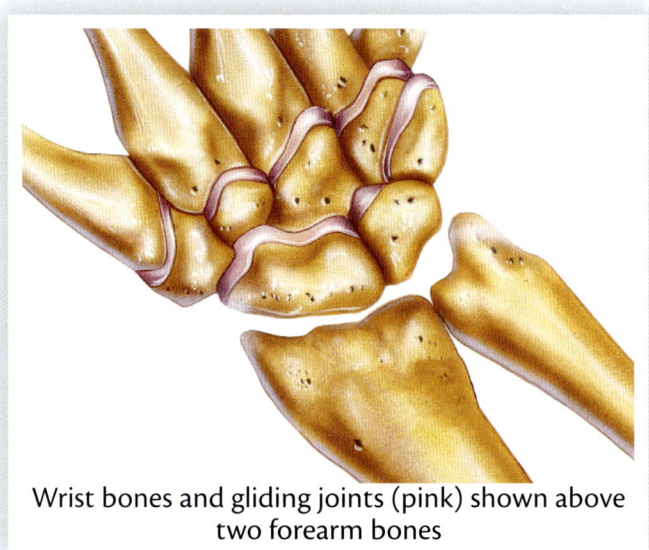

Wrist bones and gliding joints (pink) shown above two forearm bones

Your neck's pivot joint allows you to look over your shoulder.

Saddle joint

Your body has two **saddle joints** where your thumbs meet your wrists. These allow your thumbs to move in circles.

Each of your body's joints has a certain position where its bones rest together without you having to think about it. Let's try this out with your knees. Stand up and

CHAPTER 26: GOD GIVES YOU STRUCTURE

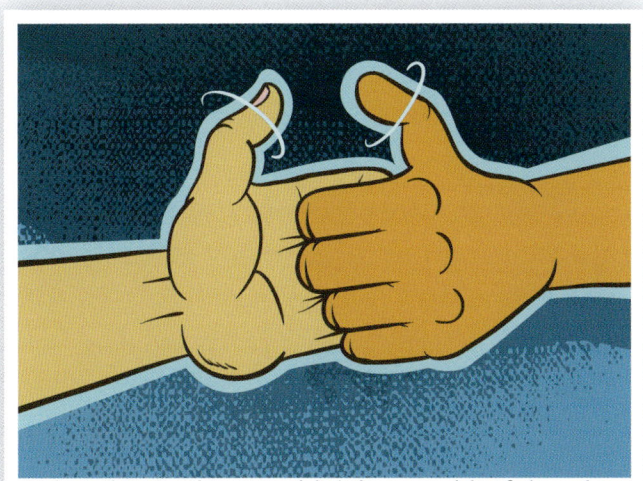

Thumb wrestling wouldn't be possible if thumbs weren't connected to the wrist by a saddle joint.

bend your knees like this boy on the right. How long can you stand this way? When your leg muscles get tired, stand up so your legs are straight. Doesn't that feel better? You could stand this way a long time. When your legs are straight, your knee joints are locked. Your leg muscles don't have to use energy to keep you from falling.

Your body has a few joints that don't move. They are called **immovable joints**. These joints connect the parts of your skull. A baby's head needs to grow, and his skull will not be completely filled in with bone until he grows up. But once the skull is all bone, the joints still will show as wiggly **sutures**.

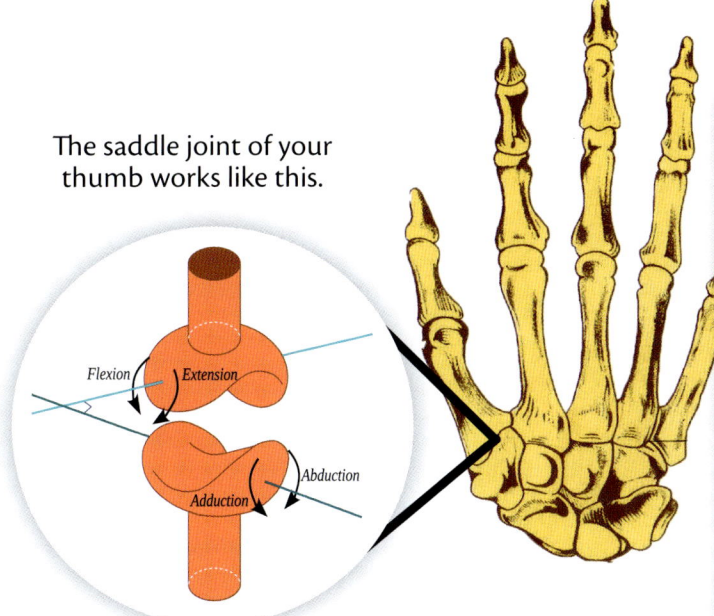

The saddle joint of your thumb works like this.

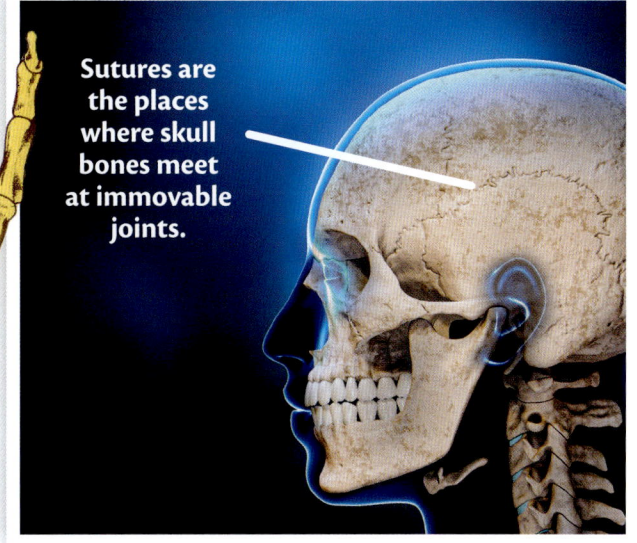

Sutures are the places where skull bones meet at immovable joints.

Time to do Activity 77 in the Activity Book!

269

Between Your Bones

You've seen how God gave you joints to separate your bones from each other. Joints make your skeleton moveable. But your bones are not completely separated at your joints. Your joints join bones together. If your bones weren't somehow attached to each other, you would wobble and collapse into zigzags. Let's look between your bones to see how they are joined together by your joints!

God connected your bones together at your joints with a strong tissue called **ligaments**. Ligaments are a kind of connective tissue that tightly connects your bones together and keeps them in place.

The area around a joint has special liquid inside. This liquid does the same job that oil does in a car. Slippery car oil surrounds your car's moving parts so they rub together smoothly. Oil keeps metal parts from wearing out where they rub together. In the same way, the slippery liquid in your joints keeps your bones from wearing out where they rub together.

God has also put cartilage between your bones. Cartilage protects your bones and joints in two ways.

- Cartilage protects by covering the ends of the bones in your joints. This smooth,

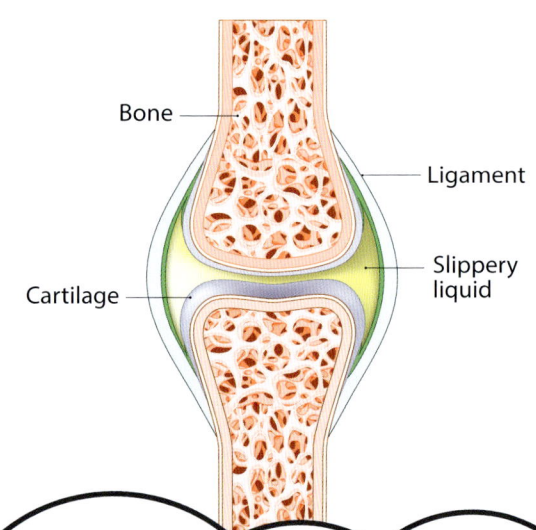

From him the whole body, joined and held together by every supporting ligament, grows and builds itself up in love, as each part does its work. (Ephesians 4:16 NIV)

In this verse, we learn that the Lord Jesus has formed the church into a body! Every member of the body acts as a ligament. The members join the body together and make it work beautifully. God gave us this verse to show that each person is important for the whole group of God's people (His church).

CHAPTER 26: GOD GIVES YOU STRUCTURE

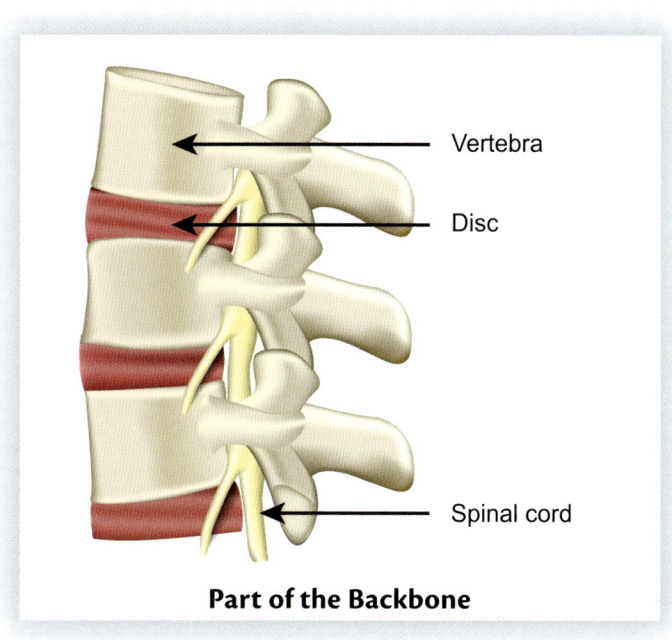

Part of the Backbone

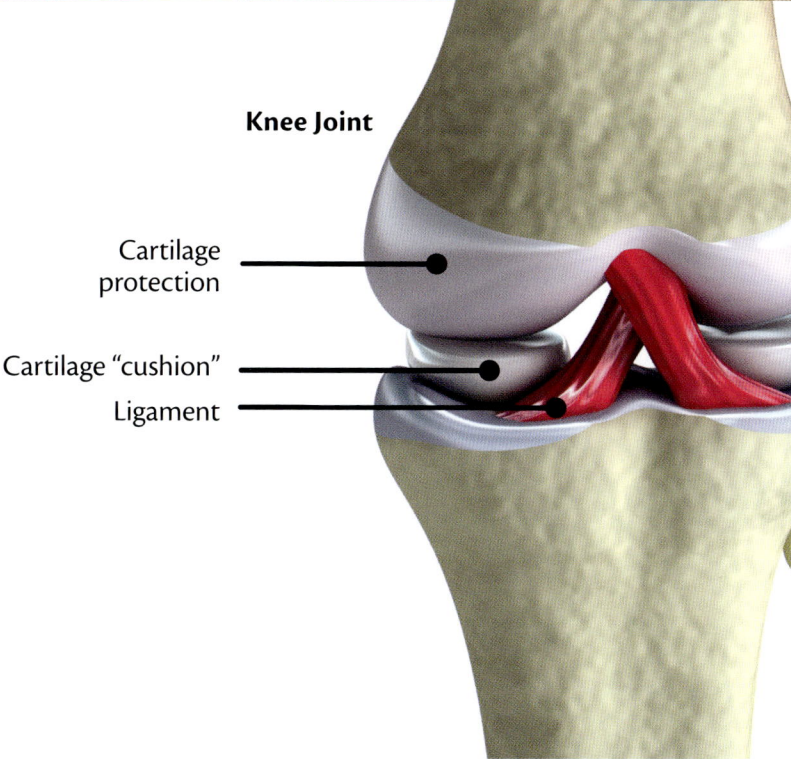

Knee Joint

tough cartilage helps your bones slide past each other. It protects your bones from rubbing and wearing out.

 Cartilage protects by absorbing bumps so certain bones don't hit each other too hard. You have cartilage "cushions" that absorb sudden bumps the way athletic shoes cushion your feet when they hit the ground. The cartilage cushions between your vertebrae are called **discs**. You also have cartilage cushions in your knees to keep your leg bones from hitting each other too hard.

Prayer

Father, I'm so glad you gave me a sturdy frame of bone. Thank You that I can move my strong skeleton at its joints. I'm thankful I don't have to hear and feel my bones grind together when I bend. You wisely gave my joints smooth, slippery protection with cartilage and special liquid. I also thank You for the cushions in my knees and between my vertebrae that help me jump and run comfortably! You are a wise Father! Amen.

 Time to do Activity 78 in the Activity Book!

CHAPTER 27
How Muscles Work

You've already learned about muscles as we talked about different parts of your body. We talked about how your nervous system sends messages to your muscles. We praised God for your heart muscle. You learned that your diaphragm muscle and the muscles between your ribs make you breathe. We talked about muscles that move liquid through your blood vessels and food through your digestive system. You also learned about the tiny goosebump muscles that make you hair stand up when you're chilly, and the little muscles attached to the three small bones inside each ear.

God gave your muscles a job. That job is to make things move! Now let's see how they do this.

How Muscles Move Bones

O LORD Gods of hosts,
Who is mighty like You, O LORD? . . .
You have a mighty arm;
Strong is Your hand. (Psalm 89:8,13)

As you grow, your muscles get stronger and you're able to do more things. God is a Spirit and doesn't have arms like you do. But this verse talks about God's "arms" to help us understand that God can do lots of amazing things.

In fact, God can do all things! He created the world and everything in it. He defeated sin and death through Jesus' death and resurrection. Our God is mighty to save!

GOD MADE ME

Even though your bones are alive, they can't move without help. Your bones must get their movement from muscles!

Puff will tell us how muscles move bones!

- One end of a muscle is attached to one bone, and the other end is attached to a different bone.
- The muscle gets shorter, which brings the bones together.
- After the muscle shortens, it can lengthen again. This moves the bones back to where they were.
- As the muscle lengthens again, a different muscle on the other side of the joint gets shorter. This also helps the bones move back to where they were. These muscle *partners* make your movements smooth and fast.

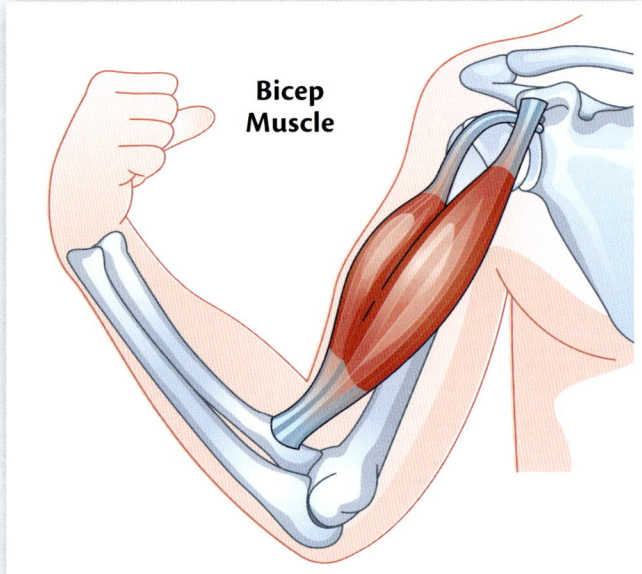

The top of the biceps muscle attaches to the shoulder blade at two different places. The bottom attaches to one of the lower arm bones. When the biceps muscle shortens, the arm bends at the elbow!

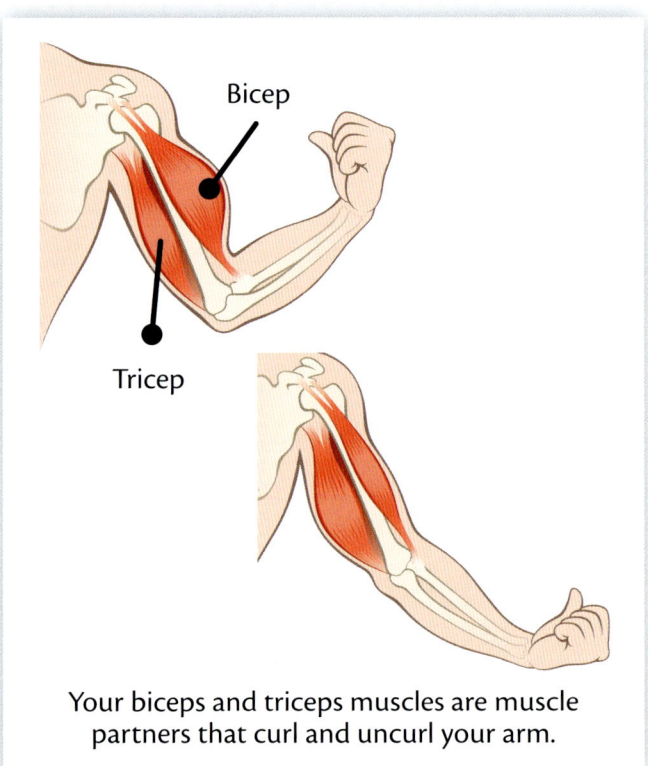

Your biceps and triceps muscles are muscle partners that curl and uncurl your arm.

Definitions

Contract (kuhn-trakt) — what a muscle does to get shorter.

Relax — what a muscle does to get longer.

Flex — to bend and swing two long bones together.

Extend — to straighten and swing two long bones apart.

CHAPTER 27: HOW MUSCLES WORK

You can feel your muscles working under your skin. To feel your biceps muscle, put your left hand around the front of your right arm just above the elbow. Curl (**flex**) and uncurl (**extend**) your right arm. Do you feel your biceps muscle working (**contracting**) as you move? Muscles feel harder when they are contracting. You will feel it working even harder if you hold something heavy in your right hand as you flex.

As your body moves, many muscles work together to make your movements smooth and to keep you balanced. Your biceps' main job is to flex your arm. But did you notice that your biceps was also working as you extended your arm? It was still contracting a bit so that your arm didn't suddenly fall. Now, without holding anything in your right hand, try feeling your biceps again as you flex your arm and then let it suddenly fall. Did your biceps stay soft as your arm fell? A muscle is soft when it's **relaxed** and not contracted.

Pushups

If you get on the floor and do pushups, your triceps muscles will be working hard. Your triceps will straighten your arms as you raise your body. Your triceps and your biceps are muscle partners on your arms. Your biceps contracts to flex your arm. Your triceps contracts to extend your arm. They do opposite jobs for the same joint.

These children are doing things with the hinge joints of their elbows and knees. Try to move into the same positions with

your body! After you try each one, look at the picture again and point to which of the child's elbows and knees are extended. Then point to the elbows and knees that are flexed, even if they are only bent a little.

Time to do Activity 79 in the Activity Book!

Zoom into the Cell: How a Muscle Cell Works

Muscle cells are long and thin. Each muscle contains many muscle cells. Muscle cells live in groups bundled together lengthwise. To understand how muscles work, it will help if we learn how just *one* muscle cell works.

Inside each muscle cell, there are several thinner strands we'll call **fibrils** (fie-brills). Muscle fibrils contain proteins that make fibrils contract. Muscles make you move because of what happens inside their fibrils. Thump is going to tell us how a muscle cell does its work:

How a Muscle Cell Works

- A muscle is resting.
- Then, a nerve sends a message to the muscle telling it to move!
- The message spreads to long proteins inside each fibril. Some of these proteins are thick and some are thin. These proteins can slide against each other.
- When the message comes, each thick protein grabs a thin protein. The thick proteins pull and slide the thin ones along next to themselves.
- This shortens the fibrils they live inside of.
- All the fibrils in each muscle cell shorten at the same time. This causes the whole cell to shorten.
- All the cells in the muscle also shorten at the same time. This causes the whole muscle to shorten and do its job.
- After a muscle does its job, certain chemicals tell it to relax again.
- This makes the sliding proteins slide back to their resting position.
- Finally, all the relaxed muscle's cells get long again, and so does the whole muscle.

CHAPTER 27: HOW MUSCLES WORK

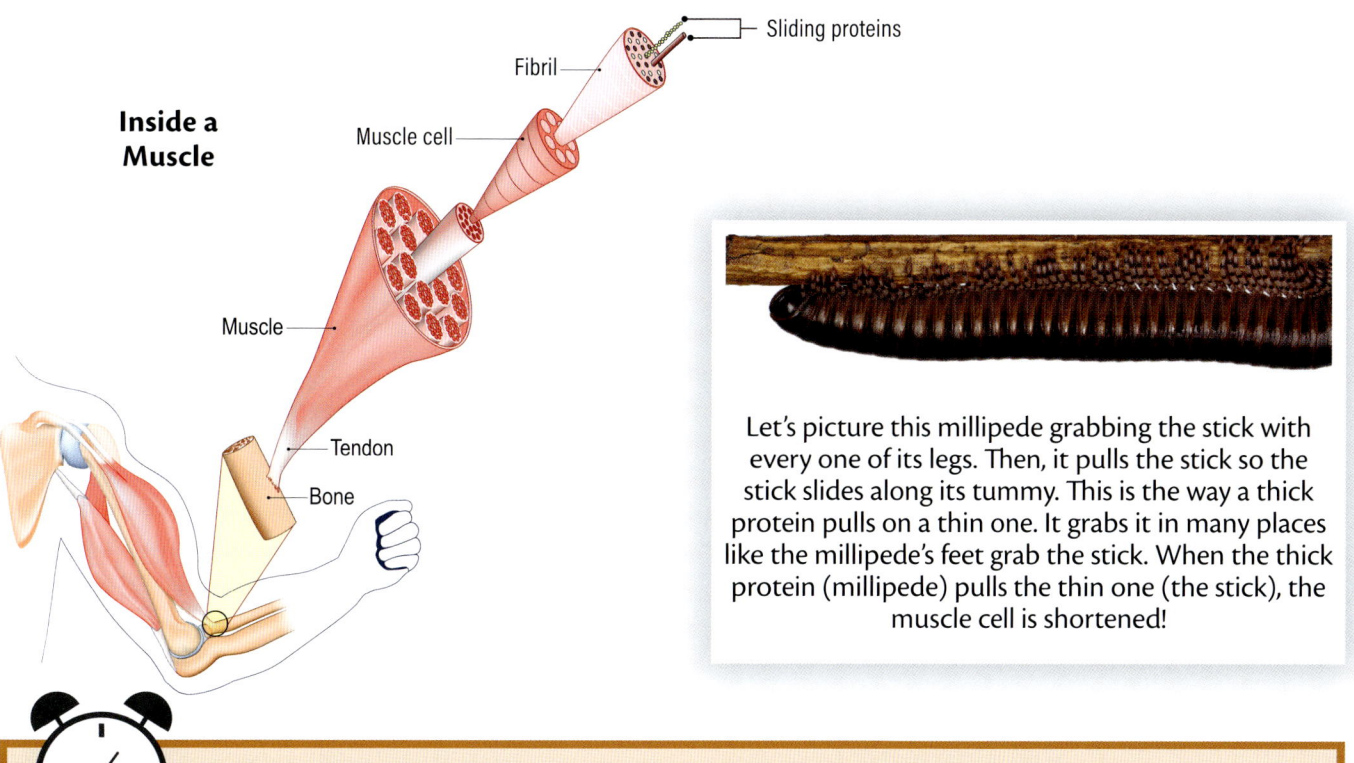

Inside a Muscle

Let's picture this millipede grabbing the stick with every one of its legs. Then, it pulls the stick so the stick slides along its tummy. This is the way a thick protein pulls on a thin one. It grabs it in many places like the millipede's feet grab the stick. When the thick protein (millipede) pulls the thin one (the stick), the muscle cell is shortened!

Time to do Activity 80 in the Activity Book!

Different Kinds of Muscles for Different Jobs

You have three kinds of muscles in your body. Let's learn about them now!

1. Skeletal muscles

Skeletal muscles are attached to bones. Their job is to move bones or to hold bones in place. You move your skeletal muscles on purpose. You use your skeletal muscles for:

- larger movements like walking, running, and swimming.
- smaller movements like writing, talking, and making expressions with your face.

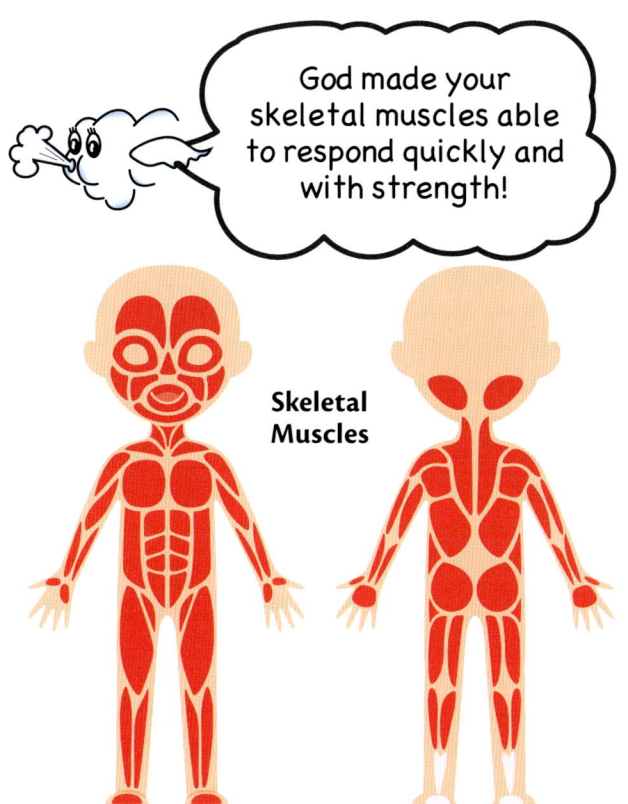

God made your skeletal muscles able to respond quickly and with strength!

Skeletal Muscles

GOD MADE ME

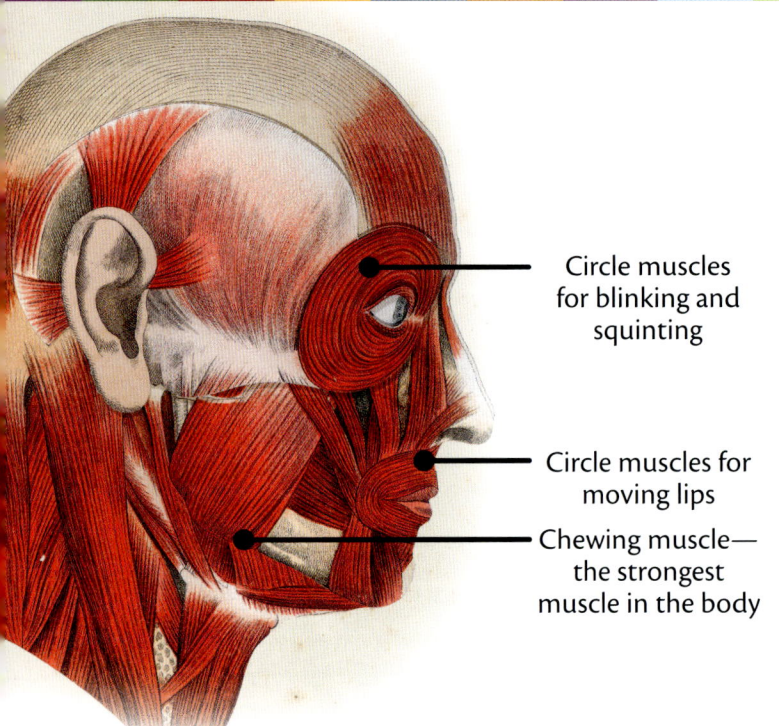

- Circle muscles for blinking and squinting
- Circle muscles for moving lips
- Chewing muscle—the strongest muscle in the body

Try to make these expressions with your face!

Most muscles in your face are attached to your skin. They were made by God so you could express yourself with your face. Face muscles also give shape over your bones. This makes everyone's face unique! Except for your chewing muscles, face muscles don't move bones.

- other small but very important movements like breathing, blinking, chewing, and swallowing.

Remember! Peristalsis is the movement that squeezes food and liquids through the digestive tract. It uses smooth muscles!

2. Smooth Muscles

Smooth muscles are not attached to bones. They are arranged in thin layers inside the walls of certain organs and tubes. Their job is to help squeeze things along. Smooth muscles contract and relax automatically. They respond slowly when they get the message to squeeze, but they can stay contracted a long time without getting tired. Smooth muscles can be found in:

- the digestive tract for peristalsis.
- the iris of the eye for opening and closing the pupil.
- blood vessels to control blood pressure.
- the bladder to squeeze out urine.
- hair follicles to make hair stand up.

3. Cardiac muscles

Cardiac muscles live in only one place in your body. They live in the walls of your heart! Their job is to contract and relax, on and off, making the perfect rhythm of your heartbeat. Cardiac muscles work automatically so you don't have to think about keeping your heart pumping.

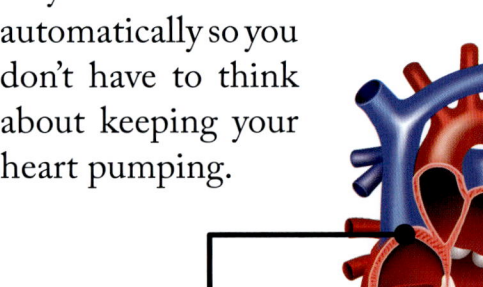

Cardiac muscle

278

CHAPTER 27: HOW MUSCLES WORK

Types of Muscle

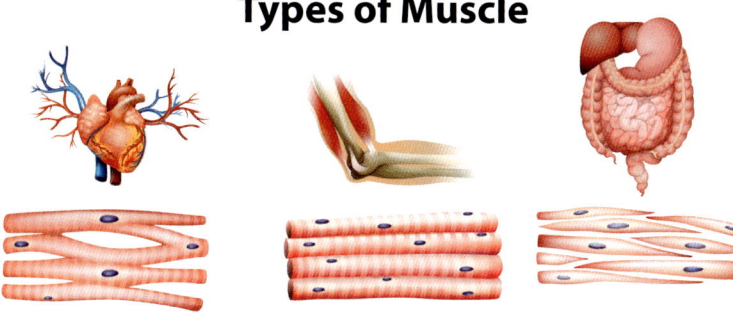

Cardiac muscle   Skeletal muscle   Smooth muscle

Zoom into the Cell

Your body's three kinds of muscles each have cells that are different from one another.

God made **cardiac muscle cells** branched and connected to each other. Because of this, the cells share their cytoplasm. (Remember, cytoplasm is the liquid inside cells.) Since electric messages travel quickly through liquid, the connected cytoplasm passes messages immediately to all the cells. This causes all the muscle cells in a chamber of your heart to contract at the same time.

When the muscle cells contract, the whole chamber squeezes out its blood. Then the cells all relax. This makes the whole chamber relax and fill with blood.

God made **skeletal muscle cells** very long and thin, and each one has many nuclei. Muscle cells often need extra protein. The extra nuclei are needed when it's time to make extra protein. (Remember, the DNA in nuclei give protein-making instructions to ribosomes.)

God made **smooth muscle cells** thick in the middle and thin at the ends. They look smooth because their sliding proteins are not arranged in stripes.

> The cells of cardiac muscles and skeletal muscles look striped because of how their sliding proteins are arranged inside.

You shall love the LORD your God with all your heart, with all your soul, and with all your strength. (Deuteronomy 6:5)

Prayer

Lord, we want to love You with all our strength! Thank You for making us to move. You thought of us when You made the muscles that move our skeletons. Help us glorify You with our bodies. Thank You for the muscles in faces! You specially created each person's face differently. Amen.

Time to do Activity 81 in the Activity Book!

CHAPTER 28
Swift, Strong, Careful Movement

"By working hard . . . we must help the weak and remember the words of the Lord Jesus, how he himself said, 'It is more blessed to give than to receive.' " (Acts 20:35 ESV)

When you first started learning about your skeleton, you learned about the bones near the center of your body that protect your organs. These are your skull, backbone, and chest bones (ribs and breastbone).

Now let's learn about the bones that take you places and do things. These bones don't stay close to the center of your body. These bones and their strong muscles can reach out and grab or step out and dance. They are your arms and legs. These **appendages** (up-end-udge-uzz) were made by God so you could do amazing things!

Legs and Feet

How beautiful upon the mountains
Are the feet of him who brings good news,

Can you use your strong arms and legs to help others who are hurt or weak?

GOD MADE ME

**Who proclaims peace,
Who brings glad tidings of good things,
Who proclaims salvation,
Who says to Zion,
"Your God reigns!" (Isaiah 52:7)**

This verse talks about "beautiful feet." What makes feet beautiful in this verse? Beautiful feet are feet that God uses to bring the good news of His salvation. There's something wonderful your feet can do. Your feet can take you to people who need to hear God's good words from you. That's what makes your feet beautiful!

Your feet and legs are the parts of your body that take you places. Let's learn about these strong friends in your musculoskeletal system.

The upper part of your leg has one long bone—the **femur**. The lower part of your leg has two long bones. When you accidentally run into something and bruise the front of your lower leg, you've whacked your sturdy **shin bone**. You have a skinnier **calf bone** deeper inside your lower leg.

If you had a twin brother exactly your weight, how hard would it be for you to lift him? It would be quite hard! But your leg muscles lift that much weight many times every day when they carry you around! Stand up with your feet flat on the floor. Now raise yourself up on tiptoes several times. The **calf muscles** on the backs of your lower legs are lifting your whole body!

Your **femur** is the longest and thickest bone in your body. The top of your femur attaches to your hip bone, and the bottom attaches to your shin bone. The part of your body made by your femur and its muscles is called your *thigh*.

For muscles to be able to move bones, they must be attached tightly to the bones. God gave you **tendons** as the perfect attachment. Tendons are a kind of connective tissue. They connect muscles to bones. Tendons are like sturdy ropes at the ends of muscles.

Tendons don't need to be as strong as ligaments, but they need to be stretchier. They must stretch when a muscle springs into action. You can easily see and feel two of your tendons under your skin above each heel. They are called your *Achilles* (uck-ill-eez) *tendons*.

Hip, Leg, and Foot Bones

Femur
Knee Cap
Calf Bone
Shin Bone

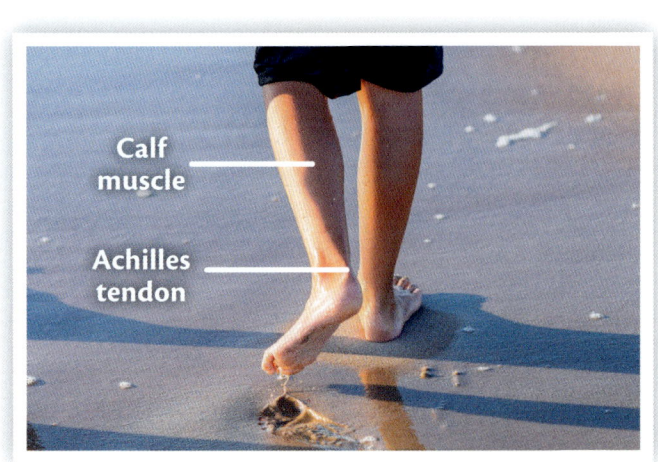

Calf muscle
Achilles tendon

CHAPTER 28: SWIFT, STRONG, CAREFUL MOVEMENT

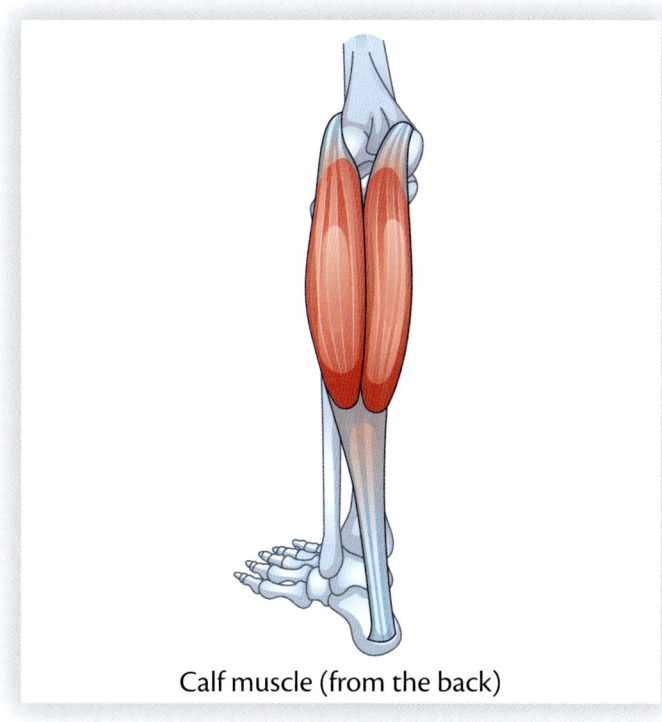

Calf muscle (from the back)

knee's hinge joint. But they have another important job. This job is to move your femur around in your hip socket. Most of these muscles have long names. For now, we'll just give them names according to where God placed them. Let's learn about two of these muscles.

The *bottom muscle* is the muscle you sit on. It makes a good cushion, but it also has important active jobs. Your bottom muscle is the largest muscle in your body. It can help you climb stairs and hills. It's the muscle that helps you stand up from a chair. It also holds you up when you stand still.

The *across-the-thigh muscle* is the longest muscle in your body. It goes from the side of your hip, across your thigh, to the inside of your knee. It helps you do four different motions. You do all four of those motions every time you get down into a cross-legged position on the ground.

Sit in a chair with your left foot resting on the floor. Bring up your right foot and lay it across your left knee. Feel your right Achilles tendon by pinching behind your ankle with your fingers. Keep feeling that tendon as you flex and extend your foot. Can you feel the tendon tighten and loosen as your move your foot? Your Achilles tendons are attached to your calf muscles. They help you walk, run, jump, and stand on tiptoe. You can also feel different tendons on the front of your ankle as you move your foot the same way.

God put an interesting, rounded bone called the **kneecap** in front of your knee joint to protect it. Your kneecaps are your body's only bones that live inside tendons. They help your thigh muscles straighten your legs at your knees.

Your upper leg muscles can work your

Isn't it wonderful how God made your legs and arms able to move slowly and gracefully sometimes, and quickly and powerfully other times?

283

GOD MADE ME

Time to do Activity 82 in the Activity Book!

Arms

You've learned that your legs can take you places. Wherever your legs have taken you, your arms and hands have the job of doing things. You will learn about the amazing hands God gave you later. Now let's learn about your arms.

Your arms can do big things. They can pull. They can push. They can carry. They can hug. They can help you balance as you move. Your arms can move your hands to the places they need to be.

The humerus is the only bone in the upper arm.

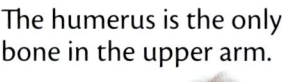

Your arms can also do small things. Your *forearm* (four-arm) is the lower part of your arm between your hand and your elbow. Its two bones and many muscles help your hand do its job. God made one whole forearm muscle for moving only the tip of your thumb! And sometimes that muscle can move one cell at a time for your careful thumb!*

You already know about the ball and socket joint at your shoulder and the hinge joint at your elbow. You know that your biceps and triceps are muscle partners that move your elbow. Now let's learn the names of your arm bones.

You have one long bone in your upper arm. It's called the **humerus** (hume-er-us). Your humerus goes from your shoulder

This boy has soapsuds on his shoulder's **deltoid** muscle. The deltoid muscle lifts your humerus bone forwards, backwards, and to the side.

Made in His Image. 2015. Dallas, TX: Institute for Creation Research, DVD.

284

CHAPTER 28: SWIFT, STRONG, CAREFUL MOVEMENT

to your elbow. Its ball-shaped *head* moves around in your shoulder socket.

Your forearm has two long bones. The bone that is attached to your wrist on your pinky's side is the **ulna** (ull-nuh). The bone that's attached on the thumb side is the **radius** (ray-dee-us). These two long bones have a pivot joint between them on their upper parts, close to your elbow. This pivot joint allows your radius to slide partway around your ulna.

You can watch this joint work if you sit with your left elbow and hand resting on a table. Slowly twist your left hand so it flops back and forth on the table. You're able to do this motion because of the pivot joint between your forearm bones.

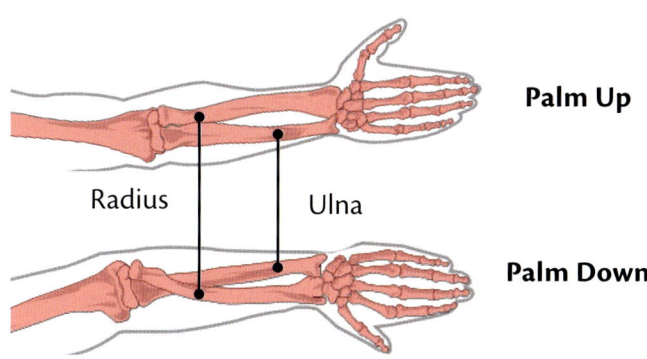

As you flop your hand back and forth on the table, the radius moves around the ulna the way a person rolls from side to side in bed.

Time to do Activity 83 in the Activity Book!

Hands and Fingers

And let the beauty of the LORD our God be upon us,
And establish the work of our hands for us;
Yes, establish the work of our hands.
(Psalm 90:17)

Your hands connect you with your world! With your hands, you can work, and you can play. You can bless others with your hands by writing a letter, painting a picture, or playing a musical instrument. Your hands can show love when you hold a baby, hold someone's hand, or hold open a door. God gave you hands to work for Him

> This verse is a prayer asking God to bless our work and make it last. We should work cheerfully and faithfully. Then, after we work hard, let's ask God to use it for His glory!

GOD MADE ME

every day by doing many different things. Let's see what God has done to make your hands so amazing.

 **Puff's Amazing Hand Facts***

 God gave your hands a *power grip* for a strong squeeze.

 God gave your hands an *endurance grip* to keep your knuckles curled a long time.

 God gave your fingers a *precision grip* for small, careful movements.

 God gave your fingers a *microforce grip* for tightly pinching small things.

 God gave you a brain with lots of special neurons made just for using your hands.

 God gave you more than 30 muscles to handle the complicated movements of each hand.

 God gave you extra touch receptors in your fingertip swirls. These receptors constantly send messages to your brain. The messages tell your brain everything your fingers feel so your hands will know what to do. Even your fingernails help you feel by sensing pressure.

Power grip

Endurance grip

It takes a precision grip to put a worm on a fishhook without poking your fingers.

Microforce grip

Made in His Image. 2015. Dallas, TX: Institute for Creation Research, DVD.

CHAPTER 28: SWIFT, STRONG, CAREFUL MOVEMENT

Your fingers can work at amazing speeds for doing things like typing and playing musical instruments. Your brain makes your fingers speedy by knowing what your fingers need to do a few movements ahead. The more you practice a skill, the more your muscles remember what to do. This makes you faster and better!

Some animals like raccoons, frogs, monkeys, and apes have hands. But no animal has hands like humans do. God has given humans special hand muscles and brains that are much more complicated than any He gave to the animals. This is why your hands are so wonderful!

**Fear not, for I am with you;
Be not dismayed, for I am your God.
I will strengthen you,
Yes, I will help you,
I will uphold you with my righteous right hand. (Isaiah 41:10)**

Prayer

Father, it's hard to imagine how many things must work together in our brains, bones, muscles, and nerves for even one tiny movement. But You understand all these things because You made us! You keep our bodies moving and working just as they were designed! Thank You for giving us bodies that function so well. Thank You for letting us move our legs and arms. You give us joy! Amen.

Time to do Activity 84 in the Activity Book!

UNIT 8
God Heals You

Have you ever skinned your knee? Do you ever catch a cold? These things happen to children quite often! Thankfully, God heals skinned knees and colds! Our memory verse reminds us that God really can heal. And, better than that, He really can save our souls. As this verse says, let's praise Him!

Our hymn is full of praises for God. We praise Him for His wisdom in how He made us. We sing of His loving care for our health. Let's keep praising God as we learn how He heals our bodies!

Memory Verse

Heal me, O LORD, and I shall be healed;
Save me, and I shall be saved,
For You are my praise.
(Jeremiah 17:14)

You can listen to this hymn by searching for "Praise to the Lord the Almighty hymn for kids" on the internet.

Hymn to Sing: Praise to the Lord the Almighty

Praise to the Lord, the Almighty,
  the King of creation!
O my soul, praise Him,
  for He is thy health and salvation!
All ye who hear,
  now to His temple draw near,
Join me in glad adoration!

Praise to the Lord,
  who with marvelous wisdom hath made thee,
Decked the with health,
  and with loving hand guided and stayed thee.

How oft in grief
  hath not He brought thee relief,
Spreading His wings to o'er shade thee!

Praise to the Lord!
  oh let all that is in me adore Him!
All that hath life and breath,
  come now with praises before Him!
Let the Amen
  sound from His people again,
Gladly for aye we adore Him!

CHAPTER 29

God Heals Your Wounds

" 'For I will restore health to you and heal you of your wounds.' " (Jeremiah 30:17)

An **injury** is damage to your body that causes pain. Do you remember that your skin and other places in your body have pain receptors? Pain receptors are one part of your nervous system that starts reflexes. Remember, reflexes are fast messages that make you pull away from pain.

We don't like getting hurt. But pain has a purpose. God made our bodies able to feel pain so we can do something to protect ourselves. If you touch something hot, the pain starts a reflex. That reflex makes you quickly move away from the danger before you end up with a severe burn.

A **wound** is an injury that breaks the skin and often causes deeper damage. A wound usually bleeds. When was the last time you needed a bandage because you were bleeding? It can be scary when you start bleeding, especially it there is a lot of blood. God wisely made your body to stop bleeding. Let's learn how that works.

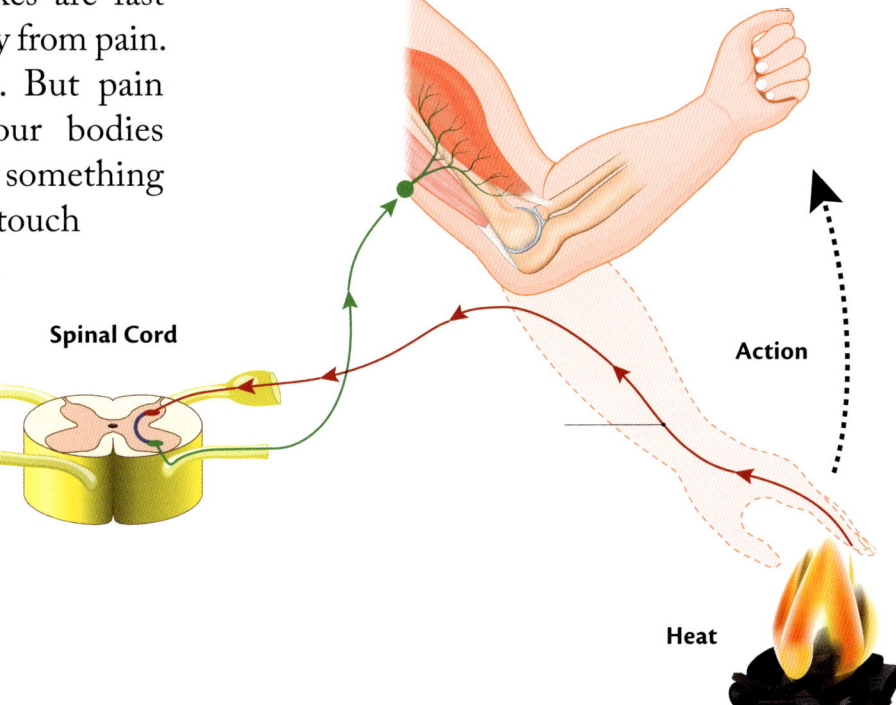

GOD MADE ME

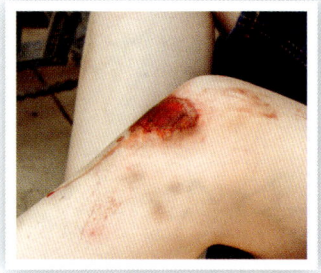
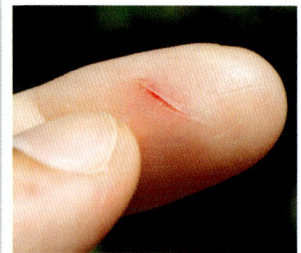

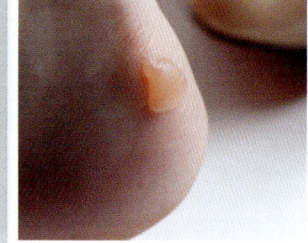

Your Blood Vessels Squeeze
Wounds can be:
- Scrapes
- Cuts
- Punctures
- Blisters

If your wound is bleeding, it means that blood vessels are broken and blood is leaking out. Your body needs its blood to keep you alive. Thankfully, God designed three jobs your body does to stop bleeding. Thump will tell us about the first job.

Job #1: Slow Down the Flow of Blood.
1. Pain receptors around your wound send chemical messages that say something is wrong.
2. The inner layer of your blood vessels also sends chemical messages saying it has been torn.
3. The smooth muscles in the walls of your damaged vessels receive both messages.
4. The damaged blood vessels **constrict**. This means the muscles squeeze to make the vessels skinnier. The skinny vessels won't let as much blood through so that you won't bleed as much.

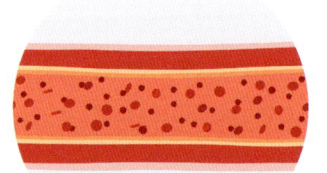
Normal vessel
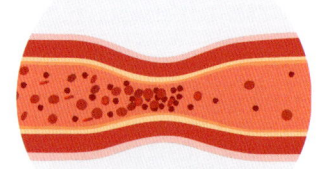
Constricted vessel

Time to do Activity 85 in the Activity Book!

Platelets Are Called In
You've learned that platelets are pieces of cells that live in your blood. Your platelets are like firefighters. Firefighters spend most of their time in a fire station. They are waiting

292

CHAPTER 29: GOD HEALS YOUR WOUNDS

and ready for an emergency. In the same way, platelets spend their time in your blood waiting for you to bleed. Puff will tell us how platelets do the second job that helps you stop bleeding.

Job #2: Plug the Hole.

1. Blood vessel walls are made of several layers. One of the inside layers is made of connective tissue. When your vessel gets damaged, the connective tissue is "showing." This inner layer puts out a chemical message so platelets will stop and help.
2. Platelets that are flowing by in your blood are attracted to this connective tissue. They begin to gather on the edges of the damage.
3. When the platelets land, they become active. They get sticky and grow a lot of branches.
4. Active platelets attract more platelets.
5. The new platelets become active and attract even more platelets.
6. Very soon, the hole in the blood vessel is plugged with platelets!

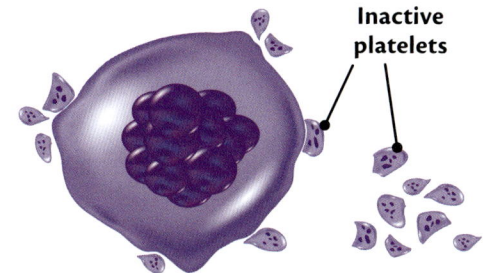

God made certain large cells that live in your bone marrow. Their very long name starts with *mega*. He gave these "mega" cells the job of breaking off little blobs of their own cytoplasm to make platelets.

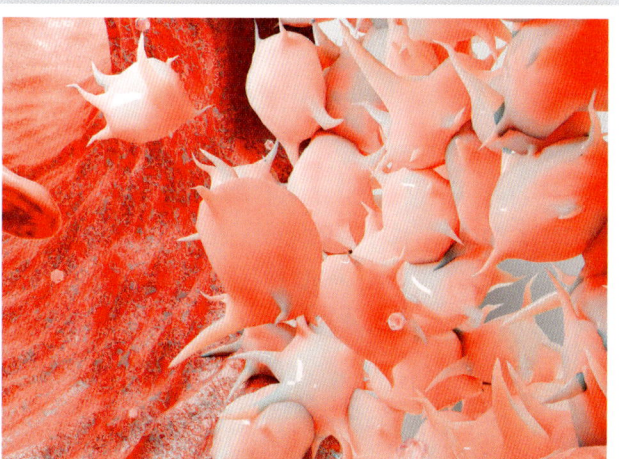

Active platelets inside a blood vessel plugging a hole

Did you know that you bleed often without noticing it? After you eat popcorn, the hulls move through your digestive tract, making little cuts. Chapped lips and fingernail cuticles may bleed without you noticing. There may be blood you can't see when your skin is stretched or barely scraped. God designed your body to stop even these small wounds from bleeding!

Time to do Activity 86 in the Activity Book!

Fibrin Makes a Net

Platelets are not strong enough to plug holes for long. They need help. The third job your body does to stop bleeding is to make a blood clot. A **blood clot** is a sturdy seal that closes a hole in your blood vessel.

A blood clot stops a vessel's bleeding. But what would happen if your blood didn't stop clotting? Too much clotting would block your blood vessel and stop the blood's flow. Then the tissues around the vessel would die because they couldn't get oxygen from your blood. But God wisely put safety steps in this system that makes clots. These steps keep clots in the right places only when they are needed.

Job #3: Make a Clot.

To make a clot, a strong **fibrin** net is formed all around the platelet plug. Fibrin is a wonderful, waterproof protein that's shaped like threads. Fibrin's threads make a net that holds the plug in place. During this process some blood cells get trapped in the net. This causes the plug to become harder. The hardened plug is now called a *clot*. Clots turn into **scabs** as they dry.

It takes about 30 different proteins to make a clot. These proteins are enzymes called *factors*. Calcium, vitamin K, and several other nutrients are also needed. Making a clot is so complicated that it would take many pages to explain all the steps. Some factors help make the clot, while other factors keep the rest of your blood liquid (not clotted). Many of these factors are made by your liver.

A good way to describe the clot-making job is to call it a *cascade*. As a mountain stream runs over a steep, rocky place, the stream becomes a cascade—it branches into smaller streams that have their own little waterfalls and pools. Sometimes water from one branch joins the water from another. Finally, the water comes together again into one stream.

A mountain stream becomes a cascade when it comes to a steep, rocky place. Your blood-clotting cascade starts when damage happens to your blood vessels. Factors from the damaged vessel and factors in your blood begin working together. They start making other factors. Some factors branch one way to make things happen.

Blood cells are trapped in a fibrin net to make a clot.

A cascade is a chain of small waterfalls.

CHAPTER 29: GOD HEALS YOUR WOUNDS

Other factors go another way to make different things happen. Just like there are many little waterfalls in each branch of a mountain cascade, there are many steps in a clotting cascade.

Often, some of the water in one branch of a mountain cascade crosses over to another branch. The same thing happens in the clotting cascade. This crossover happens because a factor is made in one branch, but it's needed in both branches.

After the mountain cascade flows through the steep place, its waters begin to come together again. The branches of your clotting cascade also come together again. Then, a few more steps are taken. When the clotting cascade is finished, your blood vessel is tightly sealed with a clot!

As we learn about clotting, we can see how wise our God is! The Lord made a system that works very well to protect our bodies. Each step will only happen if the step before it has happened. If just one step was missing, a person wouldn't stop bleeding or wouldn't stop clotting.

> Remember, enzymes are amazing chemicals that speed up jobs in your body. It takes many different clotting enzymes (factors) to change blood from a liquid to a gel. It also takes other enzymes to stop the clotting before all your blood turns to gel. If just one of these factors was missing, none of the cascade would work!

A nosebleed stops when a clot forms.

A blood clot on your skin dries and becomes a scab. However, a blood clot within your damaged vessel must be dissolved so it doesn't block your blood flow. How does a clot inside the blood vessel dissolve? Once your damaged blood vessel has healed, a certain enzyme gets busy. The enzyme cuts the fibrin net into small pieces and your clot dissolves!

Prayer

Our Lord, we praise You for how wise You are! You are so kind to give us our blood that gives us life. You also make sure we safely keep the blood that we need. Thank You for the three jobs our body does to stop blood from leaving our bodies. Only You know everything about how clotting works because You are the One who invented that amazing cascade! Amen.

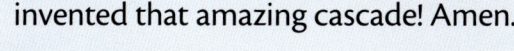

Time to do Activity 87 in the Activity Book!

CHAPTER 30

Healing: Making Things New Again

Then He who sat on the throne said, "Behold, I make all things new." (Revelation 21:5)

God makes things new!

- 🎁 Jesus gave a *new* commandment—to love each other just as He loved us.
- 🎁 God cleanses us from our sins and gives us *new* hearts.
- 🎁 One day God will give us *new* bodies in the resurrection.
- 🎁 Someday, God will make a *new* Heaven and a *new* earth.
- 🎁 During our lives in this world, God has made our bodies able to grow *new* tissue to heal our injuries.

When you have a wound that heals, remember God's goodness. He made that new skin for you!

God Gives You New Skin

Once your wound is closed with a clot, your body keeps healing with other processes God designed. Three things happen as He gives you new skin.

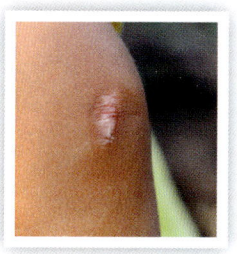

1. Inflammation

After damaged blood vessels are sealed, your wound begins to heal. Your blood platelets make special chemicals that cause **inflammation** (in-flum-may-shun). Inflammation is when:

- 🎁 Blood vessels bring in extra blood to deliver oxygen for healing. This causes the skin around your wound to get red and warm.

297

GOD MADE ME

🎁 Blood vessels allow liquid (**puss**) to leak out. Puss causes swelling. Swelling causes pain. God created puss to clean germs and dead cells out of your wounds.

🎁 White blood cells are brought to the wound. White blood cells kill germs and "eat" your damaged tissue. They also send chemical messages that tell new skin to grow.

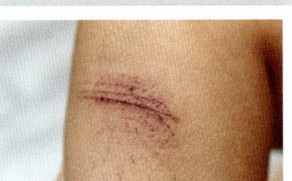

The pink skin between these scratches shows that inflammation is working to heal the wound.

2. Rebuilding

Next, it's time to **rebuild** the wounded area under your scab.

🎁 Special cells begin to make collagen (call-uh-jun). Collagen fibers make a frame for new tissue grow on.

🎁 Your wound starts to fill in with new tissue. This new tissue is speckled with the beginnings of new blood vessels.

🎁 Then, new skin begins to form over the new tissue.

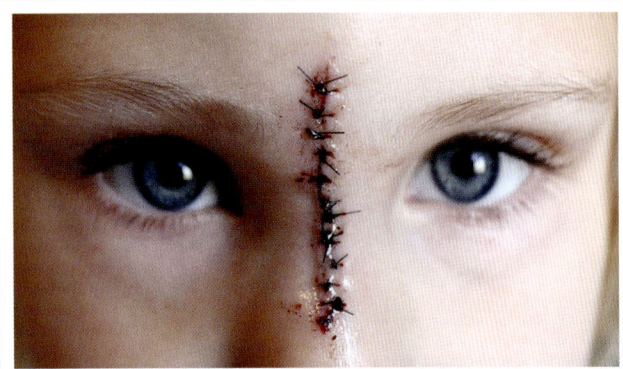

Wounds heal sooner if their edges are close to each other. A large wound may need to have its edges stitched together. Doctors have learned how to stitch wounds, so they heal with less scarring.

🎁 As your wound heals, a special kind of smooth muscle cell is busy. These cells contract to pull in the edges of your wound, making it shrink.

🎁 When your skin has healed enough, the scab will fall off.

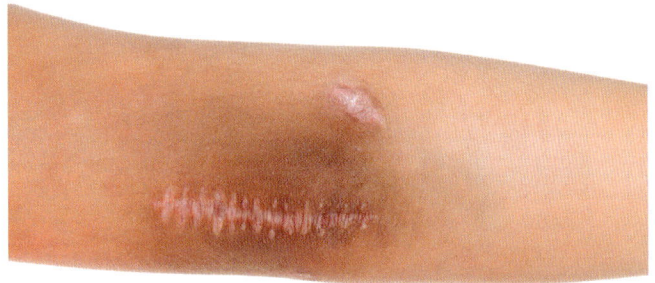

This scar is puckering as the skin strengthens. Notice there are scars from the cut and from the stitches that helped repair the cut.

3. Strengthening

Your new skin will **strengthen**.

🎁 New skin may pucker at first, until it's well attached underneath.

Definitions

Inflammation is the way our bodies begin to heal damaged tissue. When bleeding stops, blood vessels near the injuries open up to bring in extra blood. This causes redness, swelling, warmth, and pain.

Collagen is a protein that gives structure to connective tissues in the body. Collagen is the body's most plentiful protein.

A **scar** is the mark left on skin after a wound heals.

CHAPTER 30: HEALING: MAKING THINGS NEW AGAIN

- 🎁 The area might itch.
- 🎁 About three months after being damaged, your new skin will be almost as strong as it was before.
- 🎁 A **scar** will be left where the wound was. The scar will fade and may even disappear.

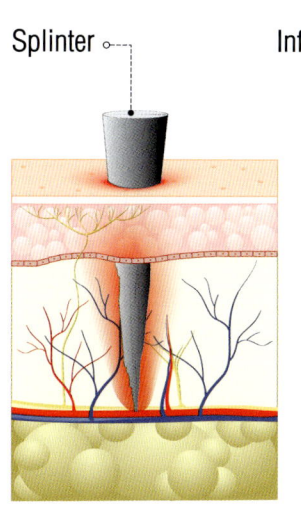
Splinter
Injured blood vessel

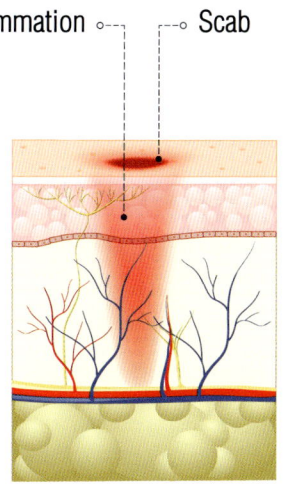
Inflammation — Scab
Splinter removed

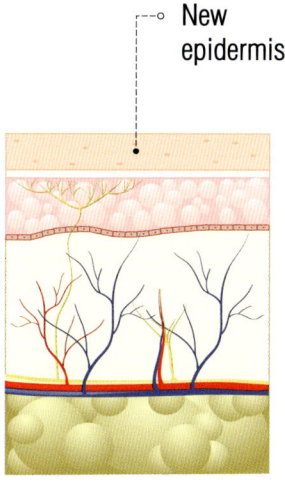
New epidermis
Freshly healed tissue

Time to do Activity 88 in the Activity Book!

God Heals Your Bumps and Bruises

Bumps and bruises are caused by injuries that bleed under your skin. They happen when blood vessels are damaged, but the skin is not broken.

- *Bruises* are caused when only the capillaries get damaged and bleed.
- *Hematomas* (heme-uh-tome-uhz) are caused when larger vessels like veins and arteries get damaged and bleed. Hematomas make bumps. They swell because a lot of blood gets trapped under the skin.

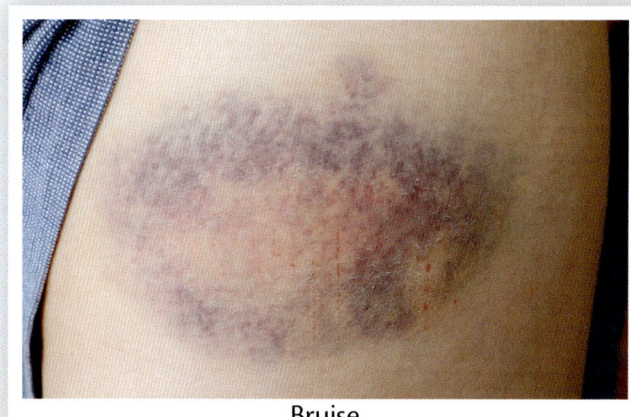
Bruise

You have probably said, "I have a bump on my head." That's easier than saying, "I have a hematoma on my head."

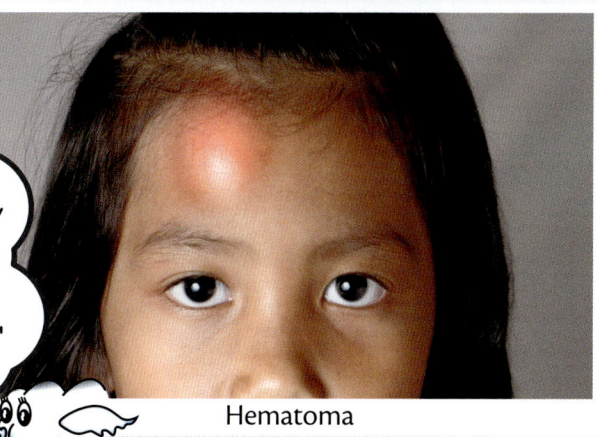
Hematoma

299

GOD MADE ME

When you first get bumps and bruises, they look red because you're seeing blood through your skin. Thankfully, this bleeding soon stops because of God's blood clotting cascade. Just as in wounds, damaged blood vessels are repaired in bruises and bumps. Soon your injury doesn't hurt as much. Your bumps go away, and the area becomes flat again.

After being injured, the trapped blood in bruises and bumps breaks down into different colored chemicals. These chemicals are carried out of the tissues one at a time. This makes the color of your injury change from red to black, blue, purple, brown, or yellow.

A doctor's word for a black eye is a fun word to say. Try saying "circumocular (sir-come-ock-you-luhr) hematoma." Circumocular means *around the eye*.

Time to do Activity 89 in the Activity Book!

God Gives You New Muscle and Bone

Sometimes, injuries happen to your muscles and bones. Bones can break from a hard fall. Muscles can tear during hard exercise or be hurt by a deep cut. Let's see how God heals these parts of your musculoskeletal system.

Muscles

- Muscle cells can get tiny tears during hard exercise. When this happens, it's time for the cell's many nuclei to get busy. They need to seal the torn cell membrane. The cell's nuclei rush over to the injury to help make a protein patch on the cell membrane. The nuclei arrive at the torn membrane in about 5 hours. The membrane is as good as new in 24 hours. That is a fast repair!

This boy and his dad are working to become stronger gradually. Training this way keeps muscle cells from tearing during hard exercise later.

Now we know why God gave each muscle cell a lot of nuclei!

300

CHAPTER 30: HEALING: MAKING THINGS NEW AGAIN

- Larger muscle injuries, like deep cuts, need different repairs. Whole muscle cells die and need to be replaced. God has given you stem cells that live between your long muscle cells. If your long muscle cells die, your stem cells replace them. Stem cells do this by dividing to make more cells. Most of the new stem cells change into long muscle cells. The stem cells that are left remain stem cells. This way, they can help if there's another injury. As muscles heal, white blood cells come and "eat" the dead muscle tissue.

The toxins from snake bites can damage muscles.

Let's look at the steps God made to heal a bone **fracture**. He makes broken bones new again with new bone material!

1. Inflammation

During inflammation, blood that leaks from the broken bone makes a clot. The clot helps hold the bone together until the next step of healing. White blood cells come in to "eat" bone and other tissue that died from the injury. This step lasts about one week.

Bones

- Have you ever broken one of your bones? If so, you know that a broken bone is very painful. You also know that, even after the pain is gone, the bone takes a long time to heal. A broken arm or leg must be used carefully until it has mostly healed. Usually, a brace or cast is put around the break to keep the ends lined up while the bone heals.

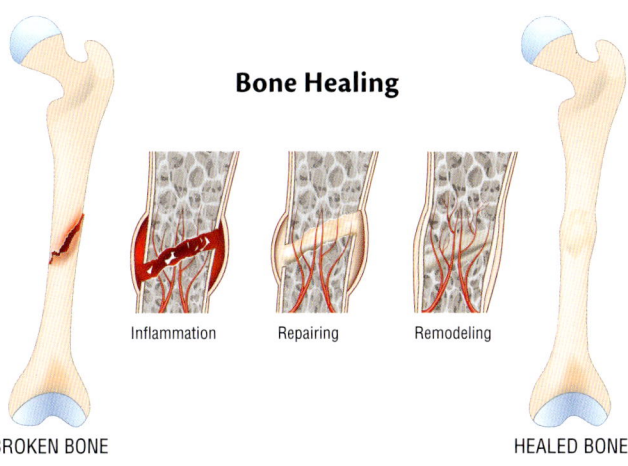

Bone Healing

BROKEN BONE — Inflammation — Repairing — Remodeling — HEALED BONE

For this girl's broken leg to heal well, she shouldn't walk on it. Instead, she's learning to swing along on crutches.

Definitions

A bone **fracture** (frack-shur) is a break in a bone.

GOD MADE ME

2. Repairing

In the repairing step, bone-building cells make a bulging frame of collagen. The collagen holds the ends of the broken bone firmly together. Minerals then harden the collagen, and it becomes spongy bone. Blood vessels begin to grow through the spongy bone. Repairing lasts two to three weeks.

3. Remodeling (ree-mod-uhl-ing)

In remodeling, spaces in the spongy bone slowly fill in with more bone. The new bone becomes harder. It's not spongy bone anymore. A bulge can still be seen around the fracture, but it's made of new, hard bone. The bulge will slowly shrink as bone-dissolving cells reshape the bone. Remodeling lasts for two to three months.

God heals damaged ligaments and

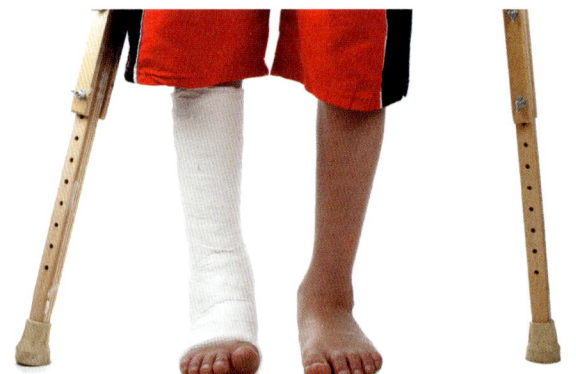

tendons with similar steps.

1. **Inflammation**: A clot forms
2. **Repairing**: Collagen forms a new frame.
3. **Strengthening**: The first collagen is replaced with the right kind of tendon collagen or ligament collagen.

Do not be wise in your own eyes;
Fear the LORD and depart from evil.
It will be health to your flesh,
And strength to your bones.
(Proverbs 3:7-8)

Prayer

Dear God, thank You for healing our bleeding wounds with inflammation and new skin. Thank You for healing the bones and muscles inside us too. We thank You for the pain of inflammation. The pain reminds us that You are kindly healing our bodies. And we praise You for the promise that all things will be made new! Amen.

CHAPTER 30: HEALING: MAKING THINGS NEW AGAIN

Time to do Activity 90 in the Activity Book!

CHAPTER 31

God Can Heal Your Illnesses

[God] forgives all your iniquities, [He] heals all your diseases. (Psalm 103:3)

When God made Adam and Eve, their bodies worked perfectly. There was no death. There was no illness. It was pleasant for them to care for God's fruitful garden where good things grew easily. But then, Adam and Eve disobeyed God. After they sinned, everything changed.

God punished Adam and Eve by bringing curses upon the world. It would no longer be easy for them to grow food. Life would become difficult. Now there would be injury, pain, illness, and even death. Adam and Eve and all the people who came after them would one day die. It's sad that Adam and Eve sinned. It's also sad that we don't always obey God either. We now live in this world that isn't perfect, and we must all die someday.

But God didn't leave mankind without hope. Jesus came to defeat sin, death, and Satan. Jesus came to die and rise again, so that our sins can be forgiven, and we can live forever with God! In this life we will have wounds and illnesses. Unless Jesus comes back first, we will also die. But these sad things won't last forever. Our eternal life with Jesus will have no more death or pain. God will wipe every

Thankfully, God gives our bodies the ability to heal many illnesses in this life. Let's learn how that works.

One curse was that thorns and thistles would grow.

GOD MADE ME

tear from our eyes. Praise God! He will make all things new! (Revelation 21:4-5)

Your Body Repels and Expels Outside Dangers

Most of the time when we get sick, it's caused by something harmful that gets into our bodies from the outside. These outside dangers are things like microbes, **parasites**, and toxins.

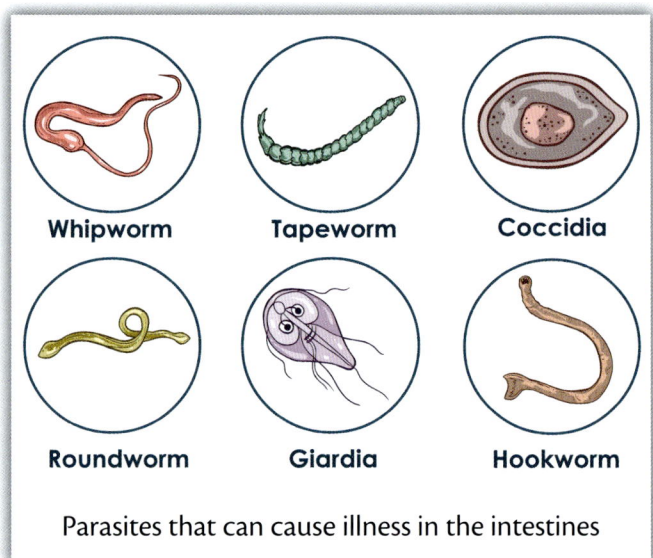

Parasites that can cause illness in the intestines

Definitions

Repel — to keep something from coming in.

Expel — to push something out.

Disease — a problem that causes the body or one of its parts to not work as it should.

Infect — to cause someone to have a disease or health problem from microbes. An **infection** is the problem people have when they are being attacked by microbes.

Parasites — creatures that live harmfully in or on someone.

Review: *Microbes* are creatures too tiny to see except with a microscope. *Toxins* are poisons made by living things like poisonous snakes and spiders.

Most microbes are not harmful, and many are helpful. We use the word *germs* for microbes that can make us sick. Germs become more harmful the more they reproduce. Your body must fight hard when germs get inside you and reproduce.

Let's review the different kinds of microbes.

Bacteria are microbes made of only one cell.

Viruses are tiny particles that can cause disease. Scientists can't decide if viruses are alive or not. They have DNA, but they can't do most of the jobs a living cell does. They must invade cells and use the cells' hard work and food to reproduce. This usually damages the invaded cells.

Fungi are microbes that act similar to plants. But they don't make their own food like plants do. Instead, they get food by absorbing it from other living things.

Thankfully, germs are not harmful if they stay outside your body. God gave you many ways to keep microbes out.

306

CHAPTER 31: GOD CAN HEAL YOUR ILLNESSES

- Skin safely covers you and **repels** germs.
- Chemicals in sweat kill germs on your skin.
- Chemicals in tears kill germs in your eyes.
- Earwax traps germs and **expels** them.
- Strong acid in your stomach kills germs.
- Mucus at your body openings contains white blood cells that kill germs coming in.
- Mucus in your nose traps germs in its stickiness. When the mucus is expelled, the germs are expelled too.
- Mucus in your respiratory system moves up your trachea. Then you cough and swallow or spit out the mucus. Germs in the mucus you swallow are pushed down your esophagus into your stomach. There, they are killed by your stomach acid.

Time to do Activity 91 in the Activity Book!

Self and Non-Self

Even though many germs are repelled and expelled from your body, some do find their way in. Soon we will learn about the amazing ways God protects our bodies from disease-causing invaders.

Your body is made of cells and proteins. Germs, parasites, and toxins are also made of cells or proteins. How does your body know which cells and proteins to fight and which ones to leave alone?

God designed your body to recognize your own cells and proteins (*self*). He also gave your body ways to recognize and fight cells and proteins that are not your own (*non-self*). Your body can "look" at each cell or protein and and say either, "This thing lives here," or "This thing is *not* supposed to be here!"

Every cell in your body is covered with hundreds of different proteins. We'll call one of these proteins the *name-tag protein*. Every person has a name-tag protein on every cell except red blood cells. But the name-tag protein looks different for

When you play a game, you need to know which pieces belong to you and which do not. White blood cells need to know which cells belong to you and which do not.

GOD MADE ME

every person. It's made of the same molecules, but the molecules are arranged differently in different people. Your cells are like a large crowd of people, where everyone has a name tag. The name tags are made the same, but they look different because they have different names written on them.

White blood cells are grayish white, not red.

Sometimes outside dangers (*non-self* things) can enter your body through cuts in your skin. Food spoiled with bacteria and their toxins may enter your digestive tract. Some viruses can invade your body cells because they have an enzyme that helps them get through mucus. When germs do get inside your body, you have another way to fight these *non-self* things. God gave you microscopic white blood cells to have microscopic fights with microscopic germs! You can't see any of this with your eyes. But there are battles happening inside you.

When you were in your mother's womb, God made your white blood cells able to recognize *self* cells. All your life, these white blood cells will be able to "read your name tag" on any kind of cell. They will know not to destroy *self* cells. Now your new white blood cells are formed in your bone marrow. There, hormones "teach" the white blood cells to recognize *self*. These new cells will know how to "read your name tag."

Germs have name tags too. God made white blood cells able to recognize many kinds of germ name tags. White blood cells have little molecules called **receptors** on the outside of their cells. A receptor will recognize and grab a germ's name tag. When the name tag becomes attached to the receptor, the white blood cell is switched on. It kills the germ.

Some white blood cells destroy anything that's not *self*. But viruses and some other diseases "hide" from white blood cells. They hide in *self* cells and secretly damage them from the inside. They can only reproduce if they are hidden inside a self cell. How will these *non-self* invaders be destroyed before too many of them spread to other cells?

God had an amazing plan. Your diseased cell takes a piece of protein from the non-self virus inside it. It places the *non-self* protein on its own name tag. This changes your cell's name into something that white blood cells don't know how to "read." Now your white blood cells know what to do. They destroy your *self* cell, including the *non-self* viruses inside it. Your body cell has given its life to destroy the germs and protect you.

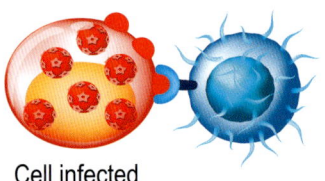

Cell infected

The infected cell (left) has put germ name tags outside on its cell membrane. The white blood cell (right) grabs the germ's name tag and will destroy the cell.

CHAPTER 31: GOD CAN HEAL YOUR ILLNESSES

Zoom in to the Cell

Proteins are long chains of amino acids made by your cells. Proteins are different from each other because of the type of amino acids they have and because of the order of those amino acids. But proteins are also different from each other because of the way they are folded.

The endoplasmic reticulum (ER) in your cells partly folds your newly made proteins. More folding happens in your cells' **Golgi** (goal-jee) **bodies**. Once proteins are folded into the right shape, they are sent off by your Golgi bodies. Golgi bodies are like little post offices sending some mail around town and other mail out of town. Golgi bodies send some proteins to places within their cell, and they send other proteins out of the cell.

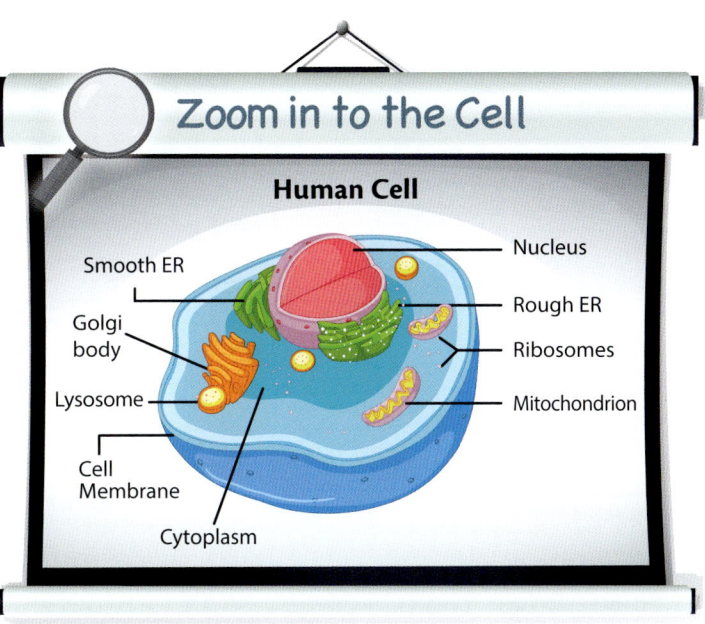

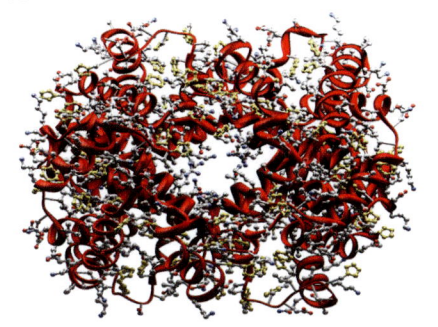

The protein that carries oxygen in your blood is folded into a square shape.

Golgi bodies also have important jobs that help your immunity. They help switch on the fighting in different kinds of white blood cells.

Time to do Activity 92 in the Activity Book!

Your Immune System

Many people throughout history have noticed:
- Wounds usually heal, even if they become infected.
- Most people who get sick do heal.
- People who heal from a disease usually don't get the same disease again.

These kinds of healing happen because of our **immune** (imm-yoon) **system**!

God has given you an immune system to protect you from germs and other outside dangers like parasites and toxins. The busiest parts of your immune system are parts that are too little to see. They are your white blood cells, alertly waiting for a

309

GOD MADE ME

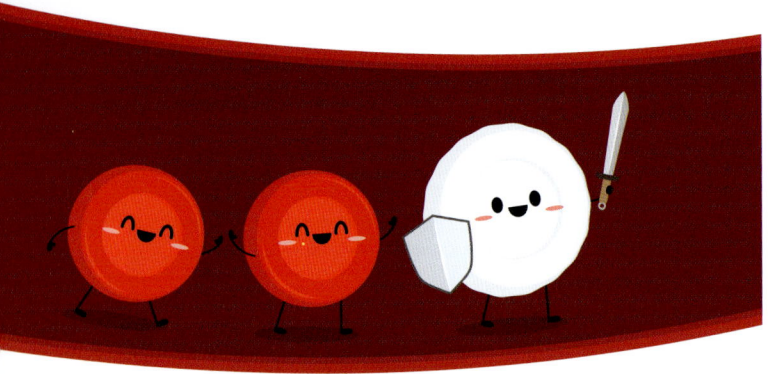

call so they can rush to defend your body.

Your immune system doesn't have any big organs like your other body systems have. Instead, your immune system is made of small areas and small parts all around your body. Even though your immune system might seem small and simple, it's really quite complicated. Scientists know they still have much to learn about the immune system.

 ## Puff's List of Immune System Structures

- Lymph nodes — small pouches in your lymph system that store white blood cells.
- **Thymus** (thigh-muss) — a small organ where certain white blood cells change to their germ-fighting form.
- Liver — a large organ that does about 500 jobs for your body. Thirty of those jobs is to make 30 different chemicals used by your white blood cells to kill germs.
- Peyer's patches — areas of mucus-making tissue that destroy germs in your small intestine.
- **Appendix** (up-end-ix) — a finger-like pouch in the large intestine that stores white blood cells and good microbes.
- **Tonsils** — areas of mucus-making tissue in your throat where germs from food and air are trapped and killed.
- Red bone marrow — the kind of bone marrow that makes white and red blood cells.
- **Spleen** — an organ that, as part of your immune system, switches on certain white blood cells. It also makes proteins that fight germs. Your spleen also helps your circulatory system by cleaning its blood and storing the mineral iron from dead red blood cells.

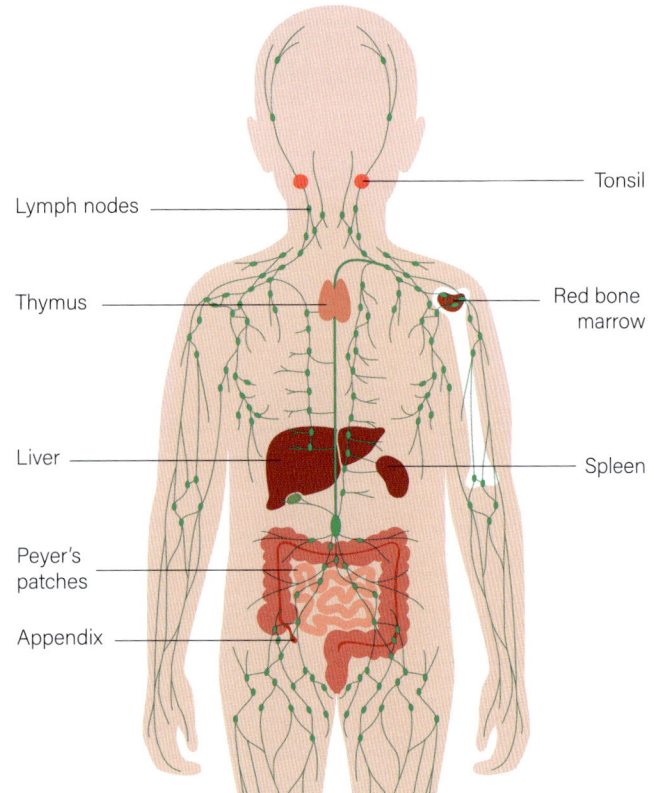

Immune System

310

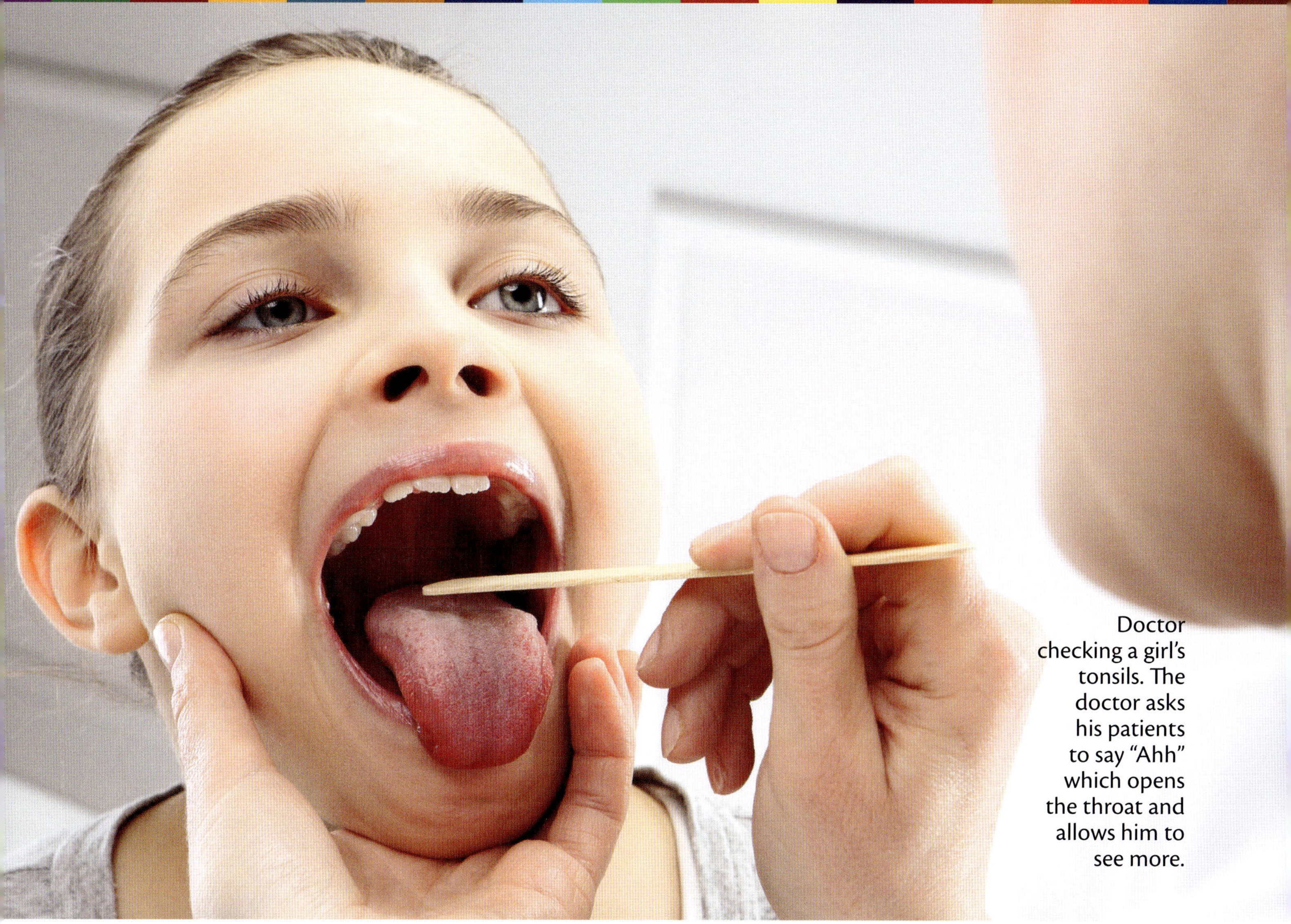

CHAPTER 31: GOD CAN HEAL YOUR ILLNESSES

Doctor checking a girl's tonsils. The doctor asks his patients to say "Ahh" which opens the throat and allows him to see more.

Prayer

Dear Heavenly Father, thank You for guarding our bodies' doorways. Thank You that, when germs do get in, we have white blood cells to fight them. We praise You for teaching our white blood cells to tell the difference between *self* and *non-self* in our bodies. It's amazing how You wisely formed our bodies' immune systems to protect us from disease! Amen.

Time to do Activity 93 in the Activity Book!

311

CHAPTER 32
Vigilant, Valiant Germ Fighters

Watch, stand fast in the faith, be brave, be strong. (1 Corinthians 16:13)

This verse tells us to be watchful (or alert) for things that would weaken our faith in Jesus. A different word for watchful is *vigilant*.

Another word, *valiant*, sounds a lot like *vigilant*, but it means to be strong and brave. We need this strength and courage to stand firm when something threatens our faith.

God has made your immune system and its white blood cells alert, standing firm against things that would weaken your body. He gave your white blood cells the strength to fight invading germs that threaten you. We can be thankful for the vigilant, valiant germ fighters He gave us!

Are you vigilant and valiant in the faith of Christ? Let's learn how your white blood cells are vigilant and valiant to protect you from disease!

Innate White Blood Cells: Ready on the Spot for Any Germ

Your body is full of white blood cells that defend you from germs. Some white blood cells live for years or even decades. Others live less than a week. Your bone marrow makes almost 100 billion white blood cells each day to replace the ones that die.

GOD MADE ME

You have several kinds of white blood cells. We can divide white blood cells into two groups: **innate** and **specific**.

First, let's learn what *innate* white blood cells are.

- Innate cells attack almost everything that is *non-self*.
- Innate cells live in tissues or blood all over your body. God put them everywhere so they would be close to any place a germ may invade.
- Innate cells respond right away.

Innate white blood cells were already with you when you were born.

Let's learn about some of your most common innate cells.

Mast cells live in tissues around your body but not in your blood. Their job is to guard your body by being alerted to danger. Mast cells are alerted to danger by "Help!" chemicals coming out of injured cells. Mast cells also recognize danger by noticing things like splinters that don't belong. They can also recognize germ name tags.

These alerts make mast cells release their own chemical messages. These messages switch on inflammation. Remember, inflammation allows liquid (puss) to leak out of your vessels near the wound. Mast cell messages also attract white blood cells from your blood. When the cells arrive to the infected area, they squeeze through the spaces in your leaky vessels and find germs to kill.

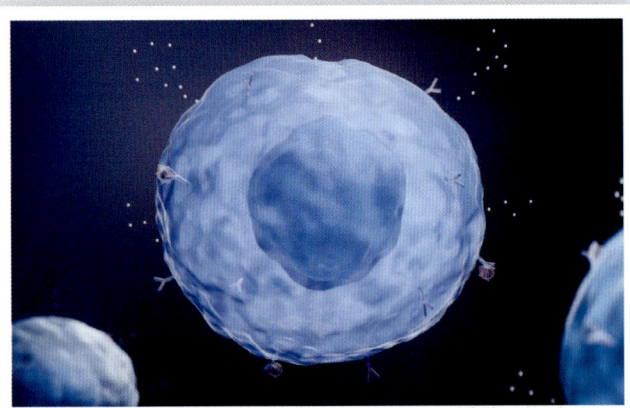
Germs became attached to some of the little Y-shaped parts of this mast cell. This caused the mast cell to release chemical messages to start inflammation.

Neutrophils (noo-troh-filz) are your body's most common white blood cell. They are also some of the largest. Neutrophils travel around in your blood, waiting for a "Help!" message. When they get that message, they stop and squeeze between your blood vessel's cells out into the damaged tissue.

God has given your innate white blood cells a way to know the name tags of many germs. They may not know exactly which germ has invaded. But they do know it is a germ, and they begin to fight! Neutrophils

CHAPTER 32: VIGILANT, VALIANT GERM FIGHTERS

kill germs by engulfing ("eating") them. Once germs are eaten, they are destroyed with packets of poison. Neutrophils also release chemicals outside their cell that begin to kill other nearby germs.

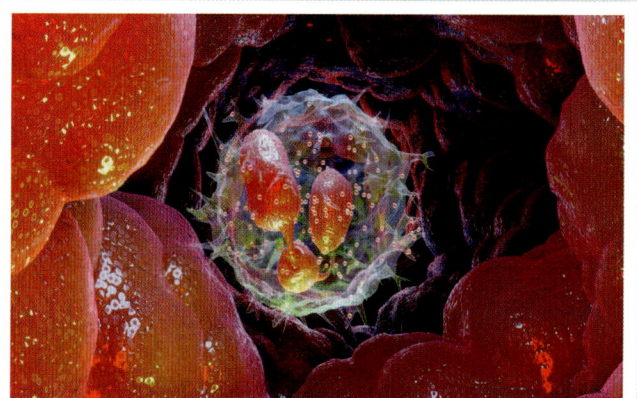
Neutrophil inside a capillary. Notice its three-lobed nucleus and tiny poison packets.

Macrophages (mack-row-fage-ezz) live mostly in your tissues instead of in your blood. Macrophage means *big eater*. Like neutrophils, macrophages are large. They also kill germs by engulfing them. Once a macrophage "eats" a germ, the germ is digested by the macrophage's lysosomes. The germ has become food!

When a macrophage finds a germ, it will save the germ's name tag and won't eat it. The macrophage sticks the name tag to its own cell membrane. This brings special white blood cells that are "trained" to kill that kind of germ. The special cells hunt for this germ, knowing it has infected that area.

Macrophages are the first white blood cells to respond to a germ. They put out chemicals that guide neutrophils to the infection. Macrophages can also switch on inflammation in the tissues.

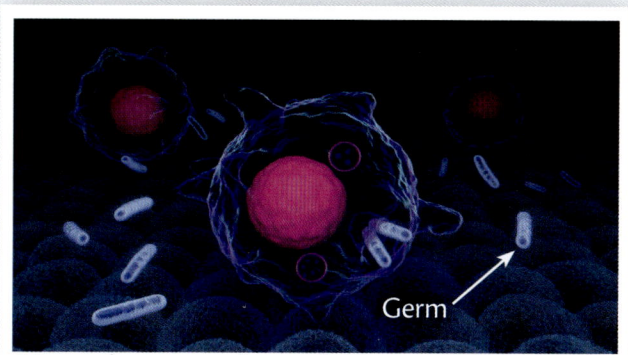
Macrophages can change shape to engulf germs and to move to a different place.

In the English language, a macrophage might have been called *big eater*. In Spanish, it might have been called *gran comedor*. Macrophages might have had many names in different languages. Instead, they have the same name in every language! That's because scientists around the world have agreed to name things using the Latin or Greek languages. It's a great way for macrophages and other science words to have the same name everywhere!

Tree-like Cells

Tree-like cells have many branches. When they grab germs, they eat them and save the germs' name tags. Then they stick the germs' name tags out on their branches.

Tree-like cells live where germs may invade (like the top of your skin). Their branches spread between your body cells, ready to grab germs.

GOD MADE ME

Then it's time for the tree-like cells to travel. They take the germs' name tags into lymph nodes. There, different white blood cells check out the name tags. When one of these cells recognizes a name tag, it divides and makes a lot more cells like itself that rush out to help kill that germ. We'll learn about these lymph-node cells later.

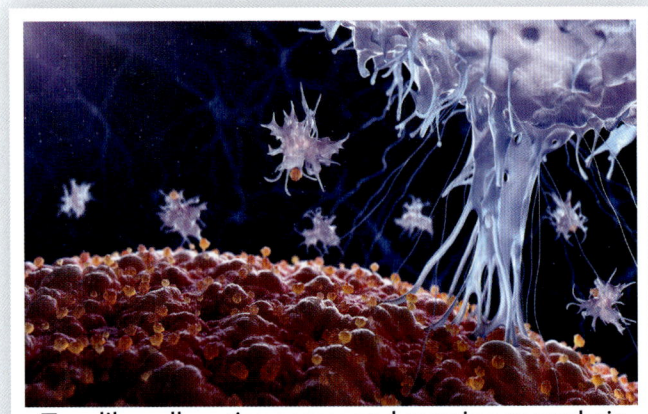
Tree-like cells eating germs and carrying away their name tags

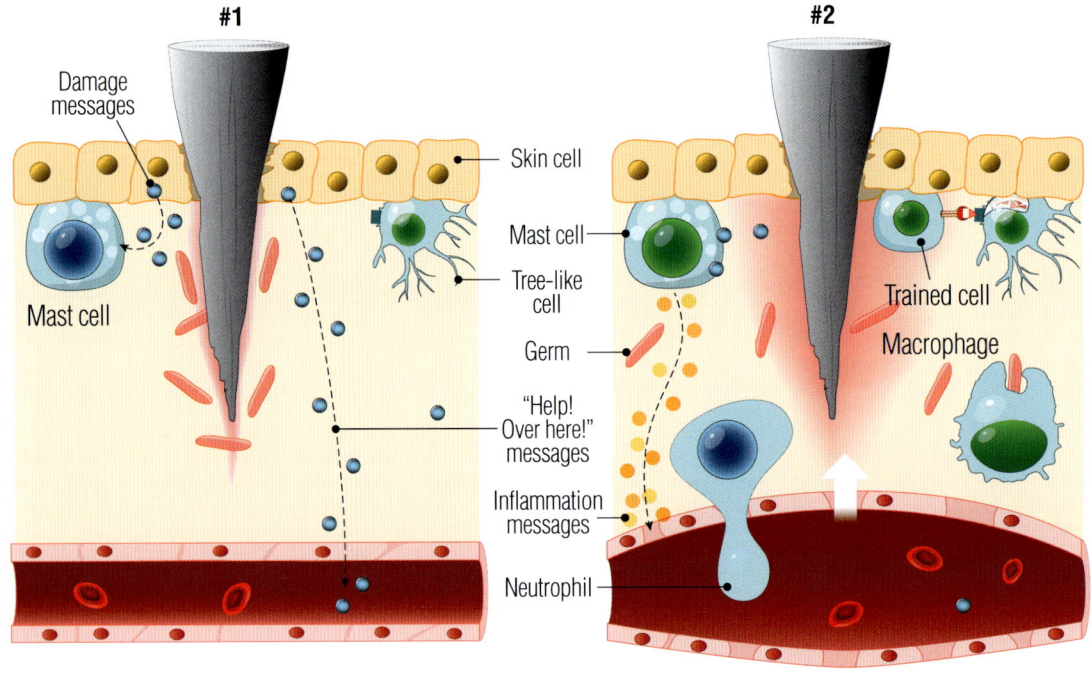

White Blood Cells Fighting Germs

Time to do Activity 94 in the Activity Book!

Specific White Blood Cells: Each Designed to Fight One Kind of Germ

Your body's second group of white blood cells is the specific group of immune cells.

- Specific white blood cells each kill only one kind of germ.
- Specific cells are switched on ("trained") in your lymph nodes where they meet

316

CHAPTER 32: VIGILANT, VALIANT GERM FIGHTERS

with germ name tags. When a specific cell recognizes a name tag, it's switched on to become active.

- Specific cells take longer to start fighting an infection. But when they do, there are a lot of them. They multiply to become many cells that are specially made for fighting that kind of germ.

Scientists still have much to learn about the immune system. As with everything in your body, God made your *specific* white blood cells work in complicated ways:

- They speedily become "trained" to fight germs.
- They remember the kind of germ they are fighting so they can fight it more quickly if that germ ever appears again.
- They are carefully tested to make sure none of them will attack *self* cells.

Sometimes, germs that come into a wound may reproduce so quickly that your innate cells can't kill them. Or flu viruses can get past your mucus and hide in your cells. This is when your specific cells become active.

Use the *Specific Immunity* picture (on page 318) when you get to #4 to learn what your specific white blood cells do when you have a bigger infection.

1. White blood cells called *T cells* are made in your bone marrow. Each T cell has been created to recognize a different germ.
2. The T cells travel to your thymus. There, they are tested to make sure they will do a good job recognizing germs.

Definitions

Antibody — a protein made and released by immune cells. Antibodies fight germs by attaching to the germs.

Antibiotic — a manmade chemical that fights germs.

They are also tested to make sure they will recognize and not kill your *self* cells. If they can't do either of these jobs, they are destroyed. Of every 10 T cells, only one will make it through testing without being destroyed.

3. The tested T cells can travel around in your lymph and blood, but they spend most of their time in lymph nodes. Each one is watching for the germ it's made to fight.
4. When germs invade, your tree-like cells bring the germs' name tags into your lymph nodes. Or a macrophage and a T

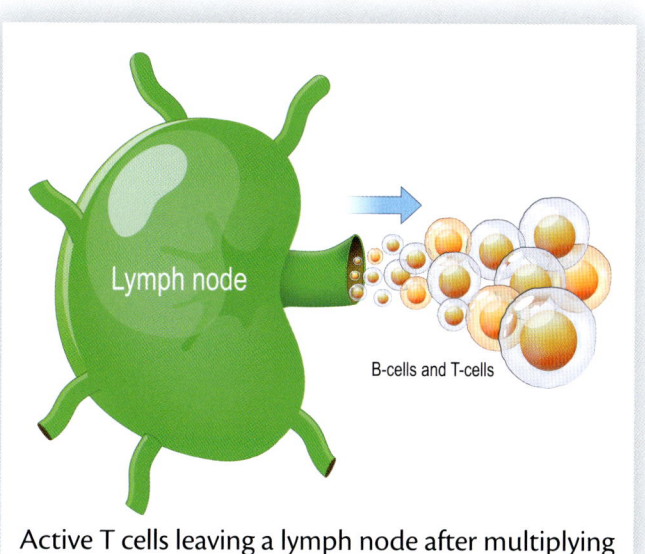

Active T cells leaving a lymph node after multiplying

GOD MADE ME

cell may meet in your lymph or blood. (**A**)

5. When a tree-like cell comes to the node, the T cells take turns trying to "read" the name tag that the tree-like cell is showing them. Being able to read the name tag means that the name tag will fit into the T cell's receptor like a puzzle piece fits into a puzzle. (**B**)

6. When the pieces fit, the T cell gets switched on. It becomes active and gets to work on a few different jobs:

7. The active T cell multiplies into many more cells that will fight the same germ. The new cells become *killer T cells*. (**C**) Killer T cells leave the lymph node and travel through the blood to the infection. There, they attach themselves

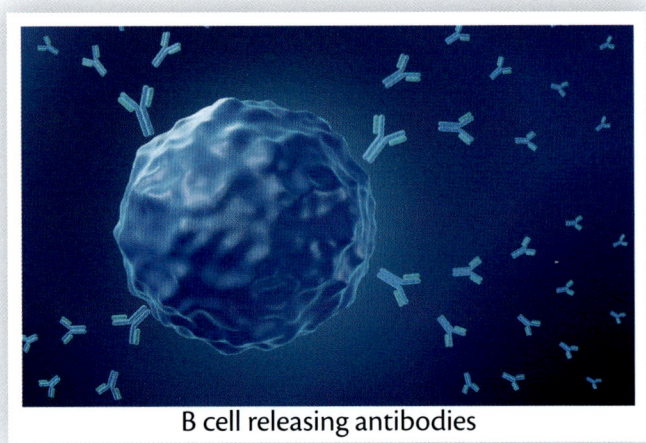

B cell releasing antibodies

to infected cells. They kill infected cells with a poisonous spray. When the cell dies, the germs inside it die too. (**D**) Immune cells eat the dead cell. (**E**)

- Active T cells switch on another kind of white blood cell—*B cells*. (**F**)

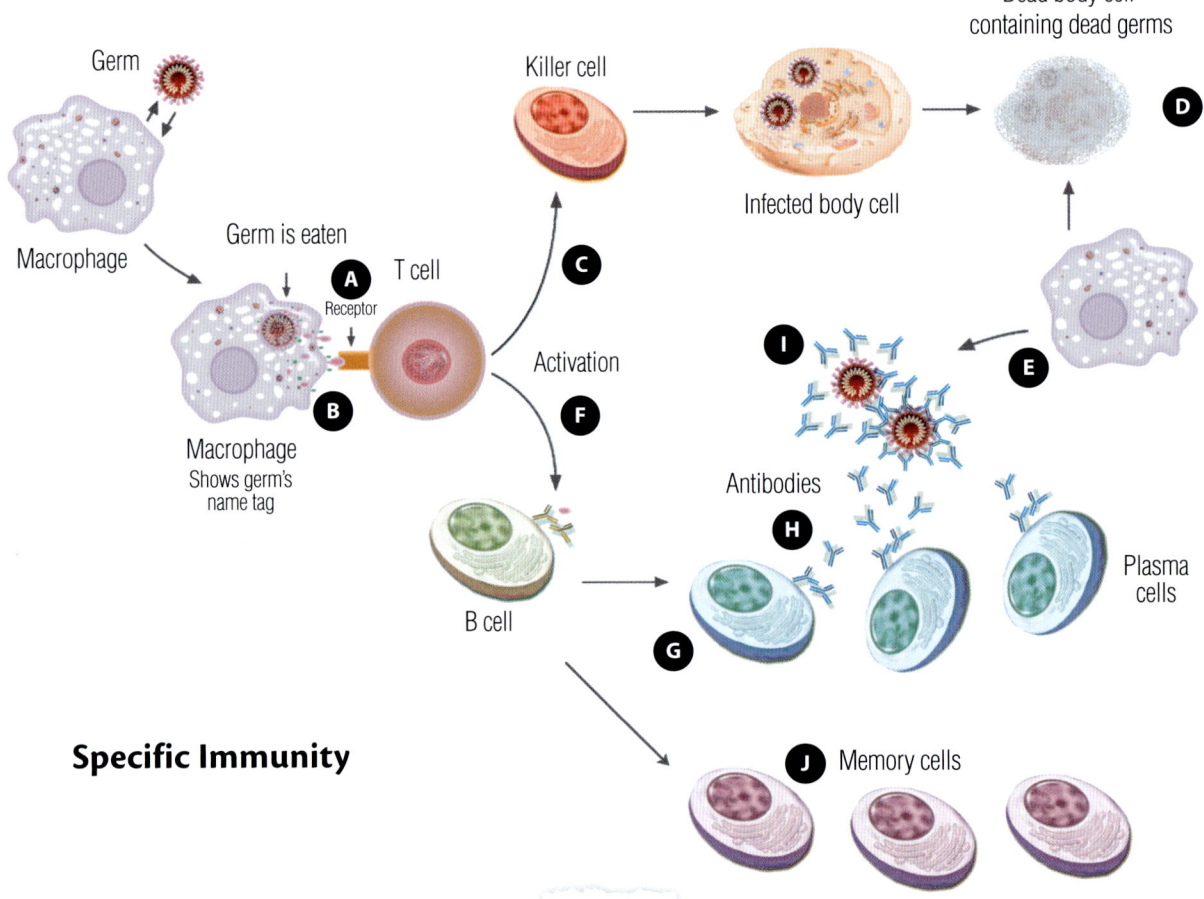

Specific Immunity

318

CHAPTER 32: VIGILANT, VALIANT GERM FIGHTERS

8. Active (switched-on) B cells divide into two new kinds of cells:
   - *Plasma cells.* (**G**) Active B cells make lots of *plasma cells* that release little Y-shaped **antibodies**. (**H**) Antibodies are germ-killing proteins. Like T cells, each antibody kills one kind of germ. God has prepared B cells to make many possible antibodies so there will be one for every possible germ! Once the right antibody is found for the germ's name tag, lots of plasma cells begin making antibodies! When antibodies attach to germs' name tags, the germs become harmless. (**I**)
   - Active B cells make *memory cells.*

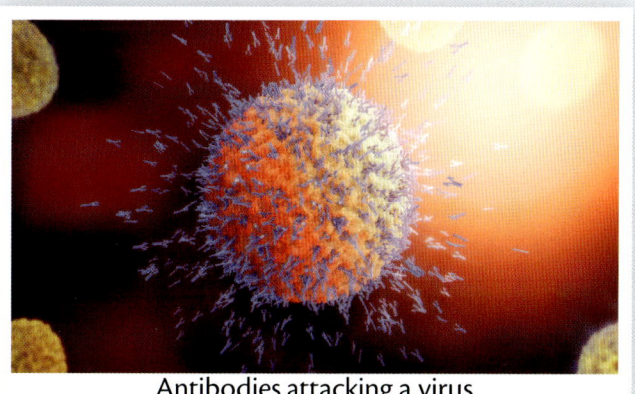
Antibodies attacking a virus

(**J**) Memory cells live a long time—maybe for the rest of your life. They keep a memory of the antibodies they made for their certain germ. If that germ ever gets inside you again, your body is ready to attack it more quickly than it did the first time. You probably won't even know the germ tried again!

Together, your T cells and B cells can fight more than a billion kinds of possible germs! That's more germs than people have even discovered. Sometimes microbes change a little, and our bodies must learn to fight the "new" germs.

Isn't God good to design our bodies with white blood cells that can fight so many germ possibilities?

T cells fight germs by killing infected cells themselves. B cells make antibodies that do the fighting!

 Time to do Activity 95 in the Activity Book!

More Ways Your Body Fights Germs

Whatever you do, do all to the glory of God. (1 Corinthians 10:31)

GOD MADE ME

Immune cells are like this mother. They have many jobs and they work hard to do them well.

You do many things. You play, work, learn, worship, talk, think, listen, clean up, eat, and sleep. Can you think of any other things you do? In the Bible, we are told that whatever we do, we should do it to the glory of God. This means that we should do things in the way that pleases God:

- We try to do things well.
- We do things kindly and cheerfully.
- We think of God while we do things, putting Him first.
- We think of ways to praise Him and talk of Him.

Everything we do should work towards the goal of glorifying God.

White blood cells do many things. Everything they do works towards their goal of keeping us healthy. White blood cells glorify God by showing how wise He is in making such an amazing design. Let's learn a few more things that white blood cells do to help your body heal.

Macrophages release chemicals that travel to your brain, telling it to make a fever. Your higher temperature can kill germs. These chemicals also tell your brain to take away your appetite for food so you don't waste energy digesting. They also make you tired and not interested in doing much. This also helps more energy go towards fighting the germs.

Mast cells notice when you eat something that may be harmful. They make you vomit or have diarrhea by adding extra water into

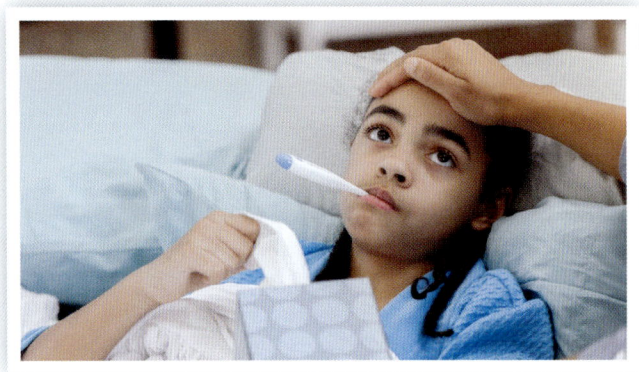

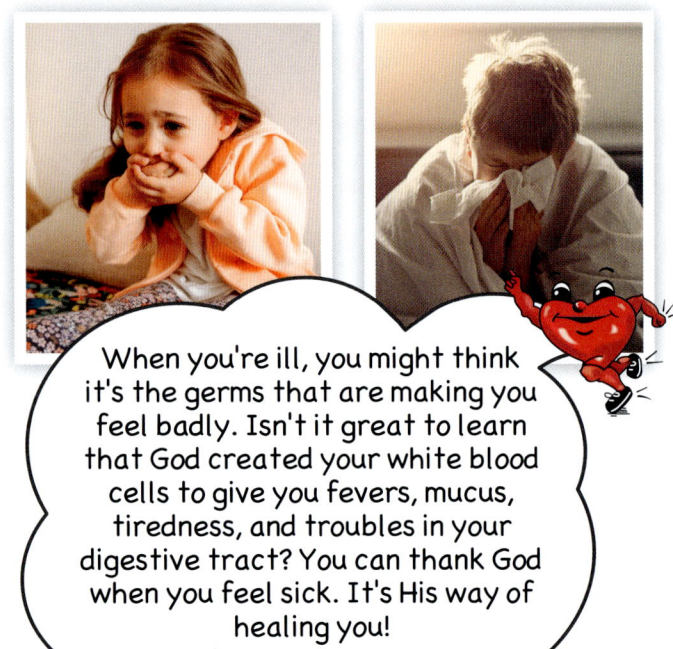

When you're ill, you might think it's the germs that are making you feel badly. Isn't it great to learn that God created your white blood cells to give you fevers, mucus, tiredness, and troubles in your digestive tract? You can thank God when you feel sick. It's His way of healing you!

CHAPTER 32: VIGILANT, VALIANT GERM FIGHTERS

your digestive tract. They also make your stomach or intestine muscles squeeze, and they move stool along more quickly. Vomiting and diarrhea expel germs from your body.

Mast cells in your respiratory tract notice when you inhale germs. They make you cough and get rid of them. These mast cells also tell you to make more mucus to trap and expel the germs.

**In everything give thanks; for this is the will of God in Christ Jesus for you.
(1 Thessalonians 5:18)**

Puff's Health Hint

You can help your immune system by washing your hands!

- Germs can live several hours or several days on places like doorknobs and toilets.
- Germs can get on your hands when you touch things.
- Germs from your hands can get inside your body when you touch your eyes, mouth, or nose.
- Washing with soap and water can kill almost all the germs. Soap destroys most germ membranes and makes the germs pop. Soap also helps lift the germs off your skin so the water can wash them away.
- It's best not to use soap that contains antibiotics. Antibiotics in soap are not any more deadly to germs than plain soap is, but they might kill the good microbes that should live on your skin. It's also not good for antibiotics to get into the water supply after we wash them down the drain.
- Soap only kills germs if your hands are soapy for at least 20 seconds. Be sure to take enough time while washing your hands.

Prayer

Dear Father, You are wonderful and wise to create our amazing immune system! When germs do get inside us, we are kept healthy by lots of microscopic defenders. You've prepared our bodies to fight many germs anywhere and anytime. Thank You for giving us germ-fighting cells that remember past invaders. They travel through our blood, keeping us ready to fight the same invaders again. Thank You, in Jesus' name, Amen.

Time to do Activity 96 in the Activity Book!

UNIT 9
God Keeps Your Body on Schedule

God created our bodies to use chemical signals that make things happen to us at the right time. These chemicals are called hormones. You sleep because hormones are switched on by the nighttime. You wake up when those hormones are switched off by the morning light. Your body is growing because of hormones, and you will stop growing when those hormones stop. Even when your body stops growing, you never have to stop growing in godliness. Our hymn reminds us that you can grow more loving at all times with God's help!

God created everything! This includes time too. Our memory verse tells us that everything happens when God plans for it to happen. If you read the verses that come after our memory verse in the Bible, you'll find that they talk about God's timing of things in our lives. They say that being born, dying, and healing happen at the right time. It's wonderful that we can know God is in charge of everything that happens!

Memory Verse

To everything there is a season,
A time for every purpose under heaven.
(Ecclesiastes 3:1)

You can listen to this hymn by searching for "Father, We Thank Thee Cedarmont Kids" on the internet.

Hymn to Sing: Father, We Thank Thee

Father, we thank thee for the night,
And for the pleasant morning light;
For rest and food and loving care,
And all that makes the day so fair.

Help us to do the things we should,
To be to others kind and good;
In all we do, in work or play,
To grow more loving every day.

CHAPTER 33
God Gives You Hormones

Watch, stand fast in the faith, be brave, be strong. (1 Corinthians 16:13)

You've learned that the electric messages of your nervous system allow you to react quickly. Your body has another system that allows you to react more gradually. This system is your **endocrine** (end-oh-krinn) **system**. Your endocrine system is made of:
- glands
- hormones made by glands
- **target cells** that are switched on by the hormones

Glands are organs that make certain chemicals (hormones) your body can use. Hormones are small amounts of chemicals that are made in one place (like a gland) and travel through your blood. Even though hormones travel everywhere in your body, they only cause certain cells to react. These *target cells* react because God gave them receptors that can only grab a certain hormone. When their receptors grab that hormone, the target cells are switched on to do the job they were created for.

Your body makes about 75 different hormones!

Hormones Keep You Balanced and on Schedule

Hormones balance things like water, salt, and calcium in your body. If your body becomes low in something, a hormone will make sure the target cells know it's time to do their job.

For example, if your blood is low in calcium, your brain tells a certain gland to release its hormone. That hormone travels in your blood to your bone-dissolving cells (the target cells for this hormone). The hormone tells these target cells that it's time to take calcium out of your bones. The bone-dissolving cells eat

GOD MADE ME

away at your bones and release calcium into your blood. When your blood has the right amount of calcium, it's balanced. The gland will stop releasing its hormone. This tells the target cells to stop dissolving your bone. Another hormone from another gland tells bone-building cells when to replace the dissolved bone.

Hormones also keep your body on schedule. Your brain knows when it's time for your body to do something. You were scheduled to be born at the right time. You are also scheduled to keep growing for a certain amount of time. When it's the right time for something, your brain tells a gland to put its hormone into your blood. Then the hormone travels through your blood, and the target cells that recognize the hormone grab it. This switches on the target cells and they start doing their job.

If you want to raise your arms, your nervous system uses electric messages. But if you want to keep holding them up, your nervous system must keep sending electric messages over and over. Your endocrine system is different. God made hormones as longer-lasting messages for things that take more time! For example, if your body needs water, the chemical messages from hormones stay in your blood until the problem is solved.

When it gets dark at nighttime, your brain tells a gland in your head to make a hormone called **melatonin** (mell-uh-tone-in). Melatonin makes you sleepy. When morning light comes, your brain tells the gland to stop making melatonin. Without melatonin, you wake up! Melatonin tells your body when it's time to sleep, and when it's time to wake up! It keeps you on a good schedule.

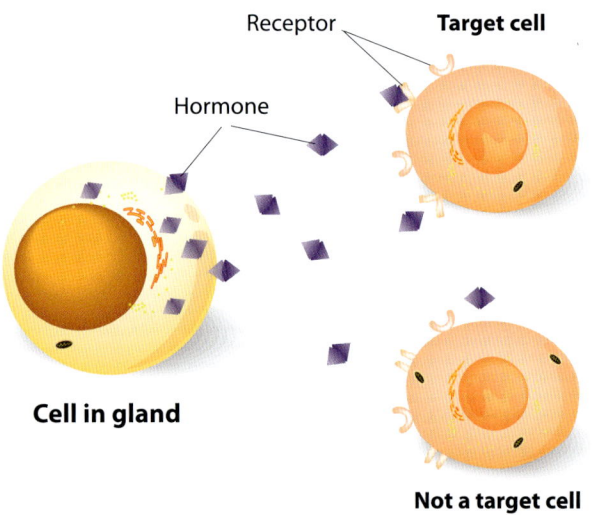

Cell in gland

Hormone

Receptor

Target cell

Not a target cell
(no receptors for this hormone)

Time to do Activity 97 in the Activity Book!

CHAPTER 33: GOD GIVES YOU HORMONES

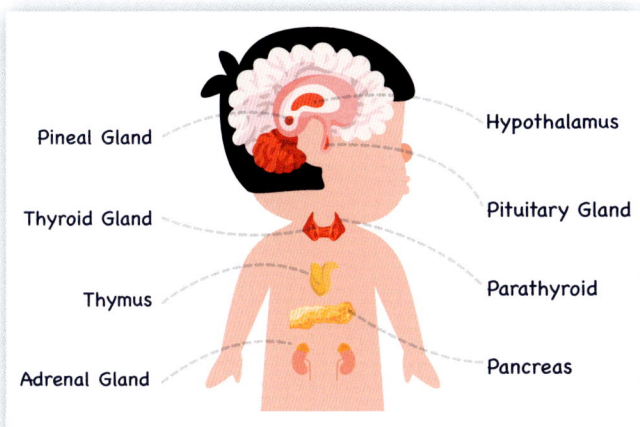

Some of your endocrine glands. Your body has other glands (not shown) that help you become an adult. God also gave you hormone-releasing cells that are not glands.

This air traffic controller tells many pilots when to take off and land their planes safely. Your T4 hormone is like an air traffic controller—responsible for a lot of things at the same time.

Glands of Your Endocrine System*

Your endocrine glands are found in different parts of your body. Each endocrine gland releases a certain kind of hormone or several hormones. You've already learned about different glands that do not release hormones. Tears and saliva are liquids made in glands, but tears and saliva are not hormones.

Tear glands and saliva glands release their liquid in large amounts. Hormones are released in much smaller amounts. Tears and saliva are released through ducts, but hormones are released from glands that don't use ducts. Instead, hormones pass out of the cells that make them into the liquid between the cells. There, your capillaries absorb the hormones so they can travel through your blood to target cells around your body.

Some hormones can switch on several different target cells in different parts of your body. For example, the hormone T4 made in your thyroid gland can:

- increase the speed of how you use energy from food.
- speed up how your cells use oxygen.
- speed up your heart rate.
- help you grow.

It takes several people to do the job of keeping a raft upright in rough water. In the same way, it takes several different hormones to do the job of keeping your blood sugar balanced.

*The teeter totter symbol ( ) will be used to show when a hormone is used for balancing the inside of the body. The clock symbol ( ) will show when a hormone helps the body do something at the right time.

GOD MADE ME

Sometimes, one function can be changed by several different hormones. For example, your blood sugar can be changed by:

- two different hormones from your pancreas.
- two different hormones from your adrenal glands.

 **Time to do Activity 98 in the Activity Book!**

Your Brain and Your Endocrine System

Let's learn how your endocrine glands know when it's time to release their hormones!

Your **hypothalamus** (high-poh-thal-uh-muss) is a part of your brain that makes sure your body stays balanced inside. It gathers information from your blood and your nervous system. Then, your hypothalamus sends messages to:

- your **autonomic nervous system**. This system sends *electric* messages to control automatic things like your heart beat and your breathing.
- your **endocrine system**. This system uses *chemical* messages to control automatic things like blood sugar.

Let's look at how your hypothalamus balances you through your endocrine system. Just as your head is in charge of your body, your hypothalamus is in charge of your endocrine system. To understand what your hypothalamus does, you must know about your **pituitary** (pit-oo-it-airy) **gland**.

Your hypothalamus is at the bottom of your brain. Your pituitary gland sits right under your hypothalamus. Although your hypothalamus is part of your brain, it also makes hormones. Two of these hormones are moved to the pituitary gland for storage. When it's time for them to be used, the hypothalamus tells your pituitary gland to release them.

Here's what your stored hypothalamus hormones do:

- One hypothalamus hormone helps you when you're thirsty. It tells your kidneys to keep water in your body

328

CHAPTER 33: GOD GIVES YOU HORMONES

instead of releasing it into your urine.

⏰ The other hypothalamus hormone helps mothers have a baby and helps them feed their milk to their babies.

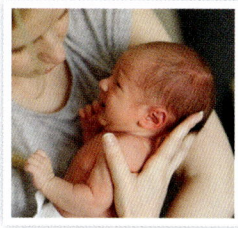

Your hypothalamus makes different hormones that are not stored in your pituitary gland. Instead, these hormones tell your pituitary gland to release its own hormones.

Together, your hypothalamus and your pituitary gland are in charge of your endocrine system. Isn't it amazing how many complicated functions and structures God has given your body? The more we learn about our bodies, the more things we find to praise Him for!

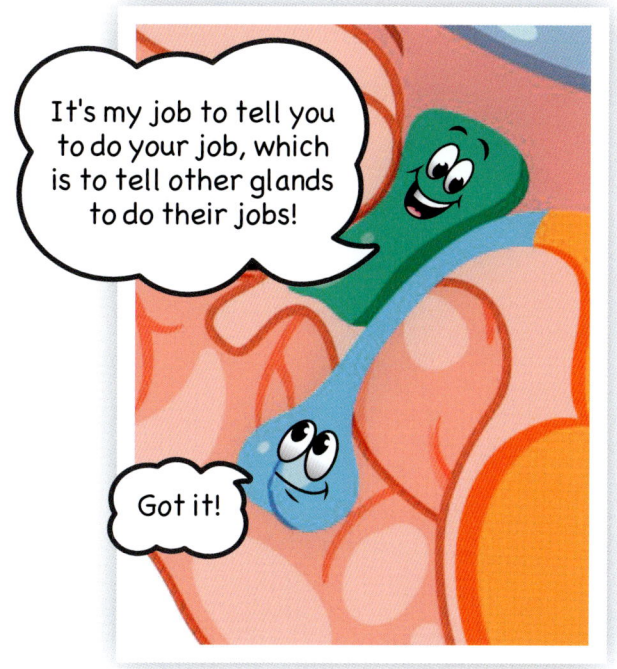

It's my job to tell you to do your job, which is to tell other glands to do their jobs!

Got it!

Prayer

Lord, we are amazed that you have made another system that does so many important things inside us! Thank You for the many chemical messengers that balance our bodies and help our bodies do things at the right time. Amen.

Time to do Activity 99 in the Activity Book!

CHAPTER 34
Pituitary, Pineal, and Thymus Glands
Keeping Your Body on Schedule

And the child Samuel grew on, and was in favor both with God and also with men. (1 Samuel 2:26 KJV)

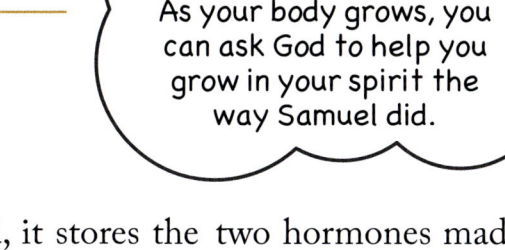

The boy, Samuel, became wiser and more godly as he grew taller and bigger. As your body grows, you can ask God to help you grow in your spirit the way Samuel did.

Several of your endocrine glands help you grow and become an adult at the right time. Let's learn about each of your endocrine glands.

Your Pituitary Gland Helps You Grow Up

Your pituitary gland is made of two parts. God placed your pituitary gland in a bony hollow of your skull behind your nose. It's attached to a little stalk, one part behind the other.

The back part of your pituitary gland doesn't make hormones. Instead, it stores the two hormones made by your hypothalamus. Your hypothalamus tells this part of your pituitary gland when to release these hormones.

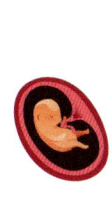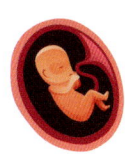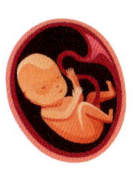

GOD MADE ME

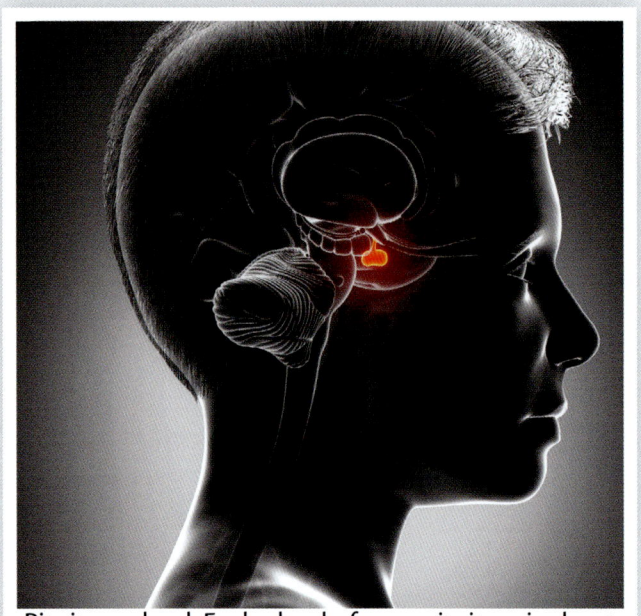

Pituitary gland. Each gland of your pituitary is about the size of a cherry pit. "**Pit**" will help you remember "**pit**uitary."

The front part of your pituitary gland makes hormones that:

- ⏰ cause you to grow taller and bigger by making your bones and muscles grow. Your body began making this hormone when you were two months old in your mother's womb.
- 👥 tell your adrenal (uhd-reen-uhl) gland to release some of its balancing hormones.
- 👥 tell your thyroid (thigh-roid) gland to release one of its hormones.
- ⏰ help mothers make milk.
- ⏰ are needed for the first human cell to begin life in the womb.

 **Time to do Activity 100 in the Activity Book!**

Your Pineal Gland Helps You Sleep

Your **pineal** (pin-ee-uhl) **gland** is about the size of a grain of rice. It lives at the end of a little stalk, tucked between the two halves of your brain.

Your pineal gland makes the hormone, melatonin, that:

- ⏰ helps you sleep at night.

When evening comes, your eyes notice that it's getting darker. The part of your eye that notices this is your retina, the lining inside your eye. But the retina neurons that notice the darkening day are not the same neurons that give you sight. Your dark-sensitive neurons live between your sight neurons. God gave these neurons the job of noticing nighttime and daytime to keep your body on schedule.

As it gets dark, these neurons send electric messages to a bundle of nerves that acts like a "clock" for your body. This bundle

CHAPTER 34: PITUITARY, PINEAL, AND THYMUS GLANDS

then tells your pineal gland to make the hormone melatonin. The melatonin is released and travels through your blood as a chemical message. Its target cells are the nerves that make you feel sleepy. These nerves make you fall asleep and stay asleep while it's dark.

When morning light comes, your retina cells notice this, even behind your eyelids. Your pineal gland is told to stop making melatonin before it's time to wake up.

Scientists are still learning about other helpful things melatonin does for your body. It seems to protect your cells, help with blood sugar, and keep your body's "clock" on schedule.

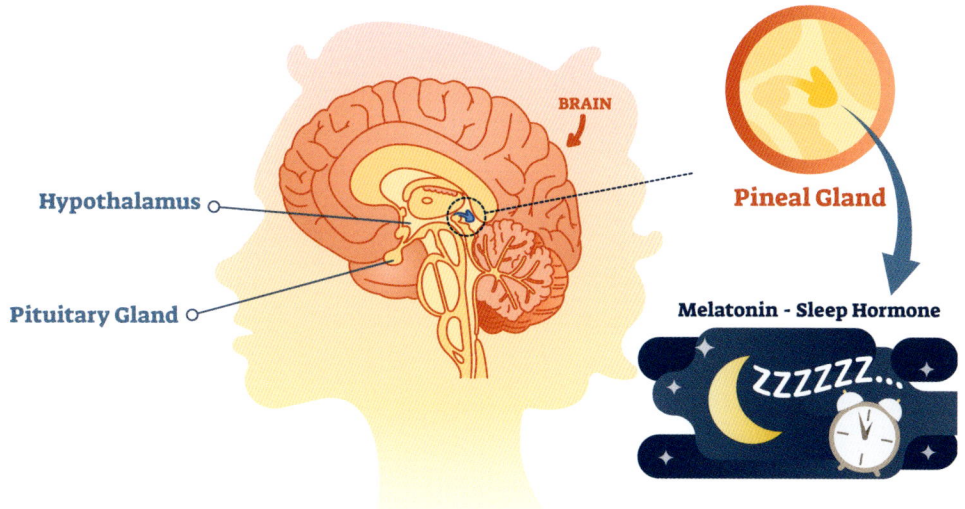

The **pine**al gland got its name because it looks like a little **pine**cone. This may help you remember its name.

Thump's Health Hint

Computer, TV, and phone screens use a certain kind of light that helps us see their screens. But this kind of light confuses our eyes and makes them feel like it's daytime even when it's not. When our eyes think it's daytime, so does our pineal gland. Our pineal gland won't make melatonin if it doesn't get nighttime messages.

It takes time for melatonin to do its job. God designed our bodies to make melatonin work slowly, as the day gets darker, so we have time to get ready for bed. So, even after we stop looking at electronic screens, it may take us a while to fall asleep. Be careful about looking at electronic screens in the evening. They may make it harder for you to fall asleep.

Sleep is important for our minds and bodies. God made sleep, so don't waste this gift!

Keep sound wisdom and discretion;
So they will be life to your soul ...
Yes, you will lie down and your sleep will be sweet.
(Proverbs 3:21-24)

 Time to do Activity 101 in the Activity Book!

GOD MADE ME

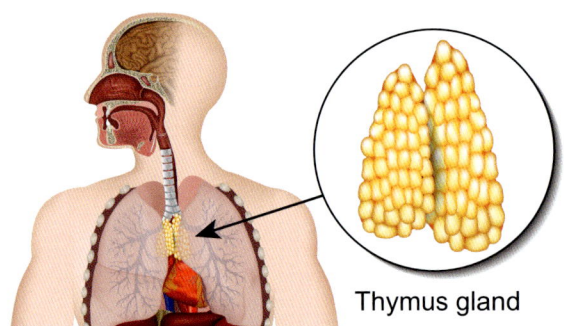

Thymus gland

Your thymus gland sits behind your breastbone and in front of your heart. This gland was named "**thym**us" because it looked like a handful of the herb, **thyme**. This may help you remember its name.

Your Thymus Gland Helps Your Immune System Grow Up

You have already learned that your thymus is the place where the T cells of your immune system are tested. Your thymus tested most of your T cells before you were even born! It was destroying the T cells that didn't have the right kind of receptors to fight germs. It was also destroying the T cells that didn't recognize "self" to keep them from attacking your body's own cells.

More of your T cells are being tested now in your childhood. By the time you are in your teens, almost all your T cells will have been tested. Then, your thymus will start shrinking. Your T cells will be ready for you to use the rest of your life. Your immune system will then be all grown up!

Your thymus is also an endocrine gland! It releases:

- ⏰ hormones that turn the first T cells into different kinds of T cells.
- ⏰ hormones that help with immunity.
- ⏰ a small amount of melatonin.
- a hormone that helps balance your blood sugar.

Prayer

Dear God, You've made our bodies do so many things at the right time. You've made us grow up. You give us sleep using melatonin from the pineal gland. And You've started out our T cells with hormones from our thymus gland. Besides changing our bodies at the right time, you have kept our bodies balanced inside. Thank You for your loving kindness! Amen.

 Time to do Activity 102 in the Activity Book!

CHAPTER 34: PITUITARY, PINEAL, AND THYMUS GLANDS

CHAPTER 35
Thyroid, Parathyroid, and Pancreas Glands
Keeping You Diligent

He who has a slack hand becomes poor, But the hand of the diligent makes rich. (Proverbs 10:4)

If I have a slack hand, it means I'm not using my muscles to do anything. My hand is limp and floppy because I'm being lazy. This Bible verse says I will become poor. But if I'm diligent, I will work hard. God will bless that work. Let's learn about how your endocrine system helps you be diligent!

Your Thyroid Gland Turns Food into Energy

Your butterfly-shaped **thyroid** (thigh-roid) **gland** sits at the front of your neck like a bow tie. It's attached to your trachea and moves along with your voice box when you swallow. God gave your thyroid gland many jobs so that you have the energy you need to be diligent.

This coin's design shows an ancient Greek shield called a **thy**os. The **thy**roid gland was named after this shield because of its shape. Maybe this will help you remember its name.

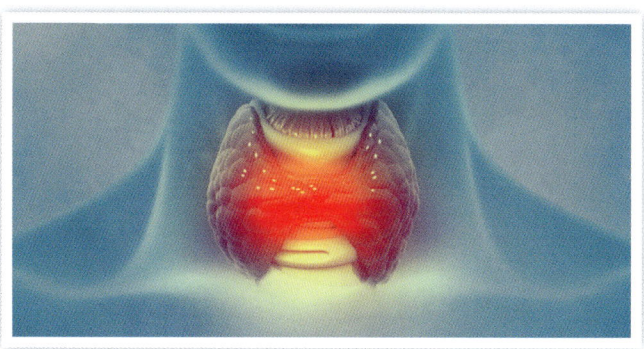

Your thyroid gland produces:

- two hormones that make things in your body work at the right speed. These hormones tell your cells how quickly to use energy from your food.

337

GOD MADE ME

They control heart rate, body temperature, the speed food travels through your digestive tract, how your muscles contract, and how quickly your skin and bones heal. Your pituitary gland keeps track of how much of these hormones are in your blood. If the amount gets low, your pituitary tells your thyroid to release these two energy hormones. If you didn't have enough of these two hormones, your body and mind would feel lazy, cold, and sluggish.

> **Thump's Health Hint**
>
> Your thyroid's two energy hormones are the only things in your body that contain the mineral **iodine** (eye-oh-dine). Your thyroid gland must have iodine to do its job.
>
> Iodine is an essential mineral—your body can't make it, so you must eat it. Earth's oceans contain a lot of iodine. Because of this, seafoods are rich in this important mineral. Soil contains less iodine. Foods raised on soil contain less iodine than seafood does.

- **one hormone that keeps the calcium in your blood from getting too high.** Your thyroid gland senses the amount of calcium in your blood. If your calcium is high, your thyroid releases a "Stop!" hormone to your bone-dissolving cells. This stops those cells from eating away at your bones so that no more calcium goes into your blood. Without this hormone, your blood would have too much calcium. Your muscles would be weak, you would feel sluggish, and your emotions would change quite often.

Foods that Supply Iodine for Your Thyroid Gland

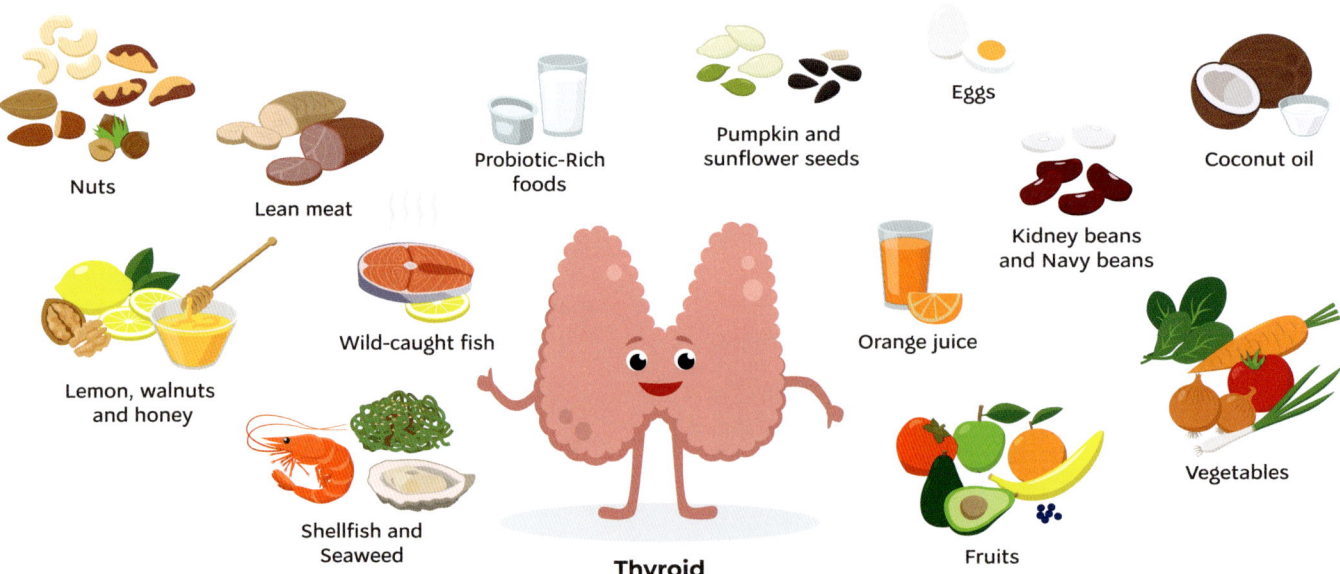

You can help your thyroid by eating foods from the sea like fish, shellfish, and seaweed. You can also eat foods that are more likely to absorb iodine from the soil.

338

CHAPTER 35: THYROID, PARATHYROID, AND PANCREAS GLANDS

Time to do Activity 103 in the Activity Book!

Your Parathyroid Keeps Your Muscles Working

People usually have four **parathyroid glands** on the back of their thyroid gland. But some people have only three parathyroid glands. A few people may have eight or more! Each parathyroid gland is small and flat like a dried lentil.

Your parathyroid gland makes a hormone that:

adds calcium to your blood. It does this by sending its chemical message into your blood. The message is picked up by your bone-dissolving cells. These target cells scrape at the bone to add calcium to your blood.

Calcium is very important in your body for:
- making your 600+ muscles contract.
- making platelets active to stop your bleeding.
- sending electric nerve messages in your body.

Your parathyroid hormone does the opposite job of your calcium-stopping thyroid hormone. It makes calcium come out of your bones until your blood has enough.

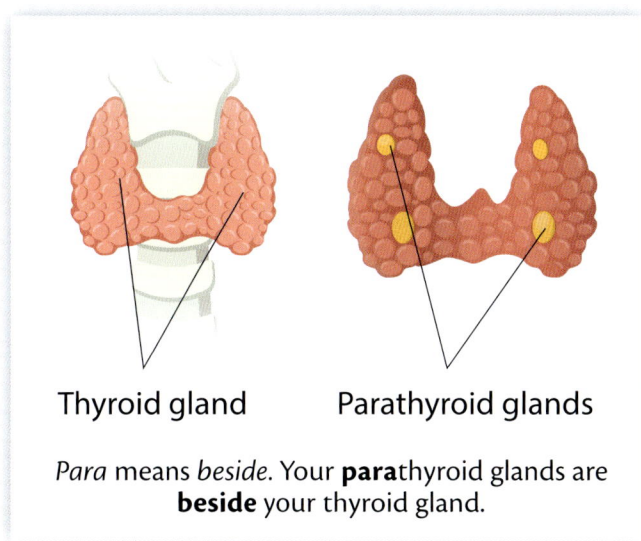

Thyroid gland   Parathyroid glands

Para means *beside*. Your **para**thyroid glands are **beside** your thyroid gland.

Time to do Activity 104 in the Activity Book!

339

GOD MADE ME

Your Pancreas Gives Your Body Sugar for Energy

Your pancreas is an organ with several jobs. We've already learned that your pancreas releases digestive juices into your small intestine to help you digest. Your pancreas is also an endocrine gland. It makes two hormones that keep the right amount of sugar in your blood.

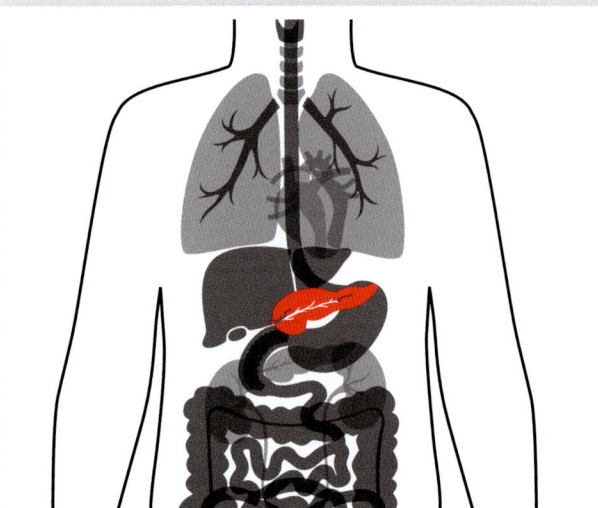

Your pancreas (red) is a digestive organ and an endocrine gland. The pancreas is dotted with little **isl**ands of cells called **isl**ets (eye-lets). Pancreas hormones are made in the islets. Digestive juices are made in the tissue between the islets.

Your two pancreas hormones have opposite jobs:

- **One hormone, insulin, keeps the amount of sugar in your blood from becoming too high.** Insulin does this by helping your body's cells absorb sugar. Your cells use this sugar as their food for making energy.

- **The other hormone keeps your blood sugar from becoming too low.** If your blood needs more sugar, this hormone tells your liver to release its stored sugar.

If you eat a sugary dessert or a starchy food like bread or potatoes, the amount of sugar in your blood will suddenly increase. It's important for your blood sugar to be balanced, so God gives you a burst of insulin to handle the burst of sugar. Insulin is like a key that unlocks your cells. It takes sugar out of your blood and puts it into your cells to give them energy. Insulin also tells your liver and muscles to store any extra sugar to keep it out of your blood. Your body will be able to use this sugar later if you need it.

If you eat a lot of sugary and starchy foods, there will be too much sugar for your body to use or to store in your liver and muscles. Since blood sugar must stay balanced, your body will store the extra sugar as fat.

CHAPTER 35: THYROID, PARATHYROID, AND PANCREAS GLANDS

Prayer

Heavenly Father, thank You for making sure our bodies have the energy we need to be diligent! You have given us food with all the things our bodies need. You also give us hormones to make sure we use food's energy at the right speed when we need it. Help us be diligent with our bodies and minds. In Jesus' name, Amen.

Time to do Activity 105 in the Activity Book!

Your adrenal glands help your body react to sudden dangers!

CHAPTER 36
Hormones for Hard Times

Your Adrenal Glands Help You React

The last endocrine glands we will learn about are your **adrenal glands**. You have two adrenal glands, each sitting on top of a kidney. Although one is pyramid shaped, and the other is more rounded, both adrenal glands release the same group of hormones. Your adrenal glands have an inside part and an outside part.

The *inner* parts of your adrenal glands release:

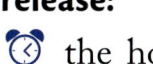 the hormone **adrenalin** (uh-dren-uh-lin) that gives you quick action in an emergency.

If you are suddenly frightened, your nervous system automatically sends "fight or flight" messages to the inner part of your adrenal glands. This causes adrenalin to be released into your blood. When this happens:
- your heartbeats speed up.
- you get a burst of sugar into your blood.
- more oxygen goes to your brain.
- your blood pressure gets higher.

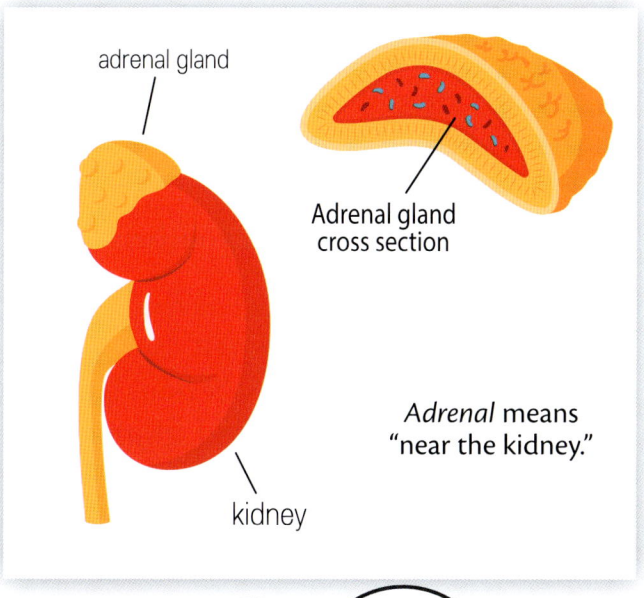

Adrenal means "near the kidney."

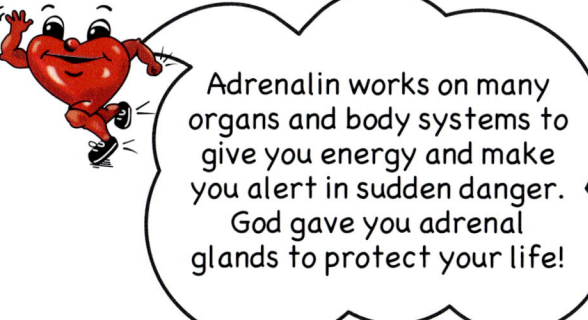

Adrenalin works on many organs and body systems to give you energy and make you alert in sudden danger. God gave you adrenal glands to protect your life!

GOD MADE ME

- less blood goes to your skin and digestion.
- more blood goes to your muscles, heart, and brain.

The *outer* layer of your adrenal glands releases:

- a hormone that tells your kidneys how much water and salt should be in your blood.
- a hormone that helps you become an adult.
- a hormone that helps your body during longer-lasting problems. Serious illnesses, exhausting exercise, and things that make you feel troubled or worried can be stressful to your body. This hormone gives your blood a little extra sugar for a longer time. It causes sugar to be made out of muscle tissue and certain organs in order to make sure enough energy goes to the brain. It also keeps the immune system from causing too much inflammation. These changes may make the body weak if stress lasts a long time.

. . . casting all your care upon Him, for He cares for you. **(1 Peter 5:7)**

It's important that you cast your cares upon God. We do this when we tell God our worries in prayer. And, as you do what needs to be done, trust Him and do not worry. Reading the Bible and praying will help you trust God and know that He cares for you.

Long-lasting dangers like wars and hungry sieges are stressful. Your adrenal glands help your body during stress that lasts a long time.

 **Time to do Activity 106 in the Activity Book!**

Other Hormones

God made other kinds of hormones. Here are some hormones that are not released from glands:

- Some cells release hormones that cause something to happen to their own cell. For example, when a T cell grabs a germ's name tag, it releases a hormone

CHAPTER 36: HORMONES FOR HARD TIMES

- that tells itself to make more T cells.
- Cells also release hormones that make things happen to other nearby cells. For example, in the womb, your early bone cells sent hormone messages to each other so your bones would not grow too big and make you a giant.
- Many organs and tissues (that are not glands) release hormones that make things happen in a different part of your body. For example, the fat cells around your body release a hormone that tells you when you have eaten enough. This makes you feel satisfied with that amount and you will then stop eating.
- God has given some animals extra hormones that He has not given to humans. Animals can't communicate with each other as people can. But God often gives animals certain hormones that send chemical messages to other animals. These hormones are released into the air and are called **pheromones** (fair-uh-moanz). For example, if a hungry bear disturbs a bee hive, the guard bees release an alarm pheromone into the air. This quickly calls the other bees to come out and sting.

 Time to do Activity 107 in the Activity Book!

How Should You Use Your Body?

I will praise You, for I am fearfully and wonderfully made. (Psalm 139:14)

Congratulations on all you have learned about your body!

Are you amazed and in awe of how God created your body? If so, you can say to the Lord, "I will praise You, for I am fearfully (awesomely) and wonderfully made."

Do you wake up each morning thinking of what you might do that day? Because God gave you a body, you can do many things. Let's see what God would like you to do.

GOD MADE ME

Be productive.

Then God blessed [Adam and Eve], and God said to them, "Be fruitful and multiply; fill the earth and subdue it; have dominion over the fish of the sea, over the birds of the air, and over every living thing that moves on the earth." (Genesis 1:28)

God created humans to be different than all of His other creatures. He gave us the responsibility of subduing and having dominion over His creation. This means we are in charge of the earth. God wants His earth to have lots of people who manage His creation and use it to produce other things. To do this, our days should be fruitful (productive). We learn about science so we can understand and be productive with the things God gave us. How can you be productive with God's creation today?

Be productive with the right attitude.

And whatever you do, do it heartily, as to the Lord and not to men. (Colossians 3:23)

To do something **heart**ily means to do it with your whole **heart**. This means you should put your whole self into your work. You should use your muscles to work quickly and carefully. But you should also practice doing things willingly and cheerfully in your heart. You'll have a good attitude about whatever you do if you remember that you are serving Jesus. What work will you do with a good attitude today?

CHAPTER 36: HORMONES FOR HARD TIMES

Think of how Jesus died for you and glorify Him with your body.

You are not your own, for you were bought with a price. So glorify God in your body. (1 Corinthians 6:19-20 ESV)

Christians belong to God. We don't belong to ourselves. This verse says God bought us. But He didn't buy us with money. Jesus bought us through His own death. What a high price Jesus paid for us!

It's wonderful to belong to the Lord who made us, and who also died to take away the sins that separated us from God. Now we have His help and comfort on Earth. After this life, He will lovingly take His people to Heaven. When Jesus returns, all that is wrong with our bodies will be fixed. Just as Jesus rose from the dead, we will also rise again with perfect bodies. Then we will glorify the Lord forever in the New Heavens and New Earth!

1 Corinthians 6 also says that, because we belong to God, we are to glorify Him in our bodies. This means that we turn away from sin. We also seek to do what pleases the Lord who bought us. We should be so thankful for what Jesus has done that we want to obey Him in everything!

One way you glorify God with your body is by using your energy to serve others. Ask your parents for ideas on how you can serve God and serve others today!

Prayer

Almighty Lord, thank You for everything we have learned about our bodies! We praise You for the way You fearfully and wonderfully made us. We see your wisdom in everything You made. We especially see it in our own bodies. Please give us wisdom to care for our health. Please also give us hearts that turn away from sin and turn toward serving You. We ask these things in the name of Jesus. Amen.

 Time to do Activity 108 in the Activity Book!

Glossary

A

Abdomen — the part of the body between the diaphragm (bottom of the chest) and the place where the legs begin.
Acid — something that is very sour.
Active — to be doing something.
Adopted — children are adopted when God has put them in a family with parents who are not the same parents who gave birth to them.
Adrenal Glands — the organ that releases the hormone called adrenalin.
Alveoli — tiny sacks in the lungs where cells trade oxygen and carbon dioxide with the body's blood.
Amino Acids — the special chemicals proteins are made of.
Ammonia — a toxic waste that the body makes when breaking down protein.
Antibiotic — a manmade chemical that fights germs.
Antibody — a protein made and released by immune cells. Antibodies fight germs by attaching themselves to the germs.
Anus — the opening at the end of the digestive tract where solid waste leaves the body.
Anvil — one of the three small bones in the middle ear.
Aorta — the large artery that takes blood from the heart to the body.
Appendages — the arms and legs.
Appendix — a finger-like pouch in the large intestine that stores white blood cells and good microbes.
Aqueous Humor — the liquid inside the eye that's found between the cornea and the lens.
Arteries — blood vessels that carry blood away from the heart.
ATP — the molecule that stores energy within a cell.
Atriums — the upper two chambers in the heart.
Autonomic — the part of the nervous system that is used for things done automatically.
Axon — the fiber on a neuron that sends messages out of the neuron.

B

Backbone — the bones that protect the spinal cord.
Bacteria — microbes made of only one cell.
Balance — a sense that receives messages from the inner ear and the rest of the body to help keep the body in the right position.
Ball and Socket Joint — a joint where one bone has a ball-shaped end which fits into a cup made of one or more different bones.
Bile — sticky, yellow-green liquid that the liver makes to digest fat.
Bile Duct — the tube that carries bile from the liver and gallbladder to the small intestine.
Bitter — when something tastes strong, like the taste of lemon peels or unsweetened chocolate.
Bladder — the organ where the body stores urine until it's time to empty it.
Blind Spot — the place on the eye's retina where there are no rods or cones for sight because the optic nerve is attached there.
Blood — the liquid that carries oxygen and good things to the body's cells and takes wastes away.
Blood Clot — a sturdy seal that closes a hole in the blood vessel.
Bolus — a blob of chewed food mixed with salvia.
Bone Marrow — the tissue inside bones where blood cells are made or fat is stored.
Brain — the organ in the head that thinks, makes decisions, and controls the body.
Brain Stem — the part of the brain that controls automatic things like blood pressure, balance, swallowing, hearing, breathing, and the heartbeat.
Breastbone — the bone that protects the chest and provides a place for the ribs to be attached to.

C

Calf Bone — the thinner bone inside the lower leg.
Calf Muscle — the muscle on the back of the lower leg.
Calories — measurement of the amount of energy food can produce.

GOD MADE ME

Calorimeter — a machine used to find out the amount of calories in food.
Capillaries — small blood vessels where oxygen, nutrition, and wastes are traded with body cells.
Carbohydrates — the sugar and starch found in food.
Carbon Dioxide (CO_2) — the waste gas made by the body.
Cardiac Muscle — the type of muscle that contracts the heart.
Cartilage — strong, bendy tissue that provides structure (nose and ears) and protection (ends of bones).
Cavity — an empty space or hole.
Cecum — the pouch at the beginning of the large intestine that contains microbes.
Cell — the smallest living part of the body.
Cell Body — the part of a neuron that contains the cell's organelles.
Cell Membrane — the skin-like covering that holds the cell together.
Cerebellum — the part of the brain that controls muscle movements.
Cerebral Spinal Fluid — the liquid surrounding the brain and spinal cord that protects them and collects waste.
Cerebrum — the part of the brain that controls thinking, decisions, and memory.
Chamber — a space inside the heart.
Chemical Reaction — a process that changes one set of chemicals into another.
Chromosome — a molecule of DNA wrapped around proteins.
Chyme — a pasty mixture of partly digested food.
Cilia — tiny, finger-like things that stick out of some cells.
Circulatory System — the heart, blood vessels, and blood in the body.
Cochlea — the inner ear's bony structure that's shaped like a snail's shell.
Collagen — a protein that gives structure to connective tissue in the body. It's the body's most plentiful protein.
Conception — the time when a mother's and father's DNA combines, and a unique person begins.
Conclusion — a decision about what is true after studying, experimenting, and thinking about a question.
Concussion — an injury to the brain caused by a hard bump or sudden shaking.
Cones — the receptor cells in the eye that help with seeing color and detail.
Connective Tissue — a tissue that connects things in the body.
Constrict — when something squeezes to make itself skinnier.
Contract — what a muscle does to get shorter.
Control — the thing in an experiment that is left as is. The purpose of a control is to be able to compare changed things with the unchanged control.
Cornea — the eye's clear covering over the iris and pupil.
Cranial Nerves — nerves that come out of the brain and go directly to places in the face, head, and neck.
Cranium — the round part of the skull at the top, sides, and back of the head.
Cross Section — the way something would look if it were cut in half.
Cuticle — one edge of the waterproof seal between a nail and the skin.
Cytoplasm — the liquid inside a cell.

D

Deltoid — the muscle that lifts the humerus bone forwards, backwards, and to the side.
Dendrites — the fibers on a neuron that receive messages into the cell.
Depth Perception — seeing and knowing how close something is.
Dermis — the skin layer below the epidermis where hair and new cells grow.
Detoxify — to remove a poison or toxin from somewhere it shouldn't be.
Develop — to grow and become able to do more.
Diaphragm — the flat muscle below the lungs used for breathing.
Digest — to break down food into pieces small enough to be used by the body.
Digestive Juices — liquids made by the digestive tract that contain chemicals to help break down food.
Digestive System — the structures, organs, and glands that work to break down and absorb food.
Discs — cartilage cushions between the vertebrae.
Disease — a problem that causes the body or one of its parts not to work as it should.
DNA — the set of instruction molecules in a cell.
Dominant — the form of a gene that will show up in the child. If a child gets a blue-eye gene from one parent and a brown-eye gene from the other, the child will have brown eyes because the brown-eye gene is dominant.
Duct — a tube that carries liquids away from a gland or organ.

E

Ear Canal — the one-inch-long tube that leads from outside the ear to the eardrum.
Ear Flap — the outer structure of the ear.
Eardrum — the ear membrane that collects sound vibrations from the air.
Earwax — the sticky stuff that traps dust and dead skin cells in the ear.
Efficient — without wasting time or energy.
Emotions — feelings in the mind, like joy and sadness.
Endocrine System — a body system of glands, hormones made by the glands, and target cells that are switched on by the hormones.
Endoplasmic Reticulum — an organelle that helps make protein and fat in the cell.
Engulf — to fold around something in order to take it in.
Epidermis — the skin's top layer.
Epiglottis — a flap of cartilage that closes off the breathing tube during swallowing.
Esophagus — the tube which carries food from the mouth to the stomach.
Essential Nutrient — a nutrient that is necessary for life but can't be made by the body.
Eustachian Tube — a tube that runs from the inside of the ear to

CHAPTER 36: GLOSSARY

the throat. It can open up to keep the right amount of air pressure inside the ear.
Exhale — to breath air out.
Expel — to push something out.
Extend — to straighten and swing two long bones apart.
Eye — the organ that allows sight.
Eyeball — the ball-shaped part of the eye.
Eyelid — a thin fold of skin that protects the eye.
Eye Socket — a hollow bony cup in the skull that forms a protective home for the eye.

F

Faint — to become unconscious for a short time.
Farsighted — to see faraway things well but to see close things poorly.
Fat — a body tissue that stores energy and protects from cold and injury.
Fatty Acid — one kind of molecule that fats are made of.
Femur — the long bone in the upper part of the leg.
Fertilization — when the DNA from both parents is joined and a unique person begins.
Fiber — a kind of carbohydrate that the body cannot break down.
Fibrils — thin strands that contract within muscle cells.
Fibrin — a thread-like, waterproof protein that partly makes up a blood clot.
Fight or Flight — the way someone's nerves automatically respond to sudden danger or fright.
First Aid — the first help someone receives for an injury or illness to keep a problem for getting worse and to help it heal.
Flex — to bend and swing two long bones together.
Focus — to bend light so a clear picture is seen.
Fontanels — the spaces between a baby's skull bones.
Food — things that are eaten to give energy and life.
Fovea — the place in the back of the eye where the sharpest vision is sensed.
Fracture — a break in a bone.
Free Nerve Endings — receptors that sense temperature and pain.
Frontal Lobe — the part of the brain (behind the forehead) that thinks, learns, stores memories, and tells muscles to move.
Fungi — living things that act kind of like plants but don't make their own food like plants do.

G

Gallbladder — the organ where bile is stored until needed.
Gamete — a cell with half of the DNA instructions for a new life.
Gastric Pits — places in the stomach where mucus and digestive juices are made.
Gene — a place in the DNA molecule that gives instructions for a certain trait.
Generation — a group people of about the same age and are grouped according to who their parents are.
Genetics — the study of genes and the traits they show.
Gland — an organ that automatically releases a hormone or other chemical.
Gliding Joint — a joint that allows the bones to slide across each other.
Glymphatic System — a system that removes waste from the brain.
Goblet Cells — the cells that make mucus in the main part of the nasal cavity.
Golgi Bodies — a cell organelle that sends proteins within and outside of the cell.
Grains — the seeds of grasses; they provide high-carbohydrate food.
Gray Matter — grayish-pink brain tissue that has a lot of neuron cell bodies.

H

Hair Cells — the hearing receptors.
Hair Follicles — deep pits that make hair in the skin's dermis.
Hammer — one of the three small bones in the middle ear.
Heart — the organ that pumps blood through the body.
Hinge Joint — a joint that allows bones to swing back and forth as a hinge does to a door.
Hip bones — bones that connect the leg bones to the body and protect some of the organs in the abdomen.
Hormone — a chemical that's made in one part of your body and causes something to happen in another part.
Human Being — a man, woman, or child.
Humerus — the bone that extends from the shoulder to the elbow.
Hyoid — the bone that helps with swallowing, coughing, making sound, moving the tongue, and preventing choking.
Hypodermis — the fatty, bottom layer of skin.
Hypothalamus — part of the brain that makes sure the body stays balanced inside. It does this by gathering information from the blood and nervous system and sending it to the autonomic nervous and endocrine systems.

I

Immovable Joints — joints that do not move.
Immune System — the body system that protects against germs and other outside dangers like parasites and toxins.
Inactive — to not be doing anything.
Infect — to cause someone to have a disease or health problem from microbes.
Infection — the problem people have when they are being attacked by microbes.
Inflammation — when blood vessels bring in extra blood to deliver oxygen for healing.
Inhale — to breathe air in.
Inhibit — to discourage or cause less of something.
Injury — damage to your body that causes pain.
Innate White Blood Cells — immune cells that recognize things that do not belong to the body and attack them.
Inner Ear — the deepest part of the ear; contains bony structures filled with liquid.
Interneurons — neurons that receive and process information from the sensory neurons and pass commands to the body through motor neurons.

GOD MADE ME

Iris — the ring of color surrounding the pupil of the eye.
Iron — an important mineral in blood that helps deliver oxygen to the body.

J
Joint — the place where separate bones come together, allowing the skeleton to bend and move.

K
Keratin — a tough protein found in skin, hair, and nails.
Kneecap — the round bone that protects the knee joint.

L
Lacrimal Gland — the gland that makes tears.
Large Intestine — the organ that absorbs useful liquid from chyme and removes solid waste from the body.
Left Atrium — the upper space in the left side of the heart.
Left Ventricle — the lower space in the left side of the heart.
Legumes — plants in the bean and pea family.
Lens — the transparent, rounded structure that focuses light onto the back of the eyeball.
Ligaments — connective tissue that tightly connects bones together, keeping them in place.
Liver — a large, complicated organ that helps your body digest, absorb nutrients, store important things, remove harmful chemicals, stop bleeding, fight germs, make hormones, and send nutrients and hormones where they are needed.
Lobes — different sections of the brain
Long-Term Memory — the type of memory that stores information a long time.
Lower Jawbone — the moving bone of the chin.
Lungs — the organs of the respiratory system that trade oxygen and carbon dioxide between blood and air.
Lunula — the white, crescent-shaped part of the nail root that can be seen through the nail.
Lymph — a small amount of plasma that leaves the capillaries and collects waste between body cells.
Lymph Duct — a tube where lymph is collected.
Lymph Node — a place where wastes are cleaned out of the lymph.
Lymph System — the system of ducts and nodes that removes certain wastes put out by body cells.
Lysosome — an organelle that acts as the digestive system for the cell.

M
Macula — the place on the retina that helps with the center of a person's vision.
Matter — anything that has mass.
Mechanical — like a machine.
Melanin — a pigment that causes tanning to protect skin from the sun.
Melatonin — a hormone that causes sleepiness.
Membrane — a skin-like covering or layer of tissue that separates one place from another in the body.
Microbe — a very small living thing that cannot be seen with the eyes only.
Microscopic — too small to be seen with the eyes only.
Microvilli — tiny "fingers" that live on the outer surface of the cells on intestinal villi.
Middle Ear — the ear's air-filled space that's surrounded by bone.
Minerals — chemicals that plants take from the soil and pass along to us in food.
Mitochondria — cell organelles that make energy.
Mitosis — the process God made for one cell to become two new cells by dividing.
Mixed Nerves — nerves that contain both sensory and motor neurons.
Molecule — the smallest piece of something that still is that thing.
Monitor — be watchful over.
Motor Nerve — a nerve that sends commands from the brain to the body.
Motor Neuron — a neuron inside a motor nerve that commands muscles and some organs to do something.
Mucus — a thick, sticky, slimy liquid that has many jobs in the body.
Muscles — groups of cells that can get shorter or longer to help control movement in the body.
Myelin — the fatty covering of axons.

N
Nail Bed — the layer under the nail plate that contains blood vessels and nerves.
Nail Plate — the part of the nail that shows.
Nail Root — the part of the nail that the nail plate grows from.
Nasal Cavity — the space inside the nose made of bones, tissue, and blood vessels.
Navel (belly button) — the place where the umbilical cord was attached to a baby while in the womb.
Nearsighted — to see near things well but to see faraway things poorly.
Nephron — a tiny structure that, along with a million others, purifies blood in the kidney.
Nerve — a group of neurons that sends messages between the brain and the body.
Nerve Fibers — the axon and dendrites of a neuron.
Nervous System — the system that sends electrical messages throughout the body.
Neurons — nerve cells.
Nuclei — more than one nucleus.
Nucleus — the organelle that acts like the brain of the cell.
Nutrient — an important thing in food, like protein and vitamins, that give what is needed for growth and life.

O
Occipital Lobes — the parts of the brain that receive messages from the eyes, figure out colors, and recognize faces.

CHAPTER 36: GLOSSARY

Odor — any smell, pleasant or unpleasant.
Oil Glands — tiny glands in the skin that make oil for keeping the hair and skin soft.
Olfactory — having to do with the sense of smell.
Olfactory Bulbs — the part of your brain where smell messages are combined.
Olfactory Glands — the glands that make mucus in the olfactory tissue of the nasal cavity.
Olfactory Tissue — the "ceiling" of each side of the nasal cavity where smell receptors live.
Optic Nerve — the nerve that connects the eye to the brain.
Oral Cavity — inside the mouth.
Organ — a structure that has a certain shape, a certain place, and a certain job to do in the body.
Organelles — the small structures with special jobs inside cells.
Otolith Organs — organs in the inner ear that help sense the body's gentle motion.
Outer Ear — the parts of the ear that can be seen (ear flap and ear canal).
Oval Window — the small opening between the middle ear and the inner ear where the air vibration of the middle ear becomes the liquid vibration of the inner ear.
Oxygen (O^2) — a gas in the air that's necessary for life.

P

Pancreas — an organ that makes digestive juices and delivers them to the small intestine.
Papillae — small bumps on the tongue; some contain taste buds and others help move food around in the mouth.
Parasites — creatures that live harmfully in or on someone.
Parathyroid Gland — the gland making a hormone that adds calcium to the blood.
Parietal Lobe — the part of the brain that understands language, figures out the skin's touch messages, and senses the body's position.
Peristalsis — the movement that squeezes food and liquids through the digestive tract.
Permanent Teeth — the teeth that replace baby teeth but will not be replaced themselves.
Perspiration — the watery liquid that skin makes for keeping the body cool and for removing certain toxins.
Pheromones — hormones that some animals release into the air that affect other animals.
Photosynthesis — the food-making process in plants.
Pigments — chemicals that have a color.
Pineal Gland — the gland making the hormone melatonin.
Pituitary Gland — the gland that stores hormones made by the hypothalamus as well as hormones it makes itself.
Pivot Joint — a joint that allows one bone to rotate against another bone.
Placenta — a special structure that takes nutrition and oxygen from the mother's blood and transfers it to the baby's blood.
Plasma — the liquid part of blood; not including the blood cells.
Platelets — pieces of cells that help blood to clot and stop bleeding.

Poison — a chemical that could harm a living thing.
Pregnant — to have a baby growing in one's womb.
Primary Teeth — the first teeth to grow in the mouth (baby teeth).
Proprioception — the sense that helps the body automatically know where the arms, legs, and other body parts are.
Proteins — large molecules made of amino acids.
Pump — a machine that uses pressure to force air or something else to flow.
Pupil — the black circle on the front of the eye where light enters the eyeball.
Puss — liquid that comes from the blood to clean germs and dead cells out of wounds.

R

Radius — the bone in the arm that is attached to the wrist on the thumb's side.
Rebuild — to put something back together that has been broken.
Receptors — places on the outside of cell membranes where certain chemicals can attach and cause something to happen to the cell.
Rectum — the last space in the large intestine where stools are stored until it's time for them to leave the body.
Red Blood Cells — disk-shaped cells that carry oxygen and give blood its red color.
Reflex — a reaction that is fast because the nerve messages travel only to and from the spinal cord and not also to the brain.
Relax — when a muscle gets softer and longer.
REM sleep — a deep stage of sleep when the eyes move around quickly (Rapid Eye Movement).
Repel — to keep something from coming in.
Reproduce — when something makes more living things of its kind.
Respiratory System — your lungs and all the tubes and openings that air must pass through to get to your lungs.
Rest and Digest — the way someone's nerves automatically calm the body so it can do its normal jobs like digesting and making urine.
Retina — the inner, light-sensitive lining of the eyeball.
Ribosomes — little, machine-like structures that make proteins inside cells.
Ribs — chest bones that protect your heart and lungs.
Right Atrium — the upper space in the right side of the heart.
Right Ventricle — the lower space in the right side of the heart.
Rods — the receptor cells in the eye that help with seeing brightness, motion, and black and white.
Round Window — a membrane-covered hole in the ear's cochlea that helps sound vibrations travel as they should.

S

Saddle Joint — a joint that allows the bone to move in many directions but is still very stable.
Saliva — the mouth's watery liquid that moistens food and begins to break it down.
Salivary Glands — glands that make saliva.

Salty — when something tastes like salt.
Savory — when something has a rich, meaty taste.
Scab — a dried blood clot.
Scalp — the skin that covers the top of the head.
Scar — the mark left on skin after a wound heals.
Sclera — the protective, white part of the eye.
Semicircular Canal — the inner ear's bony structure that senses motion of the head like nodding or shaking.
Senses — the ways the body receives information from the world; seeing and hearing are two of the senses.
Sensory Nerves — nerves that carry messages from the senses to the brain and spinal cord.
Sensory Neurons — the neurons inside the nerves that carry messages to the brain and spinal cord; sensory neurons are often receptors for the senses.
Septum — the structure that divides the right side of the nasal cavity from the left.
Shin Bone — the thick, front bone in the lower leg.
Short-Term Memory — the type of memory that only stores information for a short time because it's not important for the future.
Shoulder Bones — bones that connect the arm bones to the upper body.
Sister Chromosomes — two identical copies of the same chromosome.
Skeletal Muscles — the type of muscles that are attached to bones and cause them to move.
Skeleton — all the body's bones connected together.
Skull — the bony, hollow case that protects the brain.
Small Intestine — the part of the digestive system that absorbs nutrients from food and puts them into the blood.
Smooth Muscles — the type of muscles that are arranged in thin layers inside the walls of certain organs and tubes.
Socket — a hollow place made for something to fit inside.
Somatic — the part of the nervous system that is used for things done on purpose.
Soul — the part of a person that can know and love God.
Sound Wave — a kind of energy that's released when something vibrates.
Sour — the taste of something acidic.
Specific White Blood Cells — immune cells that can specifically attack a certain type of germ.
Sphincters — muscles that surround an opening and can close it by squeezing.
Spinal Cord — a cord of nerve tissue that runs down the length of the spine; it sends and receives messages between the body and the brain.
Spinal Nerves — nerves that go between the spinal cord and the rest of the body.
Spine — the backbone and spinal cord.
Spleen — an organ that, as part of the immune system, switches on certain white blood cells and makes germ fighting proteins.
Stem Cells — cells that can divide and make any kind of cell in the body.
Stimulate — to encourage or cause more of something.

Stirrup — one of the three small bones in the middle ear.
Stomach Acid — a strong chemical made in the stomach to digest food.
Stool — the solid waste that leaves the digestive system when someone uses the toilet.
Sutures — the places where skull bones are joined together.
Sweat — perspiration.
Sweet — something that tastes like sugar.
Synapse — a space between neurons.
System — a group of organs and structures that work together to do a certain job.

T

Target Cells — body cells that have receptors that respond to certain hormones, causing the cell to do a job.
Taste Buds — tiny, barrel-shaped structures living just under the surface of the papillae. They contain taste receptor cells.
Taste Receptor Cells — cells inside taste buds that can sense food chemicals in salvia.
Tear Duct — the tube that carries tears from the eye to the inside of the nose.
Tears — the liquid that keeps the eye wet, protects it from germs, and delivers oxygen and nutrition.
Temporal Lobe — the part of the brain that figures out messages from the senses and remembers complicated things that were seen or heard (used for art, music, and reading).
Tendon — strong connective tissue that connects muscles to bones.
Thymus — a small organ where certain white blood cells change to their germ fighting form.
Thyroid Gland — an organ that produces hormones to help the body have enough energy and to control its calcium levels.
Tissue — a group of the same kind of cells that have the same job.
Tongue — the structure in the mouth that moves food around, helps make sounds, and contains the sense of taste.
Tonsils — areas of mucus-making tissue in the throat where germs from food and air are trapped and killed.
Toxin — a poison that is made by a living thing.
Trachea — the main tube in the throat that allows air to go back and forth between the lungs.
Trait — a certain feature that depends on the genes received by the person; brown eyes and how tall someone grows are traits.

U

Ulna — the forearm bone that is attached to the pinkie side of the wrist.
Ultrasound Machine — a special camera that allows parents to see pictures and videos of their baby at certain stages in the womb.
Umbilical Cord — a tube that carries nutrition to the baby and carries waste away while the baby is in the womb. After birth, the umbilical cord falls off, leaving the navel where it had been.
Unique — something that is not like anything else.
Ureters — the long tubes which take urine from the kidneys to the bladder.
Urethra — the tube that takes urine from the bladder to the

CHAPTER 36: IMAGE CREDITS

outside of the body.
Urinary System — the kidneys, bladder, and tubes that remove urine from the body.
Urinate — to empty the bladder of urine.
Urine — the liquid put out by the kidneys that contains waste, extra water, and salt.

V

Valve — a structure that keeps liquid going in only one direction.
Veins — blood vessels that carry blood to the heart.
Ventricles — the lower two chambers in the heart.
Vertebra — an individual bone of the spine.
Vessels — tubes in the body that carry blood.
Vibrate — to move back and forth very quickly.
Villi — tiny, finger-like projections that cover the inside surface of the small intestine.
Virus — a microbe that may not be alive but uses living cells to reproduce.

Vitamins — essential molecules that help the body do its chemical reactions.
Vitreous Humor — the liquid inside the eye between the back of the lens and the retina.
Vocal Cords — folds of tissue in the voice box that stretch and make sounds as air presses against them when someone talks or sings.
Voice Box — the structure in the throat that holds the vocal cords.
Vomiting — when the stomach is irritated and causes its contents to come out through the mouth.

W

White Blood Cells — cells that help defend the body against germs.
White Matter — light-colored brain tissue that has a lot of axons.
Womb — the special place inside a mother where God makes a baby grow before it's born.
Wound — an injury that breaks the skin and often causes deeper damage.

Image Credits

All images from iStock.com with the exceptions of:

Pg. 39 | Stem cells | Wikimedia Commons
Pg. 40 | Six-week embryo | Wikimedia Commons
Pg. 41 | 20-week embryo | nlrc.org
Pg. 42 | Newborn baby | Sarah Bryant (SarahLeePhoto.com)
Pg. 76 | Cave | unsplash.com
Pg. 76 | Boy | unsplash.com
Pg. 144 | Receptors in skin | Wikimedia Commons
Pg. 148 | Parts of skin and hair | Wikimedia Commons

Pg. 160 | Frog's ear hair cell | Wikimedia Commons
Pg. 168 | Holes in bony plate | Wikimedia Commons
Pg. 169 | Tongue | Wikimedia Commons
Pg. 183 | Apples | Wikimedia Commons
Pg. 249 | Proteasome | Wikimedia Commons
Pg. 317 | Immune response | Wikimedia Commons
Pg. 337 | Coin | Wikimedia Commons